Plant Adaptation to Extreme Environments in Drylands—Series II

Plant Adaptation to Extreme Environments in Drylands—Series II

Guest Editors

Xiao-Dong Yang
Guang-Hui Lv
Nai-Cheng Wu
Xue-Wei Gong

Basel • Beijing • Wuhan • Barcelona • Belgrade • Novi Sad • Cluj • Manchester

Guest Editors

Xiao-Dong Yang
Department of Geography &
Spatial Information
Ningbo University
Ningbo
China

Guang-Hui Lv
College of Ecology and
Environment
Xinjiang University
Urumqi
China

Nai-Cheng Wu
Department of Geography &
Spatial Information
Ningbo University
Ningbo
China

Xue-Wei Gong
Institute of Applied Ecology
Chinese Academy of Sciences
Shengyang
China

Editorial Office
MDPI AG
St. Alban-Anlage 66
4052 Basel, Switzerland

This is a reprint of the Special Issue, published open access by the journal *Forests* (ISSN 1999-4907), freely accessible at: https://www.mdpi.com/journal/forests/special_issues/0R0H8WYD7S.

For citation purposes, cite each article independently as indicated on the article page online and as indicated below:

Lastname, A.A.; Lastname, B.B. Article Title. *Journal Name* **Year**, *Volume Number*, Page Range.

ISBN 978-3-7258-1188-5 (Hbk)
ISBN 978-3-7258-1187-8 (PDF)
https://doi.org/10.3390/books978-3-7258-1187-8

Cover image courtesy of Xiaodong Yang

© 2024 by the authors. Articles in this book are Open Access and distributed under the Creative Commons Attribution (CC BY) license. The book as a whole is distributed by MDPI under the terms and conditions of the Creative Commons Attribution-NonCommercial-NoDerivs (CC BY-NC-ND) license (https://creativecommons.org/licenses/by-nc-nd/4.0/).

Contents

Preface

Most of the countries in the arid regions of the world, especially in Eurasia and Africa, are developing countries. Economic development related to food security and people's livelihoods is the primary task of development for these countries. This has led to a sharp contradiction between socio-economic development and plant protection, resulting in a serious loss of biodiversity and biological resources. Therefore, explaining the adaptability of plants to environmental stress in arid areas; the current pattern of distribution, diversity, and ecosystem function of these plants; and the factors driving their changes have become a global academic issue.

With the support of MDPI, we gathered research from all relevant areas, including water uptake and use mechanisms, physiological and biochemical adaptation strategies of plants, plant–soil microbial interactions, and ecosystem services, in the Special Issue titled "Plant Adaptation to Extreme Environments in Drylands—Series II: Studies from Northwest China". This Special Issue includes thirteen innovative research papers covering diverse ecosystems such as desert sparse shrublands, natural forests, and artificial forests. The articles in this Special Issue reveal the complex interactions and adaptation differences between different ecosystems in arid areas in a multidimensional way, showing the rich results of biodiversity and ecological adaptation, as well as representing the latest advances in research on a wide range of adaptations of plants in arid areas to extreme environments.

Our Special Issue reprint provides cutting-edge theories and knowledge, especially for those interested in plant protection in arid areas, as well as scholars in universities and research institutions. Despite our best efforts to collect papers that cover as many relevant topics as possible, to provide a comprehensive introduction to the basic knowledge and scientific preamble on the adaptation of plants to environmental stress in arid areas, this Special Issue represent only a small portion of the scientific knowledge in this field. We hope that more scholars will pay attention to the research on plant adaptation to extreme environments in arid areas. More importantly, we need more scholars to invest in researching of plant protection, management, and utilization and using academic results to provide theories and technologies for the protection and utilization of plant resources in arid areas.

We are indebted to all the authors who contributed their work to this reprint and to all reviewers whose comments and suggestions helped to improve the quality of the accepted papers. We sincerely thank MDPI and *Forests* for their help and hard work in specific issue collection and reprint production.

Xiao-Dong Yang, Guang-Hui Lv, Nai-Cheng Wu, and Xue-Wei Gong
Guest Editors

Editorial

Plant Adaptation to Extreme Environments in Drylands—Series II: Studies from Northwest China

Xiao-Dong Yang [1,2,*], Sai-Qiang Li [1], Guang-Hui Lv [2], Nai-Cheng Wu [1] and Xue-Wei Gong [3]

[1] Department of Geography & Spatial Information Technology, Ningbo University, Ningbo 315211, China
[2] School of Ecology and Environment, Xinjiang University, Urumqi 830046, China
[3] Institute of Applied Ecology, Chinese Academy of Sciences, Shenyang 110016, China; gongxw@iae.ac.cn
* Correspondence: yangxiaodong@nbu.edu.cn

Citation: Yang, X.-D.; Li, S.-Q.; Lv, G.-H.; Wu, N.-C.; Gong, X.-W. Plant Adaptation to Extreme Environments in Drylands—Series II: Studies from Northwest China. *Forests* 2024, 15, 733.

https://doi.org/10.3390/f15050733

Received: 18 March 2024
Revised: 17 April 2024
Accepted: 22 April 2024
Published: 23 April 2024

Copyright: © 2024 by the authors. Licensee MDPI, Basel, Switzerland. This article is an open access article distributed under the terms and conditions of the Creative Commons Attribution (CC BY) license (https://creativecommons.org/licenses/by/4.0/).

Arid and semi-arid lands cover more than one-third of the Earth's terrestrial area [1]. The environmental conditions in these areas are very harsh, characterized by scarce rainfall, high temperatures and evaporation, soil salinization, poor soil nutrition, and low vegetation coverage [2]. Affected by global climate change, these regions will face more frequent and intense extreme weather events, such as prolonged droughts and extreme summer heatwaves, which will pose major challenges to local ecosystems and biodiversity. Revealing strategies and mechanisms of plant adaptation to extreme environments (such as drought, salinity, and high temperatures) in arid zones has become one of the research hotspots, which is of great relevance for taking appropriate measures to protect and manage local vegetation and ecosystems. Despite the efforts of academic researchers, differences in adaptation strategies between species and changes in plant adaptation across environmental gradients require more thorough and accurate research. This will provide practical solutions to the ecological challenges faced by arid and semi-arid regions.

We gathered research from all relevant areas, including water uptake and use mechanisms, physiological and biochemical adaptation strategies of plants, plant–soil microbial interactions, and ecosystem services, in the Special Issue titled "Plant Adaptation to Extreme Environments in Drylands—Series II: Studies from Northwest China". By bringing together these findings, we aim to deepen our understanding of how plants in arid regions adapt to the more widespread and frequent stresses of environmental change in the future. This Special Issue includes twelve innovative research papers covering diverse ecosystems such as desert sparse shrublands, natural forests, and artificial forests. Among them, 11 papers are located in Northwest China, while the remaining papers are based on published data using bibliometric methods to cover the global scope. We are fortunate that most of the research in the Special Issue is located in the Ebinur Wetland Nature Reserve. This reserve is an ideal site for studying biodiversity and ecological adaptability in arid and semi-arid areas of Northwest China due to its unique biodiversity and extremely arid environment [3]. This gathering of papers at the research site allows our papers to logically form a cohesive whole. The articles in this Special Issue reveal the complex interactions and adaptation differences between different ecosystems in arid areas in a multidimensional way, showing the rich results of biodiversity and ecological adaptation, as well as representing the latest advances in research on a wide range of adaptations of plants in arid areas to extreme environments.

In an arid environment, the survival and reproduction of plants are severely challenged due to the extreme lack of water. In order to adapt to this water-limited environment, plants have evolved a series of unique water absorption and regulation mechanisms, such as adjusting leaf stomata to absorb water and changing their structure to store water more effectively. Li et al. [4] first discussed the foliar water uptake (FWU) of male and female *Populus euphratica* Oliv. in different growing seasons and its relationship with photosynthetic capacity and anatomical structure from a gender perspective. The results showed that the FWU capacity of both male and female plants increased with the growing period. Although the FWU capacity of female plants was better than that of males due to their developed water storage structure, the male plant was better than the female plant in alleviating the leaf water deficit and carbon benefit. An in situ wetting experiment also found that in the middle of the growing season, wetting treatment significantly improved the pre-dawn water potential and photosynthetic capacity of *P. euphratica*. This study revealed the different water use strategies of male and female *P. euphratica* under drought conditions, providing a new perspective for understanding the foliar water uptake adaptation of plants to extreme environments. By measuring the contact angle, FWU parameters, and hydraulic parameters, Wang et al. [5] explored the differences among six desert species' FWU capacity and the effects of leaf wettability and hydraulic parameters on FWU capacity. The results suggested that all six species had FWU capacity and emphasized that stomatal regulation played a more important role in the FWU than leaf wettability and leaf structure. This study provides a theoretical basis for understanding the water utilization mechanisms of plants in arid areas.

The adaptability of plants to a drought environment is realized through a series of complex physiological and biochemical mechanisms, including not only the absorption and efficient use of water but also the adjustment of their internal physiological and biochemical traits under drought stress. In order to clarify the adaptation strategies of desert plants in different habitats, Yang et al. [6] focused on the interspecific integration of chemical traits in desert plant leaves with variations in soil water and salt habitats and found that the interspecific integration of plant leaf chemical traits in high water-salinity habitats was higher than that in low water-salinity habitats. Trait networks showed that C and Ca were the core leaf chemical traits for desert plant growth. Jiang et al. [7] compared the physiological traits of the leaves and roots of *Kalidium foliatum* (Pall.) Moq. under different saline habitats in a desert region and found that different salt environments had different effects on the leaf and root traits, including chlorophyll content, anatomical structure, nutrient composition, and the content of osmotic regulatory substances. *K. foliatum* adapts to different salt environments by regulating leaf and root structure, element allocation, ion balance, osmotic regulation, and antioxidant defense mechanisms, which provides insights into how plants respond to salt stress. Additionally, Jiang et al. [8] discussed the influence of intraspecific trait variation on plant functional diversity and community assembly processes in arid desert areas of northwest China. They pointed out that the effects of spatial and environmental factors on functional diversity varied across the different scales. Spatial factors have a greater impact on functional diversity at small scales, while environmental factors have a more significant impact at larger scales. Moreover, they emphasized the importance of intraspecific trait variation in understanding and predicting the response of plant communities to environmental change in arid regions. Zheng et al. [9] analyzed the distribution patterns of δ^{13}C and major nutrient elements in different organs of *Alhagi sparsifolia* Shap. and *Karelinia caspia* (Pall.) Less. across the growing seasons in the northern part of the Taklimakan Desert and their relationship with soil environmental factors. It was found that δ^{13}C and nutrient contents were significantly different between the different organs of the two species. Soil total phosphorus was the most important environmental factor affecting δ^{13}C and nutrient elements in the two species. They also demonstrated that these two species are able to effectively coordinate and regulate their water, N, and P use strategies in response to environmental stress. Khan et al. [10] compiled a dataset of a total of 1295 desert plant species from the literature and then used bibliometric analysis to explore whether plant adaptive strategies and ecosystem nutrient concentrations shifted across three desert ecosystems (e.g., desert, steppe desert, and temperate desert). They found that leaf N, P, and C concentrations were significantly different only from those of certain other growth forms and in certain desert ecosystems. In leaves, the C concentrations were always greater than the N and P concentrations and were greater than those in soils. The element concentrations and ratios were also greater in the organs than in the soils. Additionally, the values in the leaf versus the root N and P changed among the three desert ecosystems. However, the conclusion reached in this study was different from previous macro-ecology studies in that the mean annual precipitation and mean annual temperature had no significant influences on the leaf elemental concentrations and ratios. This study deepened our understanding of resource-related adaptive strategies, indicating that plants adapt to drought-stressed environments via divergent element concentration changes, thus maintaining the stability of desert ecosystems. Aishan et al. [11] conducted a field survey of tree cavities (hollows) of *P. euphratica* in the Tarim River and found that water source was the main environmental factor affecting the formation of hollows. Moreover, the height and width of the hollows of *P. euphratica* were significantly positively correlated with the DBH, crown loss, average crown width, and the distance from trees to rivers. This study is of great significance for understanding the death and ecological adaptation mechanisms of *P. euphratica* and similar species.

Microbial activity plays a crucial role in maintaining the functional stability of vegetation–soil ecosystems [12]. Because arid zones soil is characterized by insufficient water, loose structure, and severe salinization, the interaction between plants and soil rhizosphere microorganisms is very important for plant adaptation to an arid environment. By conducting pot experiments to control the water and nitrogen sources of *Haloxylon ammodendron* (C.A.Mey.) Bunge seedlings, Zhu et al. [13] found that water deficit and nitrogen addition significantly changed the morphological traits and rhizosphere microbial diversity. They highlighted that rhizosphere microorganisms respond significantly to water deficits and nitrogen addition, thus promoting plant adaptation to environmental stress through symbiotic relationships. Wang et al. [14] studied the soil bacterial community structure and its influencing factors in an artificial *H. ammodendron* forest in the Sand Blocking and Fixing Belt of Minqin, China. It was found that *H. ammodendron* plantation significantly increased the principal nutrient contents and the diversity and richness of soil bacteria compared to the mobile dune soil. They also found that soil organic matter, total nitrogen, available potassium, pH, and electrical conductivity were the main environmental factors affecting the structure and function of the soil bacterial community. This study provides important insights into the underlying mechanism for explaining the changes in the structure and function of soil bacterial communities in artificial sand-fixation vegetation.

Biological systems in arid areas play a vital role in maintaining ecological balance and providing ecosystem services. Due to the unique climatic conditions and resource constraints, ecosystem services have become a core area of ecological research in arid areas, involving water resource management, the carbon cycle, and climate regulation. By using the InVEST model, Yao et al. [15] studied the spatial and temporal response of vegetation to runoff changes in the Ebinur Lake Basin by using remote sensing data. The results suggested that the spatial distribution of runoff was positively correlated with altitude. Vegetation coverage and types have an obvious spatial and temporal response to runoff changes. This is of great significance in terms of theoretical guidance for water resource management and ecological restoration in arid areas. Li et al. [16] explored the ecological functions of artificial shelterbelts (composed of *H. ammodendron*, *Calligonum mongolicum* Turcz., and *Tamarix chinensis* Lour.) along the Tarim Desert Highway. They found that the average daily net photosynthetic rate, carbon sequestration per unit leaf area, and oxygen release of *T. chinensis* were significantly higher than those of the other two species, but the carbon storage was the greatest in *H. ammodendron*. According to the net photosynthetic rate method and the biomass method, they obtained shelterbelts that contributed significantly to carbon storage and oxygen release. This study highlighted the role of these shelterbelts in promoting ecological restoration and sustainable development in arid desert areas.

It should be emphasized that the studies of plant adaptation to extreme environments in the arid region collected in our Special Issue are still in the preliminary exploration stage, although they provide profound insights and rich data. Further work is needed to systematically reveal plant adaptation to extreme environments, especially in the study of processes and mechanisms based on experimental, functional-based, and modeling methods.

Author Contributions: All authors (X.-D.Y., S.-Q.L., G.-H.L., N.-C.W. and X.-W.G.) contributed to the design, writing, and editing of this Editorial article. All authors have read and agreed to the published version of the manuscript.

Funding: This research received no external funding.

Acknowledgments: We are indebted to all the authors contributing their work to this Special Issue and to all Forests editorial staff who contributed to the publication of the papers.

Conflicts of Interest: The authors declare no conflicts of interest.

References

1. Yang, X.-D.; Wu, N.-C.; Gong, X.-W. Plant Adaptation to Extreme Environments in Drylands. *Forests* **2023**, *14*, 390. [CrossRef]
2. Yang, X.; Long, Y.; Sarkar, B.; Li, Y.; Lü, G.; Ali, A.; Yang, J.; Cao, Y. Influence of Soil Microorganisms and Physicochemical Properties on Plant Diversity in An Arid Desert of Western China. *J. For. Res.* **2021**, *32*, 2645–2659. [CrossRef]
3. Jiang, L.-M.; Sattar, K.; Lv, G.-H.; Hu, D.; Zhang, J.; Yang, X.-D. Different Contributions of Plant Diversity and Soil Properties to the Community Stability in the Arid Desert Ecosystem. *Front. Plant Sci.* **2022**, *13*, 969852. [CrossRef] [PubMed]
4. Li, Z.; Chen, Y.; Wang, H.; Zhang, X. Foliar Water Uptake and Its Relationship with Photosynthetic Capacity and Anatomical Structure between Female and Male *Populus euphratica* at Different Growth Stages. *Forests* **2023**, *14*, 1444. [CrossRef]
5. Wang, H.; Li, Z.; Yang, J. Effects of Leaf Hydrophilicity and Stomatal Regulation on Foliar Water Uptake Capacity of Desert Plants. *Forests* **2023**, *14*, 551. [CrossRef]
6. Yang, J.; Zhang, X.; Song, D.; Wang, Y.; Tian, J. Interspecific Integration of Chemical Traits in Desert Plant Leaves with Variations in Soil Water and Salinity Habitats. *Forests* **2023**, *14*, 1963. [CrossRef]
7. Jiang, L.; Wu, D.; Li, W.; Liu, Y.; Li, E.; Li, X.; Yang, G.; He, X. Variations in Physiological and Biochemical Characteristics of *Kalidium foliatum* Leaves and Roots in Two Saline Habitats in Desert Region. *Forests* **2024**, *15*, 148. [CrossRef]
8. Jiang, L.; Zayit, A.; Sattar, K.; Wang, S.; He, X.; Hu, D.; Wang, H.; Yang, J. The Influence of Intraspecific Trait Variation on Plant Functional Diversity and Community Assembly Processes in an Arid Desert Region of Northwest China. *Forests* **2023**, *14*, 1536. [CrossRef]
9. Zheng, Z.; Wu, X.; Gong, L.; Li, R.; Zhang, X.; Li, Z.; Luo, Y. Studies on the Correlation between δ^{13}C and Nutrient Elements in Two Desert Plants. *Forests* **2023**, *14*, 2394. [CrossRef]
10. Khan, A.; Liu, X.-D.; Waseem, M.; Qi, S.-H.; Ghimire, S.; Hasan, M.M.; Fang, X.-W. Divergent Nitrogen, Phosphorus, and Carbon Concentrations among Growth Forms, Plant Organs, and Soils across Three Different Desert Ecosystems. *Forests* **2024**, *15*, 607. [CrossRef]
11. Aishan, T.; Mumin, R.; Halik, Ü.; Jiang, W.; Sun, Y.; Yusup, A.; Chen, T. Patterns in Tree Cavities (Hollows) in Euphrates Poplar along the Tarim River in NW China. *Forests* **2024**, *15*, 421. [CrossRef]
12. Li, W.; Wang, J.; Jiang, L.; Lv, G.; Hu, D.; Wu, D.; Yang, X. Rhizosphere Effect and Water Constraint Jointly Determined the Roles of Microorganism in Soil Phosphorus Cycling in Arid Desert Regions. *Catena* **2023**, *222*, 106809. [CrossRef]
13. Zhu, M.; Jiang, L.; Wu, D.; Li, W.; Yang, H.; He, X. Effects of Water Control and Nitrogen Addition on Functional Traits and Rhizosphere Microbial Community Diversity of *Haloxylon ammodendron* Seedlings. *Forests* **2023**, *14*, 1879. [CrossRef]
14. Wang, A.; Ma, R.; Ma, Y.; Niu, D.; Liu, T.; Tian, Y.; Dong, Z.; Chai, Q. Soil Bacterial Community Structure and Physicochemical Influencing Factors of Artificial *Haloxylon ammodendron* Forest in the Sand Blocking and Fixing Belt of Minqin, China. *Forests* **2023**, *14*, 2244. [CrossRef]
15. Yao, C.; Wang, Y.; Yang, G.; Xia, B.; Tong, Y.; Yao, J.; Chen, H. Spatial and Temporal Variation in Vegetation Response to Runoff in the Ebinur Lake Basin. *Forests* **2023**, *14*, 1699. [CrossRef]
16. Li, L.; Zayiti, A.; He, X. Evaluating the Stand Structure, Carbon Sequestration, Oxygen Release Function, and Carbon Sink Value of Three Artificial Shrubs alongside the Tarim Desert Highway. *Forests* **2023**, *14*, 2137. [CrossRef]

Disclaimer/Publisher's Note: The statements, opinions and data contained in all publications are solely those of the individual author(s) and contributor(s) and not of MDPI and/or the editor(s). MDPI and/or the editor(s) disclaim responsibility for any injury to people or property resulting from any ideas, methods, instructions or products referred to in the content.

Article

Foliar Water Uptake and Its Relationship with Photosynthetic Capacity and Anatomical Structure between Female and Male *Populus euphratica* at Different Growth Stages

Zhoukang Li [1,2,3], Yudong Chen [1,2,3], Huimin Wang [1,2,3] and Xueni Zhang [1,2,3,*]

[1] College of Ecology and Environment, Xinjiang University, Urumqi 830017, China; lzkeco@163.com (Z.L.); cyd666@stu.xju.edu.cn (Y.C.); whm@stu.xju.edu.cn (H.W.)
[2] Key Laboratory of Oasis Ecology of Education Ministry, Xinjiang University, Urumqi 830017, China
[3] Xinjiang Jinghe Observation and Research Station of Temperate Desert Ecosystem, Ministry of Education, Jinghe 833300, China
* Correspondence: xnzhang@xju.edu.cn

Abstract: Foliar water uptake (FWU) is considered to be a common phenomenon in most terrestrial plants. As a supplementary water source, it plays an important role in the growth and survival of plants in arid areas. However, there is no research to explain the water absorption of plant leaves from the perspective of gender specificity. To this end, we carried out a leaf water absorption capacity experiment and in situ wetting field experiment, respectively, in the early (Initial), middle (Mid) and end (End) of the growth season of male and female *Populus euphratica*. The results of the leaf water absorption capacity experiment showed that the FWU capacity of male and female *P. euphratica* showed an increasing trend with the growth period and reached the maximum at the End period. The FWU capacity of female *P. euphratica* was significantly greater than that of male *P. euphratica* after the Initial stage. The water absorption speed (k) of male and female leaves also increased with the growth period, but the increase was not significant. The increase in leaf water content per mg of water absorbed per unit of leaf area (LWC_A) of male *P. euphratica* was always greater than that of female *P. euphratica*. Specific leaf area (SLA), leaf water saturated deficit (WSD) and water absorption parameters (FWU capacity, k) were significantly correlated. The results of the in situ wetting field experiment show that humidification significantly increased the predawn water potential (Mid period) of female and male *P. euphratica* leaves and the net photosynthetic rate (Mid period) of male *P. euphratica* leaves, but had no significant effect on chlorophyll fluorescence parameters and anatomical structure. The MFA results show that the water status of male and female *P. euphratica* leaves was significantly correlated with photosynthetic parameters, fluorescence parameters and anatomical parameters. Our results show that the foliar water uptake capacity of female *P. euphratica* leaves was stronger than that of male *P. euphratica* and shows significant dynamic changes during the growing season. This was because female *P. euphratica* has a developed water storage structure. Foliar water uptake can effectively improve the water status and photosynthetic capacity of male and female *P. euphratica*, and this improvement was more significant during the most intense period of soil water stress. These findings will deepen our understanding of the ecological adaptation of dioecious plants to foliar water uptake.

Keywords: Tugai forest; arid area; dioecy; wet leaves; physiological

Citation: Li, Z.; Chen, Y.; Wang, H.; Zhang, X. Foliar Water Uptake and Its Relationship with Photosynthetic Capacity and Anatomical Structure between Female and Male *Populus euphratica* at Different Growth Stages. *Forests* **2023**, *14*, 1444. https://doi.org/10.3390/f14071444

Academic Editor: José Javier Peguero-Pina

Received: 9 June 2023
Revised: 9 July 2023
Accepted: 12 July 2023
Published: 13 July 2023

Copyright: © 2023 by the authors. Licensee MDPI, Basel, Switzerland. This article is an open access article distributed under the terms and conditions of the Creative Commons Attribution (CC BY) license (https://creativecommons.org/licenses/by/4.0/).

1. Introduction

The plant canopy can intercept rain, fog, dew etc., resulting in wet leaves [1–3]. The most direct benefit of leaf wetting is water entering the leaf through the absorption of the water accumulated on the leaf surface, which is called foliar water uptake. It can be said that leaf water absorption is a common phenomenon in most terrestrial plants, because this phenomenon is observed in more than 85% of studied species [1,4]. As a complementary water source, although plant leaves absorb less water, this phenomenon still increases leaf water potential [5,6], promotes plant growth [7,8] and improves photosynthetic capacity [9,10].

There are 14,620 species of dioecious plants, which are an important part of terrestrial ecosystems and play an important role in maintaining the stability of regional ecosystems [11]. Studies have shown that the specific adaptation of dioecious plants to soil water demand during the growth period leads to changes in leaf morphology [12,13], structure [14,15] and physiology [16–18]. Generally, under drought conditions, female plants show higher sensitivity than male plants, which may be related to the less conservative water-use strategy of the former [19,20], but, when water is sufficient, the physiological differences between the sexes disappear [10,21]. Therefore, we infer that the secondary sexual characteristics of dioecious plants during the growth period have different responses to leaf water absorption, which is more significant in arid areas.

The water absorbed by the leaves can enter the palisade tissue, sponge tissue and epidermal cell wall, and these anatomical characteristics profoundly affect the leaf water absorption capacity of plants [22–24]. Boanares et al. [25] found that *Leandra australis* and *Myrcia splendens* had a lower leaf water uptake, because the relatively loose sponge tissue in *Leandra australis* and the higher number of secretory glands in *Myrcia splendens* reduce the number of parenchyma cells and water storage vacuoles. Dioecious plants exhibit significant differences in their morphological structure [15,26]. For example, the epidermal thickness, palisade tissue thickness and ratio of palisade tissue to spongy tissue of male *Podocarpus macrophyllus* leaves were greater than those of female plants, showing strong water retention capacity and drought resistance [27]. Combined with the above research, we believe that the morphological structure of dioecious plants is not only an important factor restricting the water absorption capacity of leaves, but also that the differences between male and female plants may lead to different leaf water absorption strategies.

The main pathway of water absorption in plant leaves is the stomata, and the value of water absorption flux is usually consistent with the diffusion of water vapor into the stomata under environmental conditions. Therefore, the water vapor flux into leaves may be partially or completely explained by transpiration, which is often called 'reverse transpiration' [28,29]. This theory redefines the leaf water absorption process and links plant leaf water absorption with the gas exchange process. Our previous studies showed that humidification significantly increased the water status of the assimilating branches of *Calligonum mongolicum* and effectively improved the photosynthetic capacity [10]. However, in other previous studies, it was found that leaf wetting was not beneficial to plant photosynthesis, which inhibited stomatal opening and reduced the carbon assimilation rate [30] or affected the carboxylation of Rubisco [31]. In arid environments, the photosynthetic capacity of female *Populus cathayana* was more limited than that of male *P. cathayana* [17,32], but can leaf water input alleviate this photosynthetic limitation of dioecious plants? This question remains to be answered.

Populus euphratica Oliv. is a dioecious tree species with strong drought resistance. It is widely distributed in arid and semi-arid desert areas and plays an important role in water conservation, windbreaking and sand fixation, stabilizing the regional climate and promoting ecosystem functions [33]. A previous study of the research team found that the proportion of dew in the water utilization sources of *P. euphratica* reached 1% [34]. Later, some scholars found that *P. euphratica* can transport the absorbed water to the root system to alleviate root stress [35,36]. These results provide a theoretical basis for this study. In general, this paper takes the dioecious plant *P. euphratica* as the research object, based on the leaf water absorption experiment and wetting experiment at different stages of the growth period, in order to clarify the water absorption characteristics of male and female *P. euphratica* leaves and the structure and photosynthetic physiological response mechanism of *P. euphratica* leaves under humidification conditions. This study will help to clarify the strategies of plant water use in arid areas and contribute to our understanding of plant water use patterns. Therefore, we propose the following two scientific hypotheses:

(1) The water absorption capacity of *P. euphratica* leaves exhibits gender differences, and different water absorption strategies are shown in each growth stage.

(2) The response characteristics of the leaf structure and photosynthetic physiology of male and female *P. euphratica* to wetting are different and show different rules in the growth stage.

2. Materials and Methods

2.1. Study Area and Plant Materials

The Ebinur Lake Wetland National Nature Reserve (82°36′–83°50′ E, 44°30′–45°09′ N) is located in Jinghe County, Bortala Mongol Autonomous Prefecture, Xinjiang, China. The annual precipitation (100–200 mm) is much lower than the potential evapotranspiration (1500–2000 mm). The average annual precipitation distribution is uneven, with more in summer and less in winter. The annual average temperature is 6–8 °C, the extreme maximum temperature reaches approximately 44 °C, and the extreme minimum temperature is −33 °C (Figure S1). The climate is extremely arid and represents a typical temperate continental arid climate [37]. The flora in the basin belongs to the Junggar Desert sub-region of the northern desert sub-region of the Palaearctic Mengxin District. There are a large number of Tugai forest communities composed of desert plants distributed along the lakes and riverbanks [38].

2.2. Plant Material and Environmental Factors

From 28 March to 28 April 2021, according to the differences between sexes in the phenological growth of *P. euphratica*, three female and three male *P. euphratica* with similar individual size were selected from the Tugai forest along the Aqikesu River in the upper reaches of the Ebinur Lake Basin and marked as "Female" and "Male", respectively (Table S1). The geographical distance between each plant was greater than 100 m, and the distance between the marked plant and the river was approximately 200 m. Then, a VP-4 sensor (Decagon Device Inc., Pullman, WA, USA) was used to measure the environmental factors such as air relative humidity (Rh), air temperature (T_{air}) and pressure (P), and a leaf humidity sensor (LWS, DECAGON, Pullman, WA, USA) was used to measure the leaf wetness. The data were recorded every 10 min and stored in the EM 50 data collector (EM 50, Decagon, Pullman, WA, USA) (Table 1).

Table 1. The characteristics of the environmental factors during the experiment period.

	Rh/%	**T_{air}/°C**	**P/kPa**	**LWS**
Initial	0.42 ± 0.08 b	19.95 ± 3.57 c	97.85 ± 0.43 a	440.08 ± 17.11 b
Mid	0.40 ± 0.10 ab	26.05 ± 2.88 a	97.08 ± 0.42 b	459.39 ± 9.14 a
End	0.47 ± 0.07 a	22.58 ± 3.62 b	97.64 ± 0.25 a	452.31 ± 8.18 ab

Note: Different lowercase letters indicate significant differences in different growth stages (ANOVA, $p < 0.05$).

The female and male *P. euphratica* in the growing season were divided into three categories according to the growth period, namely the early growth period (Initial, 16 May 2021–23 May 2021), the middle growth period (Mid, 27 July 2021–3 August 2021) and the end growth period (End, 7 September 2021–14 September 2021). The experiment was carried out in these three time periods, and the contents of the experiments were consistent. Therefore, the first description of the experiment content was taken as an example, and the next were not repeated. The experimental schedule is shown in Table S2.

2.3. Leaf Water Absorption Experiment

According to the experimental method proposed by Liang et al. [39], the leaf water absorption capacity (FWU capacity) and water absorption speed (k) were obtained. The experiment was conducted from 17 May 2021 to 18 May 2021 on female and male *P. euphratica*.

Sampling steps: Healthy twigs with fresh leaves were removed by pruning and quickly brought back to the laboratory. The average temperature in the laboratory was about 24 °C, the air relative humidity was about 35%, and there was no wind. From each tree of the same sex, 3 twigs were selected as a repeat. Determination steps: The leaves were cut from the twigs and randomly divided into 14 parts, with 4 leaves per part. The fresh weight of each leaf was quickly weighed with a precision balance (AL204, METTLER TOLEDO, Shanghai, China), the result was recorded as the initial mass (IW, g), and each leaf was photographed to calculate the leaf area (A, cm^2). The experiment was divided into three stages: for the first two hours, leaves were weighed every 15 min; for the next two hours, leaves were weighed every 30 min; and for the last two hours, leaves were measured every 1 h. The experiment ran for a total of 6 h, collecting data from 14 moments. During the water absorption process, the petiole was sealed with paraffin to prevent water from entering, and then 14 leaves were immersed in 14 cups containing deionized water in turn, and the petiole was fixed on the cup wall to avoid contact between the petiole and water. During each weighing, the leaves removed from the water cup were dried with a dry paper towel, and re-determined the weight after water absorption (FW, g). The fresh weight of the leaves at 6 h of immersion was used as the saturated weight after water absorption and recorded as (SW, g). After the end of the water absorption experiment, the leaves were dried in the oven for 48 h, and then the weight after drying was weighed and recorded as (DW, g). Some indicators refer to Schreel et al. (2019) [40].

The total amount of leaf water absorption (TFWU, g·g^{-1}) is

$$\text{TFWU} = \frac{\text{SW} - \text{IW}}{\text{DW}} \tag{1}$$

Leaf water absorption per unit area (FWU capcacity, mg·cm^{-2}) is

$$\text{FWU}_{\text{capacity}} = \frac{\text{SW} - \text{IW}}{\text{A}} \tag{2}$$

Leaf water content (LWC, %) is calculated as follows:

$$\text{LWC}_{\text{I}} = \frac{\text{IW} - \text{DW}}{\text{DW}} \tag{3}$$

$$\text{LWC} = \frac{\text{FW} - \text{DW}}{\text{DW}} \tag{4}$$

$$\text{LWC}_{\text{S}} = \frac{\text{SW} - \text{DW}}{\text{DW}} \tag{5}$$

$$\frac{\text{LWC} - \text{LWC}_{\text{I}}}{\text{LWC}_{\text{S}} - \text{LWC}_{\text{I}}} = 1 - e^{-kt} \tag{6}$$

In the above formula, LWC_{I} is the initial leaf water content, LWC_{S} is the saturated leaf water content, LWC is the leaf water content at a certain moment, k is the water absorption speed and t is the water absorption time.

The difference in leaf water content (ΔLWC, %) is

$$\Delta\text{LWC} = \frac{\text{SW} - \text{DW}}{\text{SW}} \times 100 - \frac{\text{IW} - \text{DW}}{\text{IW}} \times 100 \tag{7}$$

The increase in leaf water content per mg of water absorbed per unit of leaf area (LWC_{A}; %·cm^2·mg^{-1}) is

$$\text{LWC}_{\text{A}} = \frac{\text{LWC}}{\text{FWU}_{\text{capacity}}} \tag{8}$$

The increase in FWU capacity per MPa decrease in Ψ_{leaf} (FWU$_i$, mg·cm^{-2}·MPa^{-1}) is

$$\text{FWU}_i = \frac{\text{FWU}_{\text{capacity}}}{\Psi_{\text{leaf}}} \tag{9}$$

The potential relative importance of FWU to alleviate drought stress at leaf level (PRI; %) is

$$\text{PRI} = \text{LWC}_A \times \text{FWU}_{\text{max}} \tag{10}$$

The actual relative importance of FWU to alleviate drought stress (ARI; %·MPa^{-1}) is

$$\text{ARI} = \text{LWC}_A \times \text{FWU}_i \tag{11}$$

2.4. In Situ Wetting Experiment

(1) Experimental design

According to the selected female and male *P. euphratica*, two parts of healthy branches with leaves were selected for each tree and marked as treatment (TR) and (CK), respectively. The in situ wetting experiment began on 16 May 2021. The leaves on TR branches were sprayed with distilled water at predawn to ensure that the surface of the leaves was moist, and the wetting effect lasted for 1 h. Branch leaves marked as CK were not treated for natural control. The above humidification process was repeated every day for the next 6 days, and the number of humidification days throughout the experimental period was 7. The test method is based on Cavallaro et al. [41].

(2) Determination of leaf water potential at predawn and midday

On 23 May 2021 (the last day of the experimental period), leaf water potential samples of female and male *P. euphratica* under TR and CK treatments were collected at predawn (4:00, local time) and midday (10:00) on that day. The specific collection method was to use branch scissors to remove the twigs with leaves of female and male *P. euphratica* under the wetting and control treatments and immediately put them in a self-sealing bag before placing them in an incubator (4 °C) for preservation, and then quickly bring them back to measure the leaf water potential with a dew point water potential meter (WP4C, Decagon Device Inc., Pullman, WA, USA). From each tree, three twigs were cut as repeats.

(3) Determination of photosynthetic and chlorophyll fluorescence parameters

In order to explore the response of the photosynthetic physiology and fluorescence physiology of female and male *P. euphratica* leaves to humidification, at midday (10:00, local time) on 23 May 2021, the net photosynthetic rate (P_n, µmol CO_2·m^{-2}·s^{-1}), stomatal conductance (g_s, mol H_2O·m^{-2}·s^{-1}), intercellular carbon dioxide concentration (C_i, µmol CO_2·m^{-2}·s^{-1}) and transpiration rate (T_r, mmol H_2O·m^{-2}·s^{-1}) of the leaves of female and male *P. euphratica* were measured using a photosynthetic measurement system (LI-6400XT, Li-COR, Inc., Lincoln, NE, USA). Water use efficiency (WUE, µmol·mmol^{-1}) is the ratio of net photosynthetic rate to transpiration rate. Three healthy leaves were selected for each *P. euphratica*, and each leaf was measured 10 times. After the determination of photosynthetic parameters, the labeled leaves were wrapped with tinfoil paper for 30 min of dark adaptation. After the dark adaptation, the chlorophyll fluorescence parameters of each *P. euphratica* were measured. These parameters include minimum initial fluorescence (F_o), maximum fluorescence under dark adaptation (F_m), maximal photochemical efficiency (F_v/F_m), minimum fluorescence under light adaptation ($F_o{}'$), maximum fluorescence under light adaptation ($F_m{}'$), light energy capture efficiency in light system II ($F_v{}'/F_m{}'$), photochemical quenching coefficient (qP) and electron transfer efficiency (ETR).

(4) Leaf anatomical sample collection and determination

After the determination of photosynthetic and chlorophyll fluorescence parameters, we collected anatomical samples, and the samples of *P. euphratica* leaves were collected using a mixed sampling method. A total of 15 female *P. euphratica* leaves were collected (5 leaves were collected for each female *P. euphratica* individual), and 15 leaves were fully mixed and divided into 3 samples as repetition; then, each sample collected was immediately placed in a sample bottle containing FAA fixed solution (70% ethanol 90 mL + formaldehyde 5 mL + acetic acid 5 mL). Then, the samples for dissection were taken to the laboratory for the determination of anatomical parameters such as leaf thickness (LT), upper and lower epidermal cell thickness (TUE; TLE), upper and lower palisade tissue thickness (TUP; TLP) and sponge tissue thickness (TS).

2.5. Statistical Analysis

Before data analysis, a normality test was carried out. If it did not conform to the normal distribution, logarithmic transformation was carried out to meet the requirements of data analysis. An independent-sample *t*-test was used to test the difference in the water relationship index between male and female *P. euphratica*, and a paired-sample *t*-test was used to test the differences in the same sex in different growth periods.

General linear mixed models (GLMMs) were built to evaluate the influence of FWU capacity on oxidative response related to leaf traits. The leaf traits and sex were used as fixed explanatory variables, and the response variables were water uptake per unit area (FWU capacity) and water uptake speed (k). The identity of each measured plant was used as a random effect variable to control for potential pseudo-replication biases in our models.

Three-way repeated-measures analysis of variance was used to evaluate gender (S), time (T), treatment (W) and their interaction effects. The Kolmogorov–Smirnov test was used to determine whether the data obeyed the normal distribution. The spherical test was used to analyze whether the interaction term exhibited any interaction. If the outcome of the spherical test was $p > 0.05$, the spherical hypothesis was satisfied. If $p < 0.05$, the spherical hypothesis was not satisfied. At this time, ε needed to be corrected. When $\varepsilon < 0.75$, the Greenhouse–Geisser method was used for correction; when $\varepsilon > 0.75$, the Huynh–Feldt method was used for correction [42], and the Bonferroni *test* was used for multiple comparisons. The differences between the 12 groups were analyzed by one-way ANOVA.

Multivariate factor analysis was performed using the FactoMineR package in R (R v.4.2.1, http://cran.rproject.org). Taking the sex of *P. euphratica* as the classification variable and using water potential data, photosynthesis parameters, the chlorophyll fluorescence parameter and anatomic parameter data as the variable set, the relationship between the variable sets of male and female *P. euphratica* are discussed.

The above analysis was performed in SPSS 19.0 (IBM, Chicago, IL, USA) and R (R v.4.2.1, http://cran.rproject.org). Charts were produced in Origin 2021 (OriginLab, Northampton, MA, USA), Excel 2013 (Microsoft, Redmond, WA, USA) and Visio 2010 (Microsoft, Redmond, WA, USA). The data are expressed as mean + standard deviation.

3. Results

3.1. Dynamics of Foliar Water Uptake

With the growth period, the FWU capacity of male and female *P. euphratica* showed an increasing trend, reaching the maximum at the End period. The FWU capacity of female *P. euphratica* showed significant differences in the three periods ($p < 0.05$). The FWU capacity of male *P. euphratica* in the End period was significantly greater than in the Initial and Mid periods ($p < 0.05$). The difference in FWU capacity between male and female *P. euphratica* occurred only during the Mid and End periods and reached the maximum at the End period (Figure 1A). The k of male and female leaves also increased with the growth period, but the increase was not significant ($p > 0.05$). In each period, there was little difference in water absorption rate between male and female leaves (Figure 1B).

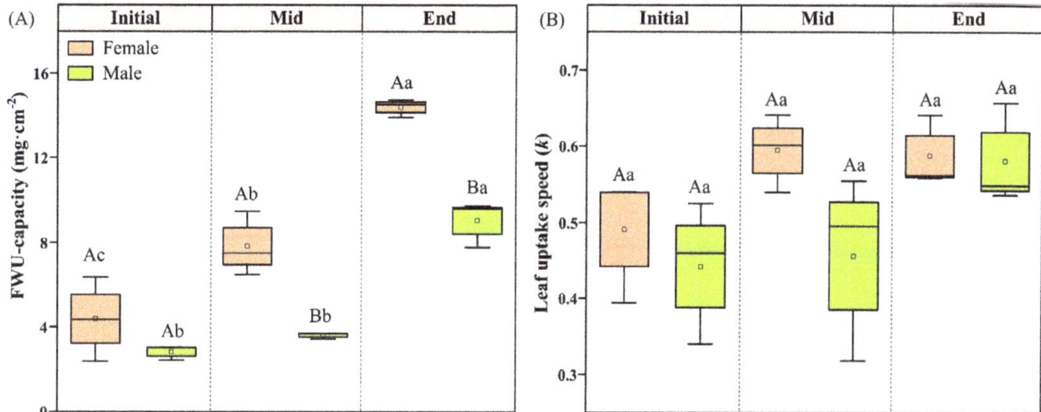

Figure 1. FWU capacity (**A**) and leaf uptake speed (**B**) of female and male *P. euphratica* in different growth stages. Note: Capital letters indicate significant gender differences between males and females at the same growth stage (independent-sample *t*-test, $p < 0.05$), and the lowercase letters indicate significance differences in the same sex in different growth stages (paired-sample *t*-test, $p < 0.05$).

The TFWU and FWU$_i$ of female *P. euphratica* were significantly higher than those of male *P. euphratica* in each period ($p < 0.05$) (Table 2). Those of both male and female *P. euphratica* in the End period were significantly higher than those in the Initial and Mid periods ($p < 0.05$), and there was no significant difference between the Initial and Mid periods ($p > 0.05$). There was no significant gender difference in ΔLWC between female *P. euphratica* and male *P. euphratica* in each period ($p > 0.05$). After the initial stage of growth, the LWC$_A$ of male *P. euphratica* was always significantly higher than that of female *P. euphratica* ($p < 0.05$). The PRI and ARI of male and female *P. euphratica* showed significant differences only in the Mid period ($p < 0.05$). The PRI of female *P. euphratica* in the Initial period was significantly lower than that in the Mid and End periods, while the PRI of male *P. euphratica* increased with the growth period, and there were significant differences between each growth period ($p < 0.05$). The ARI of male and female *P. euphratica* in the End period was significantly higher than that in the Initial and Mid periods ($p < 0.05$) (Table 2).

Table 2. The parameters of leaf water uptake in different growth stages of female and male *P. euphratica*.

Time	Sex	TFWU (g·g^{-1})	ΔLWC (%)	LWC$_A$ (%·cm^2·mg^{-1})	FWU$_i$ (mg·cm^{-2}·MPa^{-1})	PRI (%)	ARI (%·MPa^{-1})
Initial	Female	0.48 ± 0.22 Ab	0.49 ± 0.25 Ab	0.11 ± 0.03 Ab	5.54 ± 2.52 Ab	0.70 ± 0.17 Ab	0.62 ± 0.32 Ac
	Male	0.25 ± 0.03 Bb	0.35 ± 0.05 Ab	0.12 ± 0.03 Ab	3.15 ± 0.39 Bb	0.38 ± 0.10 Ac	0.38 ± 0.05 Ab
Mid	Female	0.50 ± 0.07 Ab	6.95 ± 2.58 Aa	0.87 ± 0.17 Ba	7.60 ± 0.95 Ab	8.28 ± 1.59 Aa	6.67 ± 1.72 Ab
	Male	0.29 ± 0.01 Bb	4.21 ± 0.22 Ab	1.17 ± 0.09 Aa	2.40 ± 0.12 Bb	4.68 ± 0.35 Bb	2.80 ± 0.10 Bb
End	Female	1.00 ± 0.04 Aa	10.13 ± 3.70 Aa	0.71 ± 0.27 Ba	20.58 ± 0.62 Aa	9.83 ± 3.75 Aa	14.47 ± 5.28 Aa
	Male	0.71 ± 0.08 Ba	9.00 ± 1.15 Aa	0.99 ± 0.01 Aa	11.33 ± 1.36 Ba	9.69 ± 0.11 Aa	11.25 ± 1.44 Aa

Note: Capital letters indicate significant gender differences between males and females at the same growth stage (independent-sample *t*-test, $p < 0.05$), and the lowercase letters indicate significance differences in the same sex in different growth periods (paired-sample *t*-test, $p < 0.05$).

The SLW was significantly correlated with FWU capacity and *k* ($F = 12.515$, $p = 0.003$; $F = 5.281$, $p = 0.035$); in addition to SLW, WSD was also significantly correlated with FWU capacity and *k* ($F = 38.372$, $p < 0.001$; $F = 7.472$, $p = 0.015$). SLA and LWC only had significant effects on FWU capacity ($F = 10.652$, $p = 0.005$; $F = 7.591$, $p = 0.014$) (Table 3).

Table 3. Correlation analysis between leaf traits and water absorption parameters.

Independent Variable	Dependent Variable	F	p
SLA	FWU capacity	10.652	0.005 **
	k	3.918	0.065
SLW	FWU capacity	12.515	0.003 **
	k	5.281	0.035 *
LWC	FWU capacity	7.591	0.014 *
	k	4.291	0.055
WSD	FWU capacity	38.372	0.000 ***
	k	7.472	0.015 *

Note: General linear mixed models (GLMMs) were used for the biotic independent variables evaluated. ***, $p < 0.001$; **, $p < 0.01$; and *, $p < 0.05$.

3.2. The Effects of Wetting on Water Potential

Single factors (T, S and W), two-factor interactions (T × S, T × W and S × W) and the three-factor interaction (T × S × W) had significant effects on the Ψ_{Pre} of *P. euphratica* leaves. Further analysis of the effects of various factors on the interaction term found that the interaction effect of S × W in the End period was significant ($F_{1,2} = 46.443$, $p = 0.021$), and the Ψ_{Pre} of the control and treatment was significantly different for female *P. euphratica*. Under the wetting treatment conditions, there was a significant difference in Ψ_{pre} between the Initial period and the End period ($F_{1,2} = 155.769$, $p = 0.006$). There were significant differences in Ψ_{Pre} between female and male *P. euphratica* during the Mid period. T and S had significantly separate effects on Ψ_{Mid} ($F_{2,4} = 145.004$, $p < 0.001$; $F_{1,2} = 53.353$, $p = 0.018$). The interaction between T and S during the Initial period had a significant main effect on the Ψ_{Mid} of *P. euphratica* ($F_{2,4} = 49.787$, $p = 0.001$), but there was no significant difference between the female and male *P. euphratica* treatments and the control (Table 4).

Table 4. The leaf water potential of female and male *P. euphratica* in the wetting experiment.

Time	Sex	Treatment	Ψ_{Pre} (MPa)	Ψ_{Mid} (MPa)
Initial	Female	CK	-1.25 ± 0.14 bc	-1.85 ± 0.28 bc
		TR	-1.3 ± 0.14 cd	-1.70 ± 0.23 b
	male	CK	-1.16 ± 0.19 a	-1.22 ± 0.14 a
		TR	-1.20 ± 0.19 ab	-1.26 ± 0.12 a
Mid	Female	CK	-1.76 ± 0.13 h	-1.89 ± 0.23 cd
		TR	-1.38 ± 0.15 ef	-2.01 ± 0.17 cde
	male	CK	-1.74 ± 0.12 h	-1.84 ± 0.12 bc
		TR	-1.43 ± 0.10 f	-2.01 ± 0.28 cde
End	Female	CK	-1.09 ± 0.19 l	-2.07 ± 0.44 def
		TR	-1.53 ± 0.24 g	-2.17 ± 0.31 ef
	male	CK	-1.35 ± 0.33 de	-2.21 ± 0.20 f
		TR	-1.27 ± 0.10 bc	-2.20 ± 0.43 f

Note: Different lowercase letters indicate statistically significant differences between treatments (one-way ANOVA, $p < 0.05$). T, time effect; S, sex effect; W, wetting effect; T × S, Time × Sex interaction effect; T × W, Time × Wet interaction effect; S × W, Sex × Wet interaction effect; T × S × W, Time × Sex × Wet interaction effect (three-factor repeated-measures ANOVA, $p < 0.05$) (Table S3). The same as below.

3.3. The Effects of Wetting on Plant Photosynthesis and Fluorescence

Among the variation characteristics of P_n in male and female *P. euphratica*, T and W had significant effects on P_n ($F_{2,4} = 326.156$, $p < 0.001$; $F_{1,2} = 25.374$, $p = 0.037$), and there was also an interaction effect ($F_{2,4} = 14.882$, $p = 0.014$), while the gender S factor did not show a significant effect ($F_{1,2} = 1.483$, $p = 0.347$). Further analysis of its individual effects showed that, under CK conditions, the net photosynthetic rates of female and male *P. euphratica* in the Initial period were significantly different from those in the Mid and End periods

($F_{2,4} = 34.927$, $p = 0.042$; $F_{2,4} = 86.637$, $p = 0.013$). There was no significant difference in P_n between the Mid and End period. Under the TR condition, only the Initial and Mid periods showed significant differences. Only male *P. euphratica* showed differences in wetting treatment during the Mid period ($F_{1,2} = 23.430$, $p = 0.040$) (Figure 2A).

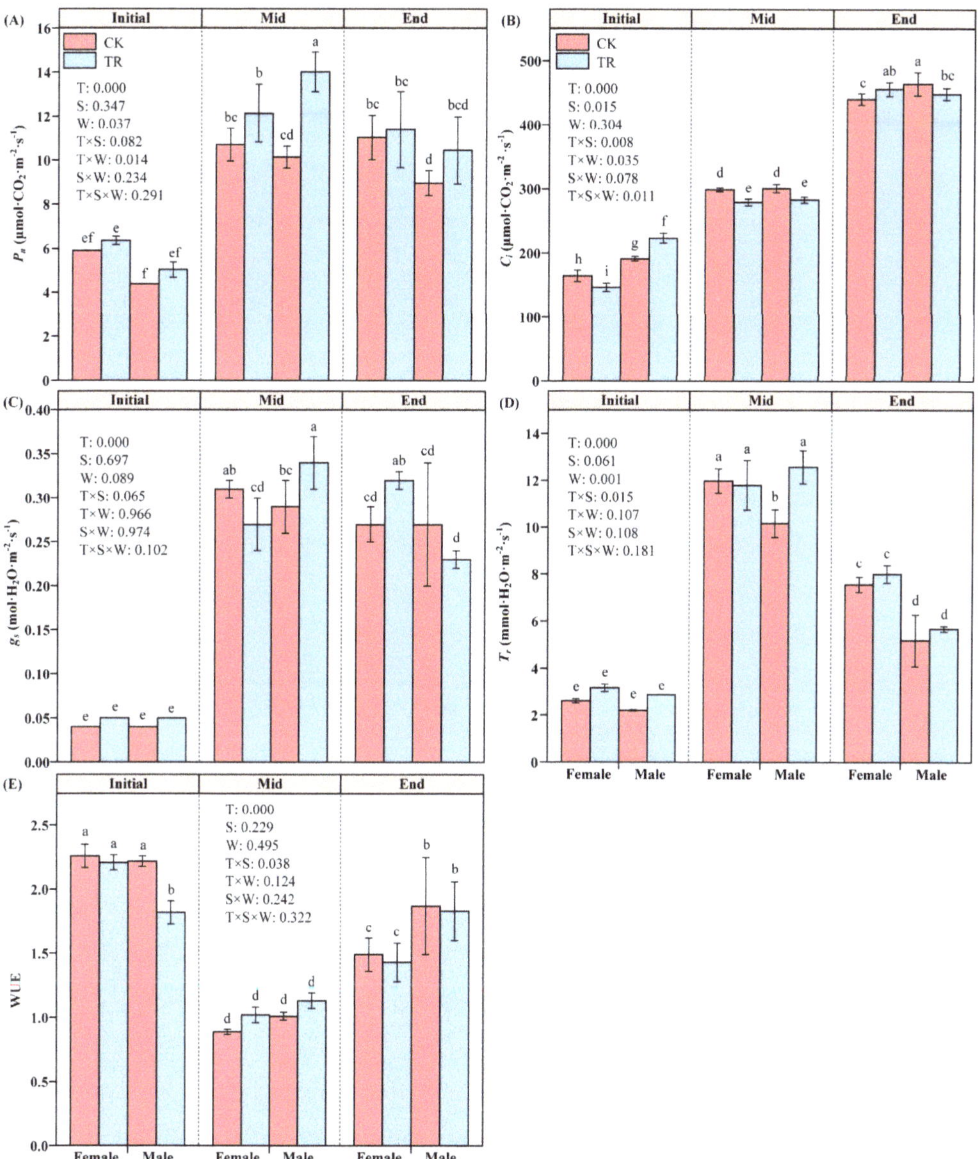

Figure 2. The photosynthetic parameters of female and male *P. euphratica* in the wetting experiment. P_n: net photosynthetic rate (**A**); C_i: intercellular carbon dioxide concentration (**B**); g_s: stomatal conductance (**C**); T_r: transpiration rate (**D**); WUE: water use efficiency (**E**). Note: Different lowercase letters indicate statistically significant differences between treatments (one-way ANOVA, $p < 0.05$) (Table S4).

Among the variation characteristics of C_i in female and male *P. euphratica*, time T and S had significant effects on C_i ($F_{2,4}$ = 1781.667, p < 0.001; $F_{1,4}$ = 65.707, p = 0.015). T × S × W had a significant three-factor interaction effect on C_i ($F_{2,4}$ = 16.967, p = 0.011). T × S and T × W had significant two-factor interaction effects on C_i ($F_{2,4}$ = 19.999, p = 0.008; $F_{2,4}$ = 8.691, p = 0.035). There were significant differences between TR and CK in each period ($F_{2,4}$ = 938.950, p < 0.001; $F_{2,4}$ = 671.835, p < 0.001). The C_i of *P. euphratica* under the TR condition had significant gender differences in the Initial and Mid periods (Figure 2B).

Among the variation characteristics of g_s in male and female *P. euphratica*, only time T had a significant effect on g_s ($F_{2,4}$ = 274.322, p < 0.001). Other factors or the interaction between factors had no significant effect on g_s (Figure 2C). Among the variation characteristics of T_r in male and female *P. euphratica*, T and W had significant effects on T_r ($F_{2,4}$ = 326.156, p < 0.001; $F_{1,2}$ = 25.374, p = 0.037). When *P. euphratica* was under the TR condition, T × S had a significant two-factor interaction effect on T_r ($F_{2,4}$ = 326.156, p < 0.001). Further analysis of its individual effects found that the T_r of male and female *P. euphratica* in the Initial period was significantly different from that in the Mid and End periods. Under the TR condition, the T_r of male and female *P. euphratica* showed a significant difference only at the End stage ($F_{1,2}$ = 82.507, p = 0.012) (Figure 2D). In the variation characteristics of male and female *P. euphratica*, T had a significant single effect on WUE ($F_{2,4}$ = 259.077, p < 0.001). There was a significant two-factor interaction effect of T × S on T_r ($F_{2,4}$ = 8.300, p = 0.038). Further analysis of its individual effect found that under both CK and TR conditions, the WUE of male and female *P. euphratica* showed differences between the Initial period and End period and Mid period (Figure 2E).

In terms of the results for the chlorophyll fluorescence parameters, the time factor T only had a significant single effect on F_o, F_m, F_o', qP and ETR (p < 0.05). The sex factor S had a significant single effect on F_v'/F_m' (p < 0.05), and the treatment factor W had no significant single effect on any fluorescence parameter (Table 5).

There was an interaction effect of T × S on the F_o' of *P. euphratica* under the condition of TR. It was found that the F_o' of *P. euphratica* in the Initial period was significantly different from that in the Mid period ($F_{2,4}$ = 7.628, p = 0.021) (Table 5).

F_m', F_v'/F_m' and qP had significant three-factor interaction effects. Among them, *P. euphratica* F_m' had an interaction with T × S under TR conditions ($F_{2,4}$ = 7.887, p = 0.041). Further analysis of its individual effects found that the F_m' of females at the Initial period was significantly different from that at the Mid period, and the F_m' of females at the Mid and End periods was significantly different, while the Initial and End period had no significant difference (Table 5).

For female *P. euphratica*, there was a significant difference between the treatment and the control only at the Initial stage ($F_{1,2}$ = 43.925, p = 0.022), while for males, there was a significant difference between the treatment and the control at the End stage ($F_{1,2}$ = 206.660, p = 0.005). After wetting treatment, the F_v'/F_m' of female *P. euphratica* was significantly different between the Initial and Mid periods (p < 0.05) (Table 5).

There was interaction between T × S in qP under TR conditions ($F_{2,4}$ = 35.698, p = 0.003). Further analysis of its individual effect showed that there was no significant gender difference in the qP of *P. euphratica* under TR conditions. Comparing the differences between the growth periods, it was found that the qP of female and male *P. euphratica* in the Initial period showed significant differences from that in the Mid period ($F_{2,4}$ = 9.854, p = 0.025; $F_{2,4}$ = 16.309, p = 0.044) (Table 5).

Table 5. The fluorescence parameters of female and male *P. euphratica* in the wetting experiment.

Time	Sex	Wetting	F_o	F_m	F_v/F_m	$F_o{}'$	$F_m{}'$	$F_v{}'/F_m{}'$	qP	ETR
Initial	Female	CK	318.57 ± 9.26 cde	1328.27 ± 54.36 cde	0.76 ± 0.01 ab	295.53 ± 12.50 bcd	886.57 ± 142.82 abc	0.66 ± 0.05 a	0.24 ± 0.05 cde	92.28 ± 15.00 bc
		TR	295.60 ± 21.82 de	1243.73 ± 159.44 de	0.76 ± 0.02 ab	281.02 ± 31.81 cd	582.14 ± 107.29 cd	0.51 ± 0.05 cde	0.43 ± 0.04 a	126.14 ± 21.27 a
	male	CK	360.98 ± 29.56 bcde	1035.47 ± 3.23 bcde	0.65 ± 0.03 c	273.60 ± 22.84 cd	665.25 ± 125.33 cd	0.58 ± 0.04 abc	0.27 ± 0.00 bcd	88.77 ± 5.87 bc
		TR	274.41 ± 21.88 e	792.05 ± 160.33 e	0.65 ± 0.06 c	260.25 ± 17.18 d	451.40 ± 81.36 c	0.42 ± 0.07 e	0.40 ± 0.08 ab	94.82 ± 6.61 bc
Mid	Female	CK	400.97 ± 101.83 abc	1580.22 ± 46.49 abc	0.75 ± 0.06 ab	325.60 ± 54.30 bcd	839.05 ± 374.39 abc	0.58 ± 0.10 abc	0.21 ± 0.08 de	51.33 ± 14.75 de
		TR	370.01 ± 21.65 bcd	1601.33 ± 162.35 bcd	0.77 ± 0.01 ab	354.59 ± 37.13 b	1036.90 ± 162.63 ab	0.65 ± 0.04 ab	0.12 ± 0.04 e	32.78 ± 9.36 e
	male	CK	325.10 ± 31.65 cde	1246.08 ± 299.30 cde	0.73 ± 0.04 b	353.63 ± 21.58 b	791.90 ± 178.99 bcd	0.54 ± 0.08 bcd	0.19 ± 0.08 de	42.80 ± 15.27 de
		TR	352.30 ± 47.69 bcde	1279.50 ± 367.98 bcde	0.71 ± 0.05 bc	302.85 ± 36.28 bcd	554.50 ± 156.22 cd	0.44 ± 0.08 de	0.29 ± 0.09 abcd	54.61 ± 13.38 de
End	Female	CK	386.26 ± 39.07 abc	1858.62 ± 46.56 abc	0.82 ± 0.04 a	332.66 ± 9.17 bc	705.69 ± 46.34 bcd	0.53 ± 0.04 cde	0.28 ± 0.06 bcd	64.90 ± 15.79 cde
		TR	419.39 ± 42.68 ab	1776.48 ± 151.05 ab	0.76 ± 0.04 ab	309.41 ± 12.30 bcd	650.10 ± 12.92 cd	0.52 ± 0.01 cde	0.43 ± 0.15 a	99.66 ± 37.58 ab
	male	CK	429.63 ± 64.06 ab	1699.51 ± 277.76 ab	0.75 ± 0.02 ab	323.68 ± 43.64 bcd	592.58 ± 139.97 cd	0.44 ± 0.07 de	0.37 ± 0.03 abc	71.15 ± 7.23 bcd
		TR	461.23 ± 50.51 a	1727.73 ± 333.38 a	0.72 ± 0.08 bc	469.98 ± 71.60 a	1138.11 ± 286.62 a	0.58 ± 0.06 abc	0.15 ± 0.08 de	37.61 ± 21.77 e

Note: Different lowercase letters indicate statistically significant differences between treatments (one-way ANOVA, $p < 0.05$) (Table S5).

3.4. The Effects of Wetting on Leaf Anatomical Structure

Time T had significant individual effects on LT, TUE, TLE, TUP, TLP and TS (Figures 3 and 4). Among the variation in the characteristics of LT for male and female *P. euphratica*, T, S and W had significant individual effects on the LT of *P. euphratica* ($F_{2,4}$ = 30.114, p = 0.004; $F_{1,2}$ = 63.500, p = 0.015; $F_{1,2}$ = 58.224, p = 0.017). Secondly, T × S × W had a significant interaction with LT ($F_{2,4}$ = 28.222, p = 0.004). Further analysis found that S × W had a significant interaction effect only in the Initial and Mid periods. In both periods, the following rules were demonstrated: the LT of female and male *P. euphratica* was significantly different under the same treatment, and the LT of *P. euphratica* was also significantly different under different treatments for the same sex (Figure 4A). The two-factor interaction effects of T × S and S × W showed significant effects on the TUP and TS of *P. euphratica* leaves (Figure 4B, E). The difference in TUP between male and female *P. euphratica* under CK conditions was manifested in the Mid and End periods. The difference in TUP between male and female *P. euphratica* under the TR condition only appeared in the End period. The interaction effect of S × W on TS only worked in the Initial period, and the difference between male and female TS was only reflected under CK conditions (Figure 4B).

3.5. Driving Factor Analysis of Wetting Treatment for P. euphratica

In Figure 5A,B, the eigenvalues of axis 1 and axis 2 are 4.03 and 2.06, respectively, and the contribution rates of each variable set in the first and second axes are 42.85% and 21.88%, respectively(Tables S7 and S8). The cumulative contribution rate of the two axes reaches 64.73% (Figure 5A). In Figure 5C,D, the eigenvalues of axis 1 and axis 2 are 3.86 and 2.57, respectively, and the contribution rates of each variable set in the first and second axes are 40.65% and 27.06%, respectively. The cumulative contribution rate of the two axes reached 67.71% (Figure 5B), so the first two axes were selected to explain the relationship between the variable sets. There were significant correlations between leaf water potential and anatomical parameters, fluorescence parameters and photosynthetic parameters (p < 0.01). Since the water potential was negative, the angle of correlation between the indicators demonstrates a negative correlation if it is an acute angle and a positive correlation if it is an obtuse angle. The analysis found that the angle between the predawn water potential of female and male *P. euphratica* and TE in the leaf anatomical parameters was the smallest, and the angle between the predawn water potential and the WUE in the photosynthetic parameters was the smallest. It can be seen that the predawn water potential has a strong correlation with these parameters. Comparing the distribution of the fluorescence parameters of male and female *P. euphratica*, it can be seen that the angle between predawn water potential and F_v/F_m is an obtuse angle, indicating that the two are significantly positively correlated (Figure 5B,D).

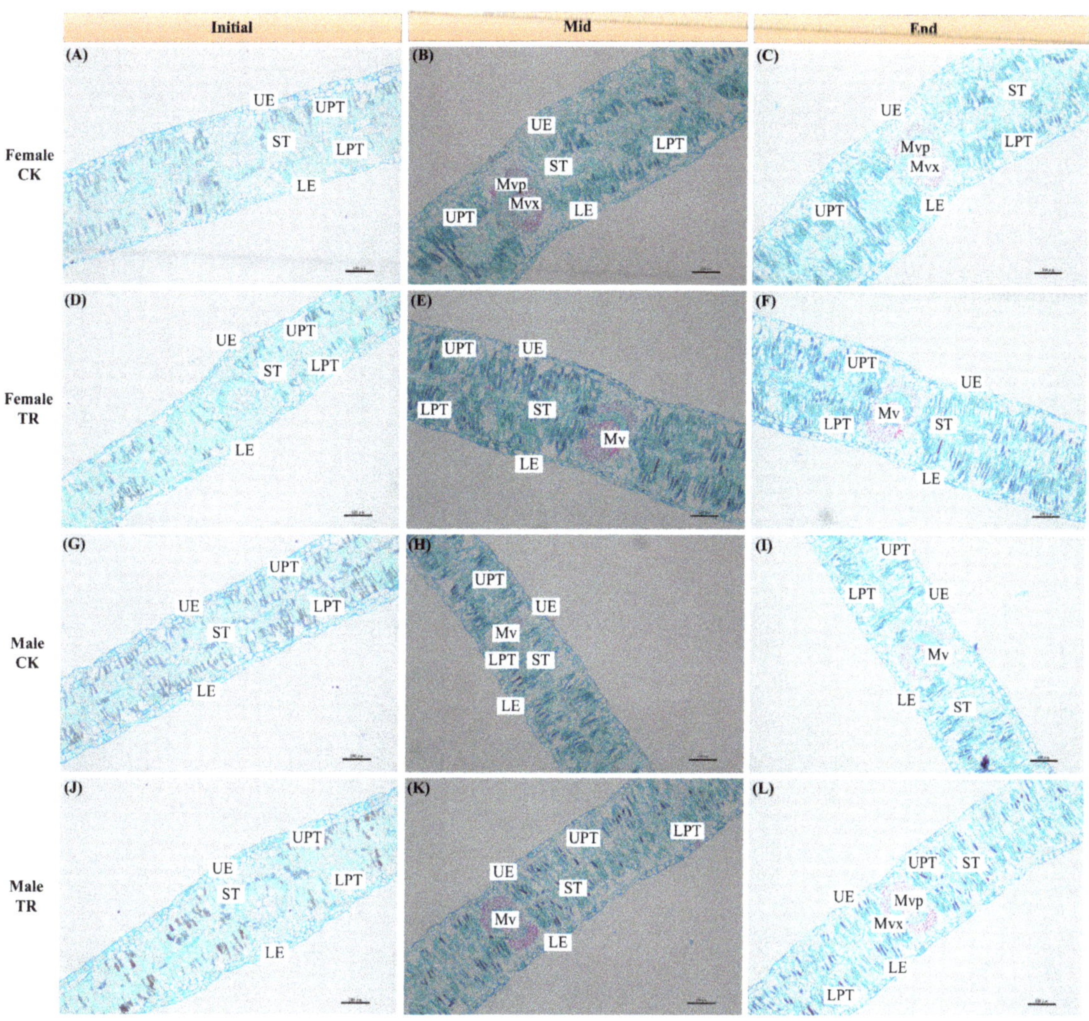

Figure 3. Leaf anatomical structure of female (**A–C**) and male (**G–I**) *P. euphratica* under CK conditions; leaf anatomical structure of female (**D–F**) and male (**J–L**) *P. euphratica* under TR conditions. Ue: Upper epidermal; LE: Lower epidermal; ST: Spongy tissue; UPT: Upper palisade tissue; LPT: Lower palisade tissue; Mv: Main vein; Mvp: Main vein phloem; Mvx: Main vein xylem.

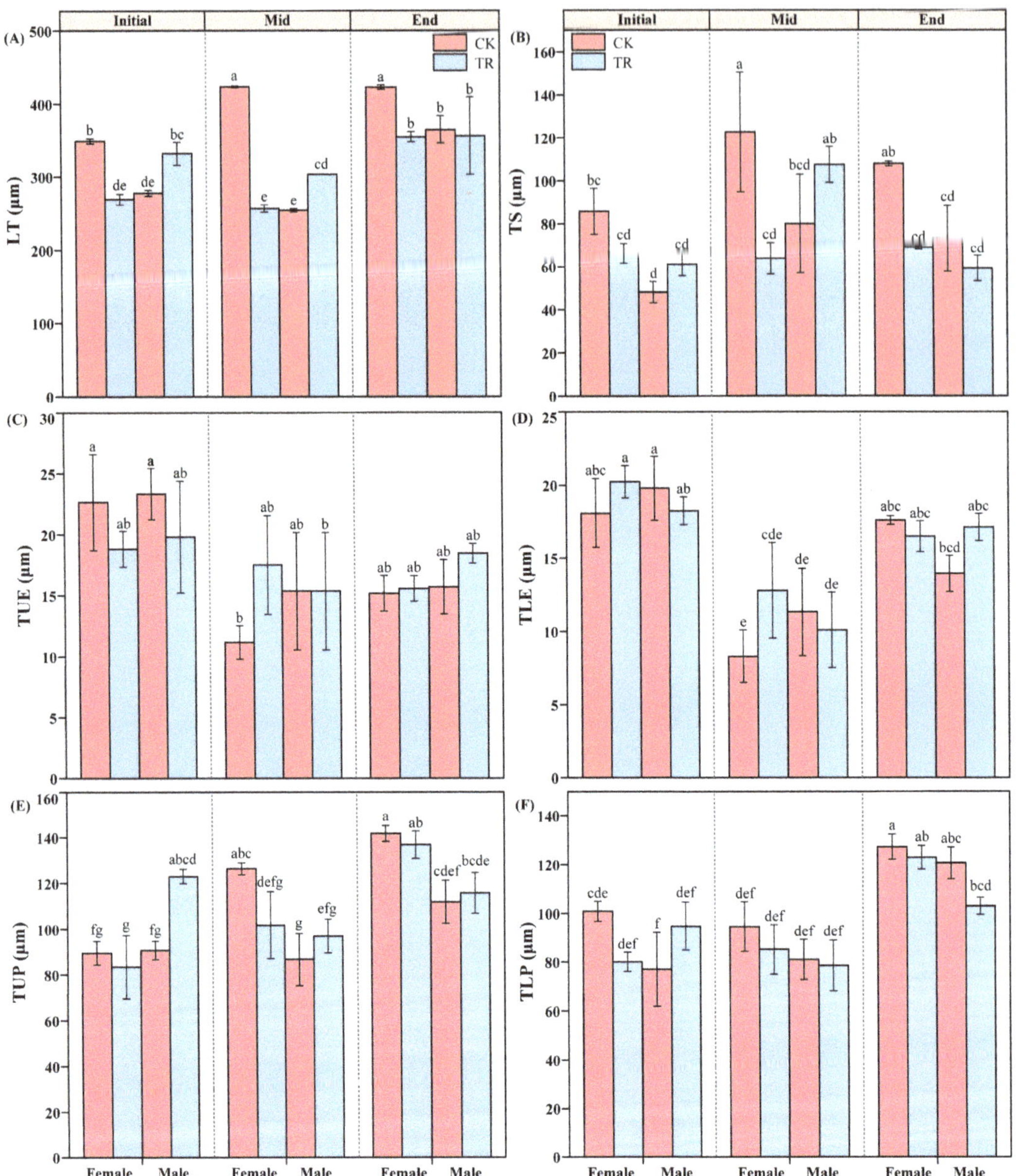

Figure 4. The anatomic parameters of female and male *P. euphratica* in the wetting experiment. LT: leaf thickness (**A**), TS: sponge tissue thickness (**B**), TUE: upper epidermal cell thickness (**C**), TLE: lower epidermal cell thickness (**D**), TUP: upper palisade tissue thickness (**E**), TLP: lower palisade tissue thickness (**F**). Note: Different lowercase letters indicate statistically significant differences between treatments (one-way ANOVA, $p < 0.05$) (Table S6).

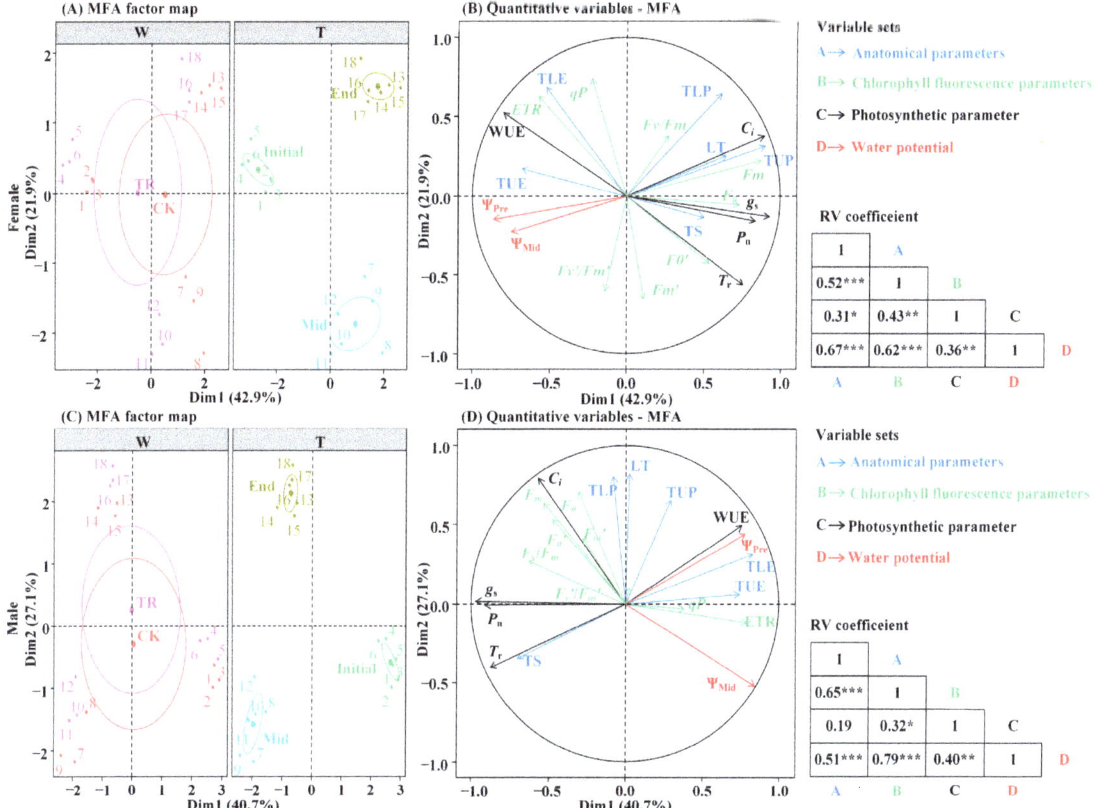

Figure 5. Multivariate factor analysis of variable sets in female and male *P. euphratica*. Note: The individual map of MFA (**A,C**) and the quantitative variables map of MFA (**B,D**). ***, $p < 0.001$; **, $p < 0.01$; and *, $p < 0.05$.

4. Discussion

4.1. Foliar Water Uptake Capacity

As long as the external water potential of the leaf is greater than the internal water potential, water will enter the leaf from the outside [1,2]. The water absorption test conducted for *P. euphratica* leaves sets an ideal condition, which reflects the maximum potential of the water absorption and water absorption rate of *P. euphratica*, so this index is often used to evaluate the capacity of leaf water absorption. The results of the water absorption experiment show that the water absorption capacity of *P. euphratica* leaves increased with the growth period, and the FWU capacity of female *P. euphratica* exhibited a significant time difference. After the Initial period, the FWU capacity of female *P. euphratica* increased significantly, becoming much larger than that of male *P. euphratica* at approximately twice the value; *k* also showed a gradual increase during the growth period, although the difference was not significant. This shows that the water absorption potential of *P. euphratica* leaves increases during the growth period, and the water absorption potential of female *P. euphratica* is stronger than that of male *P. euphratica* in the middle and later growth stages. There are two kinds of water absorption strategies for plants, namely, the storage efficiency of high water absorption and a low water absorption rate, and the speed efficiency of low water absorption and a high water absorption rate [6,25,43]. However, the results from *P. euphratica* presented in this paper show that storage efficiency and speed efficiency can coexist, which is consistent with the results of Boanares et al. [44] on *Senna reniformis*.

The difference in foliar water uptake also leads to significant changes in the degree of drought stress (LWC_A), which reduces FWU at the leaf level [40] (Table 2). Although the water absorption capacity of female *P. euphratica* leaves was stronger than that of male *P. euphratica*, from the change in LWC_A, the water absorption of male *P. euphratica* leaves alleviates the water deficit at the leaf level more strongly than that of female *P. euphratica*. During the Mid period, the atmospheric temperature was the highest and the precipitation was relatively less (Table 1; Figure S1), so the plants suffered the greatest risk of drought stress during this period. The LWC_A of male and female *P. euphratica* in this period was larger than that in the other two periods, indicating that the water absorption strategy of male and female *P. euphratica* leaves showed a high degree of environmental adaptability. Coincidentally, the advantage of male *P. euphratica* in alleviating leaf water deficit was the largest in the Mid period, indicating that male *P. euphratica* had a stronger adaptability to arid environments [45]. After further evaluating the potential (PRI) and actual relative importance (ARI) of FWU in alleviating plant water deficit, it was found that these two parameters of female *P. euphratica* were significantly greater than those of male *P. euphratica*, but this difference was only shown in the Mid period. The results show that FWU was more important than male *P. euphratica* in alleviating the leaf water deficit of female *P. euphratica*, that is, the potential value and actual value of FWU for female *P. euphratica* were higher.

Leaf traits profoundly affect leaf water absorption capacity [7,46,47]. SLW is one of the important indicators used to reflect the morphological characteristics of plant leaves. Plants with a higher SLW have a higher mesophyll density than those with a lower SLW [47], while WSD reflects the state of the leaf water saturation deficit. These two indicators are significantly correlated with FWU capacity and k, respectively. Boanares et al. [43] believed that a higher RWC was the result of a higher water absorption rate. The results reflected by WSD were actually consistent with their conclusions, that is, the higher the degree of leaf water deficit, the greater the water absorption and water absorption rate of leaves. This may be a strategy to avoid the effects of drought.

4.2. Response of Leaf Ecophysiology to Wetting

Studies have shown that FWU improves plant water status [5,48–51]. The effect of wetting on plant Ψ_{Pre} is more significant than for Ψ_{Mid} [10,52], and our results confirm this. Humidification significantly increased the predawn water potential of *P. euphratica* leaves during the Mid period, minimizing the risk of water stress caused by scarce precipitation during this period [10]. Boanares et al. [43] found that there is a relationship between leaf water uptake and plant response to wetting. Specifically, the water uptake rate is negatively correlated with Ψ_{Pre}, and the water uptake is positively correlated with Ψ_{Pre}, because plants with higher leaf water uptake slowly absorb more water, lose less water than plants that absorb it faster and keep more water in their tissues for longer. In this paper, the Ψ_{Pre} of *P. euphratica* did not change due to the gender differences in FWU capacity after humidification, which indicated that the process may be dominated by water absorption rate. This is because rapid water absorption can improve water status more quickly, which is very important for plants in arid areas.

In addition to C_i, there was no significant gender difference in the photosynthetic parameters of *P. euphratica* in each growth stage, but humidification treatment improved the photosynthetic capacity of *P. euphratica* in each growth stage, and this effect reached the maximum in the Mid period. Male *P. euphratica* had a higher carbon benefit in the Mid period. Although the C_i of male *P. euphratica* was smaller in this period, male *P. euphratica* maintained higher P_n through larger stomatal conductance [40,53]. The transpiration of *P. euphratica* was very strong in the Mid period, and WUE was significantly lower than that in the other two periods. It can be seen that the WUE of male and female *P. euphratica* leaves was improved after humidification induction, although the change was not significant. Humidification has a compensatory effect on the carbon assimilation process of *P. euphratica*. Improved leaf water status can effectively alleviate the water loss caused by transpiration [43], which is beneficial to *P. euphratica*.

Plant chlorophyll fluorescence parameters are helpful to understand the photosynthetic mechanism of plants against environmental stress, and are good indicators reflecting the relationship between plant photosynthesis and environmental stress [54,55]. F_v/F_m is the conversion efficiency of intrinsic light energy, which can reflect the ability of light system II to use light energy. The higher the value, the less likely it is that photoinhibition occurs. Under non-stress physiological conditions, the F_v/F_m of most plants is between 0.8 and 0.85, and the lower the value, the higher the degree of stress [52,56]. In addition to the End period, the F_v/F_m value of *P. euphratica* under the control condition reached 0.82, and the F_v/F_m values of *P. euphratica* in other growth stages were less than this interval, indicating that the photosynthetic organs of both were subject to different degrees of stress. After comparing the results for the fluorescence parameters, it was found that humidification did not significantly improve the light energy utilization process of plants.

Although humidification could not significantly induce a change in leaf anatomical structure, the effect of time on each anatomical parameter was very significant. During the growth period, the development of leaf thickness, spongy tissue and upper palisade tissue of female *P. euphratica* was stronger than that of male *P. euphratica*. Generally, plants with thicker leaves often have more developed water storage tissues, such as neat palisade tissue, parenchyma cells and water-containing parenchyma at the vascular end [57]; this may be one of the reasons why the FWU capacity of female *P. euphratica* leaves is greater than that of male *P. euphratica*, that is, the water storage structure contributes to leaf water absorption [58].

The change in water absorption was inseparable from cell structure support and physiological regulation (Figure 5). During the whole growth period, the response of the leaf structure and physiology of male and female *P. euphratica* to humidification exhibited both similarities and differences. In addition to WUE, there was a positive correlation between predawn water potential and photosynthetic capacity in male and female *P. euphratica*, indicating that an improvement in leaf water status due to humidification was beneficial to the process of carbon assimilation [59]. After comparing the fluorescence parameters, we found that qP and ETR were negatively correlated with the leaf predawn water potential. The larger the qP value, the greater the electron transport activity of photosystem II, which can not only reflect the redox state of the original electron acceptor QA, but also reflect the ability of the photon energy captured by photosystem II to be used for photochemical reaction. ETR reflects the apparent electron transfer efficiency under actual light intensity conditions [60,61]. Through the relationship between qP, ETR and predawn water potential, we believe that humidification inhibits the electron transfer process of the photosynthetic apparatus of *P. euphratica*, which may explain why humidification does not significantly improve chlorophyll fluorescence parameters. The predawn water potential of female *P. euphratica* was significantly positively correlated with TUP, while the predawn water potential of male *P. euphratica* was significantly positively correlated with TS, which may be due to the demand for water storage in the palisade tissue of female *P. euphratica*, while the sponge tissue of male *P. euphratica* provides a larger intercellular space and promotes gas exchange [62]. The studies have shown that male *P. euphratica* shows stronger drought resistance than female *P. euphratica* [45], and the stronger leaf water absorption capacity of female *P. euphratica* may benefit it at the individual or population level [63]. At present, we do not have sufficient data to prove that climate change has a greater impact on female or male *P. euphratica*. Especially in the past 50 years, the climate in Xinjiang has shown a trend of warming and wetting [64], which may make the mechanism of climate influence on plants more complicated. Therefore, in the future, we will focus on the role of long-term changes in climate on the structural, physiological and ecological adaptability of dioecious plants.

5. Conclusions

During the growing season, the foliar water uptake capacity of male and female *P. euphratica* leaves increased with the growth period. Among them, the water absorption capacity of female *P. euphratica* leaves was stronger than that of male *P. euphratica*, and showed a strategy of coexistence between storage and rate efficiency, which was inseparable from the developed water storage tissue of female *P. euphratica*. Leaf water absorption can effectively improve the leaf water status of male and female *P. euphratica*, and FWU effectively enhances the ecological adaptability of male and female *P. euphratica*. The leaves of female *P. euphratica* are more dependent on FWU, and the leaf water potential and net photosynthetic rate are effectively improved, but the ability of male *P. euphratica* to alleviate leaf water deficit and carbon benefit was stronger than that of male *P. euphratica*. This advantage was the largest in the middle of growth, indicating that male *P. euphratica* was more adaptable to the drought environment.

Supplementary Materials: The following supporting information can be downloaded at: https://www.mdpi.com/article/10.3390/f14071444/s1, Figure S1: Monthly precipitation and mean monthly temperature at study site in 2021; Table S1: The characteristics of female and male *P. euphratica*; Table S2: Experimental Schedule; Table S3: The leaf water potential of female and male *P. euphratica* in the wetting experiment; Table S4: The photosynthetic parameters of female and male *P. euphratica* in the wetting experiment; Table S5: The fluorescence parameters of female and male *P. euphratica* in the wetting experiment; Table S6: The anatomical parameters of female and male *P. euphratica* in the wetting experiment; Table S7: The extracted axis features of MFA for female *P. euphratica*; Table S8: The extracted axis features of MFA for male *P. euphratica*.

Author Contributions: Investigation, Z.L. and Y.C.; conceptualization, Z.L.; methodology, Z.L. and H.W.; software, Z.L. and Y.C.; writing—original draft preparation, Z.L.; writing—review and editing, Z.L.; supervision, X.Z.; funding acquisition, X.Z. All authors have read and agreed to the published version of the manuscript.

Funding: This research was supported by the National Natural Science Foundation of China (42171026; 31700354), and the Xinjiang Uygur Autonomous Region Graduate Research and Innovation Project (XJ2021G04).

Institutional Review Board Statement: Not applicable.

Informed Consent Statement: Not applicable.

Data Availability Statement: Not applicable.

Conflicts of Interest: The authors declare that they have no conflict of interest.

References

1. Berry, Z.C.; Emery, N.C.; Gotsch, S.G.; Goldsmith, G.R. Foliar Water Uptake: Processes, Pathways, and Integration into Plant Water Budgets. *Plant Cell Environ.* **2019**, *42*, 410–423. [CrossRef]
2. Dawson, T.E.; Goldsmith, G.R. The Value of Wet Leaves. *New Phytol.* **2018**, *219*, 1156–1169. [CrossRef]
3. Schreel, J.D.M.; Steppe, K. Foliar Water Uptake in Trees: Negligible or Necessary? *Trends Plant Sci.* **2020**, *25*, 590–603. [CrossRef]
4. Goldsmith, G.R.; Matzke, N.J.; Dawson, T.E. The Incidence and Implications of Clouds for Cloud Forest Plant Water Relations. *Ecol. Lett.* **2013**, *16*, 307–314. [CrossRef] [PubMed]
5. Gong, X.-W.; Lü, G.-H.; He, X.-M.; Sarkar, B.; Yang, X.-D. High Air Humidity Causes Atmospheric Water Absorption via Assimilating Branches in the Deep-Rooted Tree Haloxylon Ammodendron in an Arid Desert Region of Northwest China. *Front. Plant Sci.* **2019**, *10*, 573. [CrossRef] [PubMed]
6. Wang, H.; Li, Z.; Yang, J. Effects of Leaf Hydrophilicity and Stomatal Regulation on Foliar Water Uptake Capacity of Desert Plants. *Forests* **2023**, *14*, 551. [CrossRef]
7. Yang, X.-D.; Lv, G.-H.; Ali, A.; Ran, Q.-Y.; Gong, X.-W.; Wang, F.; Liu, Z.-D.; Qin, L.; Liu, W.-G. Experimental Variations in Functional and Demographic Traits of *Lappula Semiglabra* among Dew Amount Treatments in an Arid Region. *Ecohydrology* **2017**, *10*, e1858. [CrossRef]
8. Steppe, K.; Vandegehuchte, M.W.; Van De Wal, B.A.E.; Hoste, P.; Guyot, A.; Lovelock, C.E.; Lockington, D.A. Direct Uptake of Canopy Rainwater Causes Turgor-Driven Growth Spurts in the Mangrove Avicennia Marina. *Tree Physiol.* **2018**, *38*, 979–991. [CrossRef]

9. Carmichael, M.J.; White, J.C.; Cory, S.T.; Berry, Z.C.; Smith, W.K. Foliar Water Uptake of Fog Confers Ecophysiological Benefits to Four Common Tree Species of Southeastern Freshwater Forested Wetlands. *Ecohydrology* **2020**, *13*, e2240. [CrossRef]

10. Li, Z.-K.; Gong, X.-W.; Wang, J.-L.; Chen, Y.-D.; Liu, F.-Y.; Li, H.-P.; Lü, G.-H. Foliar Water Uptake Improves Branch Water Potential and Photosynthetic Capacity in Calligonum Mongolicum. *Ecol. Indic.* **2023**, *146*, 109825. [CrossRef]

11. Renner, S.S. The Relative and Absolute Frequencies of Angiosperm Sexual Systems: Dioecy, Monoecy, Gynodioecy, and an Updated Online Database. *Am. J. Bot.* **2014**, *101*, 1588–1596. [CrossRef] [PubMed]

12. Li, C.; Xu, G.; Zang, R.; Korpelainen, H.; Berninger, F. Sex-Related Differences in Leaf Morphological and Physiological Responses in *Hippophae rhamnoides* along an Altitudinal Gradient. *Tree Physiol.* **2007**, *27*, 399–406. [CrossRef]

13. Yu, L.; Dong, H.; Li, Z.; Han, Z.; Korpelainen, H.; Li, C. Species-Specific Responses to Drought, Salinity and Their Interactions in Populus Euphratica and *P. Pruinosa* Seedlings. *J. Plant Ecol.* **2020**, *13*, 563–573. [CrossRef]

14. Olano, J.M.; González-Muñoz, N.; Arzac, A.; Rozas, V.; von Arx, G.; Delzon, S.; García-Cervigón, A.I. Sex Determines Xylem Anatomy in a Dioecious Conifer: Hydraulic Consequences in a Drier World. *Tree Physiol.* **2017**, *37*, 1493–1502. [CrossRef] [PubMed]

15. Liu, M.; Korpelainen, H.; Li, C. Sexual Differences and Sex Ratios of Dioecious Plants under Stressful Environments. *J. Plant Ecol.* **2021**, *14*, 920–933. [CrossRef]

16. Xu, X.; Peng, G.; Wu, C.; Korpelainen, H.; Li, C. Drought Inhibits Photosynthetic Capacity More in Females than in Males of Populus Cathayana. *Tree Physiol.* **2008**, *28*, 1751–1759. [CrossRef]

17. Xu, X.; Yang, F.; Xiao, X.; Zhang, S.; Korpelainen, H.; Li, C. Sex-Specific Responses of Populus Cathayana to Drought and Elevated Temperatures. *Plant Cell Environ.* **2008**, *31*, 850–860. [CrossRef]

18. Juvany, M.; Munné-Bosch, S. Sex-Related Differences in Stress Tolerance in Dioecious Plants: A Critical Appraisal in a Physiological Context. *J. Exp. Bot.* **2015**, *66*, 6083–6092. [CrossRef]

19. Leigh, A.; Nicotra, A. Sexual Dimorphism in Reproductive Allocation and Water Use Efficiency in Maireana Pyramidata (*Chenopodiaceae*), a Dioecious, Semi-Arid Shrub. *Aust. J. Bot.* **2003**, *51*, 509–514. [CrossRef]

20. Rozas, V.; DeSoto, L.; Olano, J. Sex-Specific, Age-Dependent Sensitivity of Tree-Ring Growth to Climate in the Dioecious Tree Juniperus Thurifera. *New Phytol.* **2009**, *182*, 687–697. [CrossRef]

21. Li, C.; Ren, J.; Luo, J.; Lu, R. Sex-Specific Physiological and Growth Responses to Water Stress in *Hippophae rhamnoides* L. Populations. *Acta Physiol. Plant.* **2004**, *26*, 123–129. [CrossRef]

22. Munné-Bosch, S.; Nogués, S.; Alegre, L. Diurnal Variations of Photosynthesis and Dew Absorption by Leaves in Two Evergreen Shrubs Growing in Mediterranean Field Conditions. *New Phytol.* **2002**, *144*, 109–119. [CrossRef]

23. Gouvra, E.; Grammatikopoulos, G. Beneficial Effects of Direct Foliar Water Uptake on Shoot Water Potential of Five Chasmophytes. *Can. J. Bot.* **2011**, *81*, 1278–1284. [CrossRef]

24. dos Santos Garcia, J.; Boanares, D.; França, M.G.C.; Sershen; López-Portillo, J. Foliar Water Uptake in Eight Mangrove Species: Implications of Morpho-Anatomical Traits. *Flora* **2022**, *293*, 152100. [CrossRef]

25. Boanares, D.; Isaias, R.R.M.S.; De Sousa, H.C.; Kozovits, A.R. Strategies of Leaf Water Uptake Based on Anatomical Traits. *Plant Biol.* **2018**, *20*, 848–856. [CrossRef]

26. Liu, Y.; Li, X.; Chen, G.; Li, M.; Liu, M.; Liu, D. Epidermal Micromorphology and Mesophyll Structure of Populus Euphratica Heteromorphic Leaves at Different Development Stages. *PLoS ONE* **2015**, *10*, e0137701. [CrossRef]

27. Yang, F.; Deng, D.; Zhao, L.; Zhu, L. Comparative study on leaf morphological and structural characteristics of male and female *Podocarpus macrophyllus*. *Acta Agric. Jiangxi* **2021**, *33*, 42–47. [CrossRef]

28. Vesala, T.; Sevanto, S.; Grönholm, T.; Salmon, Y.; Nikinmaa, E.; Hari, P.; Hölttä, T. Effect of Leaf Water Potential on Internal Humidity and CO_2 Dissolution: Reverse Transpiration and Improved Water Use Efficiency under Negative Pressure. *Front. Plant Sci.* **2017**, *8*, 54. [CrossRef]

29. Binks, O.; Coughlin, I.; Mencuccini, M.; Meir, P. Equivalence of Foliar Water Uptake and Stomatal Conductance? *Plant Cell Environ.* **2020**, *43*, 524–528. [CrossRef] [PubMed]

30. Ishibashi, M.; Terashima, I. Effects of Continuous Leaf Wetness on Photosynthesis: Adverse Aspects of Rainfall. *Plant Cell Environ.* **1995**, *18*, 431–438. [CrossRef]

31. Hanba, Y.; Moriya, A.; Kimura, K. Effect of Leaf Surface Wetness and Wetability on Photosynthesis in Bean and Pea. *Plant Cell Environ.* **2004**, *27*, 413–421. [CrossRef]

32. Li, Z.; Wu, N.; Liu, T.; Chen, H.; Tang, M. Effect of arbuscular mycorrhizal inoculation on water status and photosynthesis of *Populus cathayana* males and females under water stress. *Physiol. Plant* **2015**, *155*, 192–204. [CrossRef]

33. Keyimu, M.; Halik, Ü.; Betz, F.; Dulamsuren, C. Vitality Variation and Population Structure of a Riparian Forest in the Lower Reaches of the Tarim River, NW China. *J. For. Res.* **2017**, *29*, 749–760. [CrossRef]

34. Li, W.; Lü, G.; Zhang, L.; Wang, H.; Li, Z.; Wang, J.; Ma, H.; Liu, Z. Analysis of Potential Water Source Differences and Utilization Strategies of Desert Plants in Arid Regions. *Ecol. Environ. Sci.* **2019**, *28*, 1557–1566. [CrossRef]

35. Zhang, Y.; Hao, X.; Sun, H.; Hua, D.; Qin, J. How Populus Euphratica Utilizes Dew in an Extremely Arid Region. *Plant Soil* **2019**, *443*, 493–508. [CrossRef]

36. Fan, X.; Hao, X.; Zhang, S.; Zhao, Z.; Zhang, J.; Li, Y. *Populus euphratica* Counteracts Drought Stress through the Dew Coupling and Root Hydraulic Redistribution Processes. *Ann. Bot.* **2023**, *131*, 451–461. [CrossRef] [PubMed]

37. Yang, X.; Lü, G.; Wang, Y.; Zhang, X. Water use efficiency of halophytes in Ebinur Lake Wetland Nature Reserve of Xinjiang. *Chin. J. Econ.* **2010**, *29*, 2341–2346. [CrossRef]
38. Gong, X.; Lü, G. Species Diversity and Dominant Species' Niches of eremophyte communities of the Tugai forest in the Ebinur basin of Xinjiang, China. *Biodivers. Sci.* **2017**, *25*, 34–45. [CrossRef]
39. Liang, X.; Su, D.; Yin, S.; Wang, Z. Leaf Water Absorption and Desorption Functions for Three Turfgrasses. *J. Hydrol.* **2009**, *376*, 243–248. [CrossRef]
40. Schreel, J.D.M.; von der Crone, J.S.; Kangur, O.; Steppe, K. Influence of Drought on Foliar Water Uptake Capacity of Temperate Tree Species. *Forests* **2019**, *10*, 562. [CrossRef]
41. Cavallaro, A.; Carbonell Silleta, L.; Pereyra, D.A.; Goldstein, G.; Scholz, F.G.; Bucci, S.J. Foliar Water Uptake in Arid Ecosystems: Seasonal Variability and Ecophysiological Consequences. *Oecologia* **2020**, *193*, 337–348. [CrossRef] [PubMed]
42. Maxwell, S.; Delaney, H. *Designing Experiments and Analyzing Data: A Model Comparison Perspective*; Routledge Academic Inc. Publisher: London, UK, 2003; ISBN 978-1-4106-0924-3.
43. Boanares, D.; Kozovits, A.R.; Lemos-Filho, J.P.; Isaias, R.M.S.; Solar, R.R.R.; Duarte, A.A.; Vilas-Boas, T.; França, M.G.C. Foliar Water-uptake Strategies Are Related to Leaf Water Status and Gas Exchange in Plants from a Ferruginous Rupestrian Field. *Am. J. Bot.* **2019**, *106*, 935–942. [CrossRef]
44. Boanares, D.; Ferreira, B.G.; Kozovits, A.R.; Sousa, H.C.; Isaias, R.M.S.; França, M.G.C. Pectin and Cellulose Cell Wall Composition Enables Different Strategies to Leaf Water Uptake in Plants from Tropical Fog Mountain. *Plant Physiol. Biochem.* **2018**, *122*, 57–64. [CrossRef] [PubMed]
45. Yu, L.; Huang, Z.; Tang, S.; Korpelainen, H.; Li, C. Populus Euphratica Males Exhibit Stronger Drought and Salt Stress Resistance than Females. *Environ. Exp. Bot.* **2023**, *205*, 105114. [CrossRef]
46. Chin, A.R.O.; Guzmán-Delgado, P.; Görlich, A.; HilleRisLambers, J. Towards multivariate functional trait syndromes: Predicting foliar water uptake in trees. *Ecology*, 2023; *online version of record*. [CrossRef] [PubMed]
47. Chin, A.R.O.; Guzmán-Delgado, P.; Kerhoulas, L.P.; Zwieniecki, M.A. Acclimation of interacting leaf surface traits affects foliar water uptake. *Tree Physiol.* **2023**, *43*, 418–429. [CrossRef]
48. Losso, A.; Dämon, B.; Hacke, U.; Mayr, S. High potential for foliar water uptake in early stages of leaf development of three woody angiosperms. *Physiol. Plant* **2023**, *175*, e13961. [CrossRef]
49. Ru, C.; Hu, X.; Chen, D.; Wang, W.; Song, T. Heat and Drought Priming Induce Tolerance to Subsequent Heat and Drought Stress by Regulating Leaf Photosynthesis, Root Morphology, and Antioxidant Defense in Maize Seedlings. *Environ. Exp. Bot.* **2022**, *202*, 105010. [CrossRef]
50. Eller, C.B.; Lima, A.L.; Oliveira, R.S. Foliar Uptake of Fog Water and Transport Belowground Alleviates Drought Effects in the Cloud Forest Tree Species, D*rimys brasiliensis* (Winteraceae). *New Phytol.* **2013**, *199*, 151–162. [CrossRef]
51. Holanda, A.E.R.; Souza, B.C.; Carvalho, E.C.D.; Oliveira, R.S.; Martins, F.R.; Muniz, C.R.; Costa, R.C.; Soares, A.A. How Do Leaf Wetting Events Affect Gas Exchange and Leaf Lifespan of Plants from Seasonally Dry Tropical Vegetation? *Plant Biol.* **2019**, *21*, 1097–1109. [CrossRef]
52. Wang, H.; Zhou, Y.; Qin, L.; Gong, X.; Lü, G. Influence of Dew on Fluorescence Parameter and Water Use Efficiency of *Halostachys caspica* in Different Salinity Habitats. *Arid Zone Res.* **2017**, *34*, 1124–1132. [CrossRef]
53. Yu, L.; Tang, S.; Guo, C.; Korpelainen, H.; Li, C. Differences in Ecophysiological Responses of *Populus euphratica* Females and Males Exposed to Salinity and Alkali Stress. *Plant Physiol. Biochem.* **2023**, *198*, 107707. [CrossRef] [PubMed]
54. Krause, G.H.; Weis, E. Chlorophyll Fluorescence and Photosynthesis: The Basics. *Annu. Rev. Plant Biol.* **2003**, *42*, 313–349. [CrossRef]
55. Wu, M.; Deng, P.; Zhao, Y.; Zhao, S.-H.; Chen, J.-N.; Shu, Y.; Huang, T.-F. Effects of drought on leaf growth and chlorophyll fluorescence kinetics parameters in *Cyclobalanopsis glauca* seedlings of Karst areas. *J. Appl. Ecol.* **2019**, *30*, 4071–4081. [CrossRef]
56. Howarth, J.F.; Durako, M.J. Diurnal Variation in Chlorophyll Fluorescence of *Thalassia testudinum* Seedlings in Response to Controlled Salinity and Light Conditions. *Mar. Biol.* **2013**, *160*, 591–605. [CrossRef]
57. Bryant, C.; Fuenzalida, T.I.; Zavafer, A.; Nguyen, H.T.; Brothers, N.; Harris, R.J.; Beckett, H.A.A.; Holmlund, H.I.; Binks, O.; Ball, M.C. Foliar Water Uptake via Cork Warts in Mangroves of the *Sonneratia* Genus. *Plant Cell Environ.* **2021**, *44*, 2925–2937. [CrossRef] [PubMed]
58. Hayes, M.A.; Chapman, S.; Jesse, A.; O'Brien, E.; Langley, J.A.; Bardou, R.; Devaney, J.; Parker, J.D.; Cavanaugh, K.C. Foliar Water Uptake by Coastal Wetland Plants: A Novel Water Acquisition Mechanism in Arid and Humid Subtropical Mangroves. *J. Ecol.* **2020**, *108*, 2625–2637. [CrossRef]
59. Ferreira, J.L.D.; Daniela, B.; Eliodoro, V.C.; Pinheiro, A.S.F. Do photosynthetic metabolism and habitat influence foliar water uptake in orchids? *Plant Biol.* **2022**, *25*, 257–267.
60. Richardson, A.D.; Berlyn, G.P. Spectral Reflectance and Photosynthetic Properties of *Betula Papyrifera* (*Betulaceae*) Leaves along an Elevational Gradient on Mt. Mansfield, Vermont, USA. *Am. J. Bot.* **2002**, *89*, 88–94. [CrossRef]
61. Taiz, L.; Zeiger, E. *Plant Physiology*, 5th ed.; Sinauer Associates, Inc.: Sunderland, MA, USA, 2010; ISBN 978-0-87893-866-7.
62. Chartzoulakis, K.; Patakas, A.; Kofidis, G.; Bosabalidis, A.; Nastou, A. Water Stress Affects Leaf Anatomy, Gas Exchange, Water Relations and Growth of Two Avocado Cultivars. *Sci. Hortic.* **2002**, *95*, 39–50. [CrossRef]

63. Li, Z.-K.; Chen, Y.-D.; Wang, J.-L.; Jiang, L.-M.; Fan, Y.-X.; Lü, G.-H. Foliar water uptake and its influencing factors differ between female and male Populus euphratica. *Environ. Exp. Bot.* **2023**, *213*, 105419. [CrossRef]
64. Yao, J.-Q.; Chen, Y.-N.; Guan, X.-F.; Zhao, Y.; Chen, J.; Mao, W.-Y. Recent climate and hydrological changes in a mountain–basin system in Xinjiang, China. *Earth-Sci. Rev.* **2022**, *226*, 103957. [CrossRef]

Disclaimer/Publisher's Note: The statements, opinions and data contained in all publications are solely those of the individual author(s) and contributor(s) and not of MDPI and/or the editor(s). MDPI and/or the editor(s) disclaim responsibility for any injury to people or property resulting from any ideas, methods, instructions or products referred to in the content.

Article

Effects of Leaf Hydrophilicity and Stomatal Regulation on Foliar Water Uptake Capacity of Desert Plants

Huimin Wang [1,2], Zhoukang Li [1,2] and Jianjun Yang [1,2,*]

[1] College of Ecology and Environment, Xinjiang University, Urumqi 830046, China
[2] Key Laboratory of Oasis Ecology of Education Ministry, Xinjiang University, Urumqi 830046, China
* Correspondence: yjj@xju.edu.cn

Abstract: Foliar water uptake (FWU) is one of the primary water sources for desert plants. Desert plants' water uptake capacity is essential in maintaining the balance of carbon and water. However, there are few studies on FWU capacity in desert plants and the physiological and ecological characteristics that lead to differences in FWU capacity. In order to clarify FWU strategies and the influencing factors of plants in desert ecosystems, this study measured the contact angle, FWU parameters, and hydraulic parameters to explore six desert plants' FWU capacity and the effects of leaf wettability and hydraulic parameters on FWU capacity. The results showed that all six plants had FWU capacity, among which the leaves of *Nitraria sibirica* Pall. and *Halimodendron halodendron* (Pall.) Voss had a high foliar water uptake rate (k) and high foliar water uptake accumulation (FWU storage), and the leaves of *Glycyrrhiza uralensis* Fisch. had a high k and low FWU storage. The leaves of *Populus euphratica* Oliv., *Apocynum hendersonii* Hook. f., and *Alhagi sparsifolia* Shap. had a low k and low FWU storage. Additionally, FWU capacity was mainly affected by stomatal regulation compared with leaf wettability and leaf structure. The results of this study will help to improve the understanding of the physiological and ecological adaptability of desert plants.

Keywords: arid area; foliar water uptake strategies; isohydry; leaf wettability; turgor loss point

1. Introduction

During leaf-wetting events, a water film with a water potential close to zero can be formed on the leaf surface. Water flows bidirectionally inside the leaves through pores. At this time, the external water potential of the leaf is higher than the internal water potential, thus producing the foliar water uptake phenomenon (FWU) [1]. FWU is a common physiological phenomenon in natural ecosystems [2,3] and one of the critical survival strategies for plants to maintain their water balance [4]. Previous studies have found that in the process of FWU, plants with physiological and structural differences have different FWU strategies [5–8]. In some plants, there is a trade off between the foliar water uptake rate (k) and foliar water uptake accumulation (FWU storage). The species with high FWU storage have a low k, while those with a high k have low FWU storage [7–10]. Some scholars have also pointed out that a high k and high FWU storage can coexist in the same species [7], leading to higher water use efficiency. However, the current research on FWU strategies mainly focuses on Brazil's high-fog and high-humidity habitats [7,10]. Due to the difference in water input, plants living in different habitats will adopt different water adaptation strategies [5]. Therefore, it is necessary to research the FWU strategies in other habitats.

In arid desert ecosystems, leaf-wetting events also occur frequently with water as the main limiting factor [11]. Previous studies found that most plants are capable of FWU, and that the absorbed water could be transported downward through the xylem and improve the branch water status and photosynthetic capacity, promoting the carbon and water balance of plants in arid areas [11,12]. FWU also plays an important role in the survival and

Citation: Wang, H.; Li, Z.; Yang, J. Effects of Leaf Hydrophilicity and Stomatal Regulation on Foliar Water Uptake Capacity of Desert Plants. *Forests* **2023**, *14*, 551. https://doi.org/10.3390/f14030551

Academic Editor: Claudia Cocozza

Received: 29 January 2023
Revised: 3 March 2023
Accepted: 8 March 2023
Published: 10 March 2023

Copyright: © 2023 by the authors. Licensee MDPI, Basel, Switzerland. This article is an open access article distributed under the terms and conditions of the Creative Commons Attribution (CC BY) license (https://creativecommons.org/licenses/by/4.0/).

growth of trees in drought periods, helping to repair xylem embolism and increase water conduction efficiency [13,14]. In recent years, there has been an increase in the amount of research on FWU in arid areas [11–14], but there are few studies on FWU strategies.

The differences in the physiological and structural characteristics of plants lead to different FWU strategies [5,15,16]. For example, the differences in leaf anatomical structure, water status, stomatal conductance, and oxidative metabolites between species significantly impacted FWU strategies [5–7]. Due to the diversity of species and habitats, numerous factors lead to different FWU strategies for plants. Some plants will promote FWU by evolving unique FWU structures, such as cork warts and trichomes [17,18], and maintain leaf water balance. Therefore, it is necessary to further study the influencing factors of FWU strategies and explore the water utilization strategies of plant leaves, especially in arid areas.

Leaf wettability is a common phenomenon in various habitats, showing the adhesive capacity of leaves to water [19,20]. Water adherence to the plant surface is a prerequisite for FWU to occur [14,21]. Fernández et al. found that leaf morphology and structure have an impact on FWU based on the leaf's hydrophilicity or hydrophobicity (wettability) [22]. Plants with high leaf wettability have higher FWU storage [23]. In arid areas, in order to adapt to harsh habitats and prevent water loss, plants generally form thicker cuticles on their leaves. The cuticle is the main factor affecting the wettability of the leaves, and the cuticle is hydrophobic [24–27]. Whether the wettability difference caused by the cuticle will affect FWU storage and whether it is related to k besides FWU storage remain to be elucidated [28]. Furthermore, Goldsmith and others [29] found that leaf wettability will also have a certain adaptability to changes in the environment. Under different water gradients (from tropical rainforests to cloudy mountain forests), leaf wettability will change accordingly. Areas with low precipitation and low temperature have stronger leaf hydrophobicity. Therefore, will leaf wettability change with changes in the environment? Will the difference in hydrophobicity and hydrophilicity generated by this adaptation of the environment affect the FWU strategies?

The stomata is the most important structure for water and gas exchange between leaves and the atmosphere [30], and it is the prediction index of plant response to drought. Facing the potential risk of hydraulic imbalance caused by drought, plants have evolved different stomatal regulation strategies, namely, isohydric regulation and anisohydric regulation. When the atmospheric water vapor pressure deficit (VPD) increases or the soil water content decreases, isohydric plants maintain the minimum leaf water potential relatively constant through strict stomatal regulation [31]. On the contrary, the stomata of anisohydric plants always maintains a certain opening [32,33]. Previous studies on mangrove and temperate species found that plants with anisohydric regulation had higher FWU storage, while plants with isohydric regulation had lower FWU storage [34,35]. However, at present, there is little research on FWU strategies through iso-/anisohydric regulation, especially for plants in arid areas, which needs to be discussed.

The Ebinur Lake Basin has a typical temperate continental arid climate. Due to the scarcity and uneven distribution of precipitation, water resources are incredibly scarce [36,37]. *P. euphratica*, *N. sibirica*, *H. halodendron*, *A. sparsifolia*, *A. hendersonii*, and *G. uralensis* are typical psammophytes in the Ebinur Lake Basin. They can not only improve the ecological environment in desert areas, but also weaken the ground evaporation and reduce the degree of soil drought to some extent. At the same time, the growth of psammophytes can reduce the amount of floating sand on the ground, which can create more suitable living space for other plant species and create more opportunities for the better growth and development of psammophytes [38]. Therefore, this study includes six typical desert plants in the Ebinur Lake Basin, namely, *P. euphratica*, *N. sibirica*, *H. halodendron*, *A. sparsifolia*, *A. hendersonii*, and *G. uralensis*, which were used as research objects to explore the following questions: (1) Do these six desert plants show the FWU phenomenon, and what FWU strategies do they adopt? (2) Do leaf wettability and iso/anisohydric regulation affect the FWU strategies of plants in arid areas? The purpose of this study was to explore FWU strategies from the perspective of leaf wettability and stomatal regulation, so as to enrich the understanding of the water use strategies of desert plants and provide a theoretical basis for the effective restoration and protection of degraded vegetation in desert areas.

2. Materials and Methods

2.1. Research Sites

The Ebinur Wetland Nature Reserve, which is situated in the Xinjiang Uygur Autonomous Region of northwest China (44°43′–45°12′ N, 82°35′–83°40′ E), is the lowest depression and the center of saltwater collection in the Ebinur Wetland western boundary of the Gurbantonggut Desert (Figure 1a). The reserve belongs to a typical temperate continental arid climate. The average yearly temperature ranges from 6.6 to 7.8 °C, and the annual evaporation exceeds 1600 mm [39]. The average annual precipitation is 90.8 mm [40], and the distribution is uneven throughout the year, with summer precipitation accounting for 34.0% to 66.6% of the annual precipitation. The spatial distribution of the soil in this area shows the law of meridional zonality and vertical zonality. The soil types are mainly clay, silt, and sand [41]. The main plant species in the study area are *P. euphratica*, *H. ammodendron*, *H. halodendron*, *A. sparsifolia*, *R. soongarica*, *N. sibirica*, *A. hendersonii*, *G. uralensis*, *S. terrae-albae*, *P. australis*, etc.

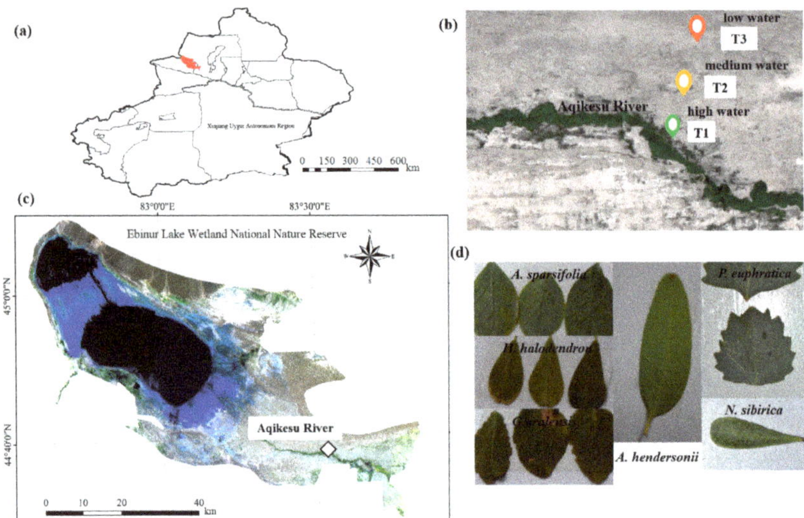

Figure 1. The study area and plots. (**a**) Location map of the Xinjiang Uygur Autonomous Region. Where the red area represents Ebinur Lake; (**b,c**) location of the study area; (**d**) leaves of *A. sparsifolia*, *G. uralensis*, *H. halodendron*, *A. hendersonii*, *P. euphratica*, and *N. sibirica*.

2.2. Field Experiments

2.2.1. Sample Plot Design

In July 2022, a 100 m × 2000 m sample strip was set up from south to north near Dongqiao Station in the reserve, along the direction perpendicular to the Aqikesu River. A 100 m × 100 m quadrat was set every 1000 m on the sample strip, with a total of 3 quadrats, which were set as high water (T1), medium water (T2), and low water (T3) (Figure 1b,c).

Soil from the 0–20 cm, 20–40 cm, and 40–60 cm soil layers was collected at three randomly selected locations within each sample square, packed into aluminum boxes, and brought back to the laboratory to determine the soil water content for water gradient division for backup. The soil water content was measured using the drying method by first measuring the fresh weight (SFW) and then drying at 105 °C for 48 h to a constant weight using an electric blast oven (DHG-9055A, CHN), and then the dry weight (SDW) of the soil samples was recorded to calculate the soil water content (SWC) using the following formula:

$$\text{SWC (\%)} = \frac{\text{SFW} - \text{SDW}}{\text{SDW}} \times 100\%$$

2.2.2. Plant Selection and Sample Collection with Determination

Plant selection: the plants selected were *P. euphratica*, *N. sibirica*, *H. halodendron*, *A. sparsifolia*, *A. hendersonii*, and *G. uralensis* (Figure 1d). The field investigation found that *H. halodendron* was only distributed in T1, and *G. uralensis* was only distributed in T1 and T2. Therefore, the plants selected in T1 were *P. euphratica*, *N. sibirica*, *H. halodendron*, *A. sparsifolia*, *A. hendersonii*, and *G. uralensis*, with a total of 6 species. The plants selected in T2 were *P. euphratica*, *N. sibirica*, *A. sparsifolia*, *A. hendersonii*, and *G. uralensis*, with a total of 5 species. The plants selected in T3 were *P. euphratica*, *N. sibirica*, *A. sparsifolia*, and *A. hendersonii*, with a total of 4 species. Four plants were randomly selected from each species in each quadrat, and their crown width and diameter at breast height/base diameter were recorded (Table S1).

Determination of morphological indicators: A total of 4 plants were randomly selected within the sample plot, and 3 leaves were randomly selected from each plant in different directions, with a total of 12 leaves. The fresh weight was recorded using a one-ten-thousandth precision balance (AL204, METTLER TOLEDO, CHN). The leaf thickness (LT) was recorded using an electronic vernier caliper with an accuracy of 0.01 mm and photographed and uploaded to IMAGE J software to calculate the leaf area. Then, the leaves were brought back to the laboratory for drying, and the dry weight was recorded. Finally, the specific leaf area (SLA) and leaf dry matter content (LDMC) were calculated using the above measurements.

2.3. Laboratory Experiments

2.3.1. Leaf Wettability

To characterize leaf wettability, the contact angle of a droplet of water was measured on the surface of a leaf. A total of 4 healthy plants were selected as replicates for each species in the sample plot, and 12 healthy and intact leaves were collected. The leaves were placed horizontally on the sampling table using double-sided adhesive tape, and then water droplets were titrated on the leaf surface with a micropipette (range 5~100 μL). Finally, the droplet side view was taken with a macro camera (Nikon Corp, Tokyo, Japan). Two equal water droplets per leaf were used as a parallel experiment of this treatment [29].

The contact angle (also referred to as θ) was measured as the angle between the horizontal line of contact of the water droplet on the leaf surface and the line tangent at the edge of the water droplet (Figure 2). A larger contact angle indicates higher leaf water repellency [29,42–44]. Before determining the contact angle, the water droplet was outlined as an ellipse to aid in a more accurate identification of the tangent. Analysis was conducted in IMAGE J v.1.47 (NIH, Bethesda, MD, USA).

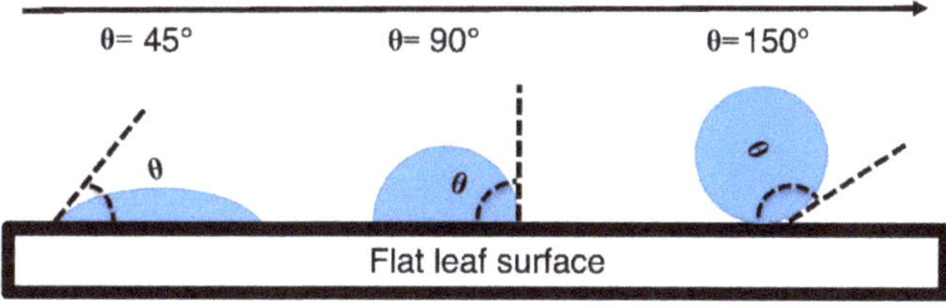

Figure 2. Sketch of contact angle of water droplet and leaf surface.

2.3.2. Determination of FWU Parameters

According to the method proposed by Liang et al. [45], the randomly selected leaves were cut from the plants, weighed immediately, and recorded as the initial weight (CW). They were then soaked in distilled water with the leaves covered with filter paper to prevent them from floating above the water surface. During absorption, the leaves were weighed every 15 min in the first 2 h, and then every 30 min in the next two hours. Finally, the leaves were immersed in water for 1 h again and weighed to obtain the saturated weight (SW) (a total of 13 measurements). The leaf area (LA) was measured before the experiment. After the experiment, the leaves were dried at 108 °C for 48 h, and the dry weight (DW) was recorded. The main FWU parameters were as follows:

Foliar water uptake accumulation per unit area ($FWU_{capacity}$; mg cm^{-2}):

$$FWU_{capacity} = \frac{SW - CW}{LA}$$

The increase in leaf water content (ΔLWC; %):

$$\Delta LWC = \frac{SW - DW}{SW} \times 100 - \frac{CW - DW}{CW} \times 100$$

Foliar water uptake accumulation (ΔM; mg·mg^{-1}):

$$\Delta M = \frac{SW - CW}{CW}$$

Foliar water uptake rate (k; mg·cm^{-2}·min^{-1}):

$$k = \frac{FWU_{capacity}}{t}$$

where SW is the leaf saturation weight, CW is the fresh weight, LA is the leaf area, DW is the dry weight, and t is time.

2.3.3. Determination of Pressure–Volume Curves

The pressure–volume (P–V) curves were determined using the method proposed by Tyree et al. [46]. Due to the small leaves and short petioles of *N. sibirica*, *H. halodendron*, *A. sparsifolia*, *A. hendersonii*, and *G. uralensis*, a pressure chamber (PMS model 1505 D, Albany, OR, USA) could not be used to determine the water potential. Therefore, the above plants' P–V curves measured about 10 cm of leafy twigs; only *P. euphratica* was directly measured using the leaves. The P–V curves were plotted following the procedures described by Sack and Pasquet-Kok [47]. The leaves were weighed and rehydrated in distilled water for 12 h at 4 °C and then gradually dried in a well-ventilated room at 25 °C. During the drying process, the leaves were weighed, and the water potential was measured using a pressure chamber (PMS model 1505 D, Albany, OR, USA) until a complete P–V curve with at least ten points was established. No 'plateau effect' was observed for any sample. Leaf dry mass was determined after at least 72 h at 70 °C. The full turgor and turgor loss points (Ψ_{tlp}) were established by considering the highest R^2 of a linear fit for the linear portion of the $-1/\Psi$ vs. 1-RWC relationship (RWC: relative water content). The following parameters were obtained from the P–V curves: osmotic potential at turgor loss point (π_o); capacitance at turgor loss point (C_{tlp}); capacitance at full turgor (C_{ft}); bulk modulus of elasticity (ε, calculated from the total relative water content); and osmotic adjustment ability (π_o-Ψ_{tlp}).

2.4. Statistical Analyses

ANOVA was performed to compare the variability of the three water gradients, and the results are shown in Table 1. Two-way ANOVA and one-way ANOVA explored the differences in the leaf wettability, FWU parameters, leaf structure, and hydraulic parameters of the plants to be tested across the water gradients as well as between species, using LSD for comparison when the variances were equal and Games–Howell for comparison when the variances were not equal. Secondly, logistic and exponential equations were used to fit the model of FWU with time, and the confidence level was 0.95. Multiple factor analysis (MFA) was used to reveal the multiple relationships among the following groups: (1) species and water gradient (high water, medium water, low water); (2) hydraulic parameters (Ψ_{tlp}, π_0, ε, π_0-Ψ_{tlp}); (3) FWU parameters (ΔM, ΔLWC, $FWU_{capacity}$, k_{45}); (4) leaf structure (LDMC, LT); (5) leaf wettability (CA). Two-way ANOVA was performed in SPSS 19.0 (IBM SPSS Inc., Chicago, IL, USA). Logistic and exponential model fitting and graphical production were completed in Origin 2023 (OriginLab, Northampton, MA, USA). MFA and RV coefficient calculations were completed using R '*FactoMineR*' and '*coffRV*' packages (R version 4.2.2 ucrt, http://cran.rproject.org accessed on 31 October 2022). The significance level was uniformly set to 0.05.

Table 1. Analysis of variance of soil water content and division of water gradients.

Water Gradient	Soil Sample Depth	Soil Water Content
High water	0–20 cm	0.131 ± 0.026 a
	20–40 cm	0.144 ± 0.013 a
	40–60 cm	0.137 ± 0.009 a
Medium water	0–20 cm	0.093 ± 0.013 b
	20–40 cm	0.119 ± 0.022 b
	40–60 cm	0.128 ± 0.018 a
Low water	0–20 cm	0.028 ± 0.007 c
	20–40 cm	0.026 ± 0.005 c
	40–60 cm	0.035 ± 0.003 b

Note: different lowercase letters indicate significant differences in soil water content in different squares (mean + SD; N = 3), $\alpha = 0.05$.

3. Results

3.1. Leaf Wettability

Among the six desert plants, the leaf contact angle (CA) of *G. uralensis* was significantly lower than that of the other plants ($p < 0.05$), and its hydrophilic ability was the strongest. *N. sibirica* came in second, and its CA was less than 90°, which is hydrophilic. The CAs of *P. euphratica*, *H. halodendron*, *A. sparsifolia*, and *A. hendersonii* were all above 90°, and the surfaces of the leaves were hydrophobic. While under different water gradients, there was no significant difference in CA among species ($p > 0.05$) (Table 2).

Table 2. Determination of contact angles of the leaves of six desert plant species.

Species	5 µL Contact Angle (°)		
	High Water	**Medium Water**	**Low Water**
P. euphratica	101.69 ± 2.22 Aa	101.62 ± 2.28 Aa	99.33 ± 3.7 Aa
N. sibirica	83.56 ± 4.62 Ab	81.82 ± 5.57 Ab	80.58 ± 3.67 Ab
A. sparsifolia	122.08 ± 6.96 Ac	127.27 ± 6.12 Ac	125.82 ± 4.48 Ac
A. hendersonii	129.22 ± 3.63 Ad	131.97 ± 4.33 Ac	131.23 ± 6.16 Ad
G. uralensis	59.42 ± 4.28 Ac	61.37 ± 3.44 Ad	-
H. halodendron	123.58 ± 4.17 e	-	-

Note: lowercase letters indicate significant differences between species, and uppercase letters indicate significant differences between water gradients (mean + SD; N = 6), $\alpha = 0.05$.

3.2. FWU Characteristics of Six Plants

3.2.1. Patterns of FWU Characteristics over Time

The six plants' foliar water uptake accumulation per unit area ($FWU_{capacity}$) showed differences. Among them, the $FWU_{capacity}$ of *N. sibirica* was significantly higher than that of the other species ($p < 0.05$), followed by *H. halodendron*, and the lowest was found for *G. uralensis*, *A. hendersonii*, *P. euphratica*, and *A. sparsifolia* (Figure 3a–c).

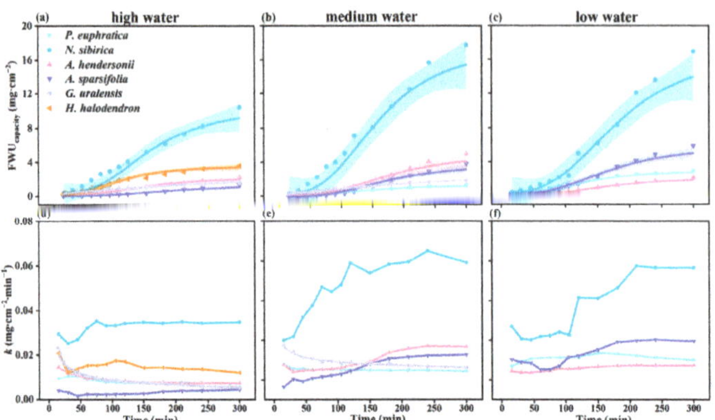

Figure 3. The variation in foliar water uptake accumulation per unit area (FWU$_{capacity}$) and foliar water uptake rate (k) over time. (**a–c**) FWU$_{capacity}$ of six plants under high water, medium water, and low water gradients over time; (**d–f**) k of six plants under high water, medium water, and low water gradients over time. Colored regions show 95% confidence intervals for the mean predicted value estimated using logistic and exponential models, and the parameters fit using each model are presented in Table S2 (N = 4).

The foliar water uptake rate (k) was also different among species. The k of *G. uralensis* decreased gradually with time. The k of *N. sibirica*, *A. sparsifolia*, and *A. hendersonii* increased initially and tended to be stable with time. The k of *P. euphratica* and *H. halodendron* increased first and then decreased with time (Figure 3d,e).

The initial k of *G. uralensis* was significantly higher than that of the other species ($p < 0.05$), followed by *N. sibirica* and *H. halodendron*. The lowest rates were found for *P. euphratica*, *A. sparsifolia*, and *A. hendersonii*. When the water was absorbed for 300 min, the k of *N. sibirica* was the highest, followed by that of *H. halodendron* (Figure 3d,e).

3.2.2. Patterns of FWU Characteristics under Water Gradients with Time

The FWU$_{capacity}$ of the six plants was different under different water gradients. The FWU$_{capacity}$ of *N. sibirica* and *A. sparsifolia* under low water was higher than that under high water. However, the FWU$_{capacity}$ of *P. euphratica* and *A. hendersonii* had no apparent regularity with the increase in the drought degree (Figure 4a–f).

Under different water gradients, with the increase in the drought degree, the k of *N. sibirica*, *G. uralensis*, and *A. sparsifolia* showed an increasing trend. There was no apparent regularity between the k of *P. euphratica* and *A. hendersonii* (Figure 3g–i).

The FWU$_{capacity}$ of the six plants showed two trends with time. One is exponential growth with time (*G. uralensis*), which is divided into a fast water absorption period, a slow water absorption period, and a saturation period (Figure 3l). The other is the logistic curve growth mode (*P. euphratica*, *H. halodendron*, *A. sparsifolia*, *N. sibirica*, and *A. hendersonii*), divided into a slow water absorption period, fast water absorption period, and saturation period, taking *A. sparsifolia* as an example (Figure 4f).

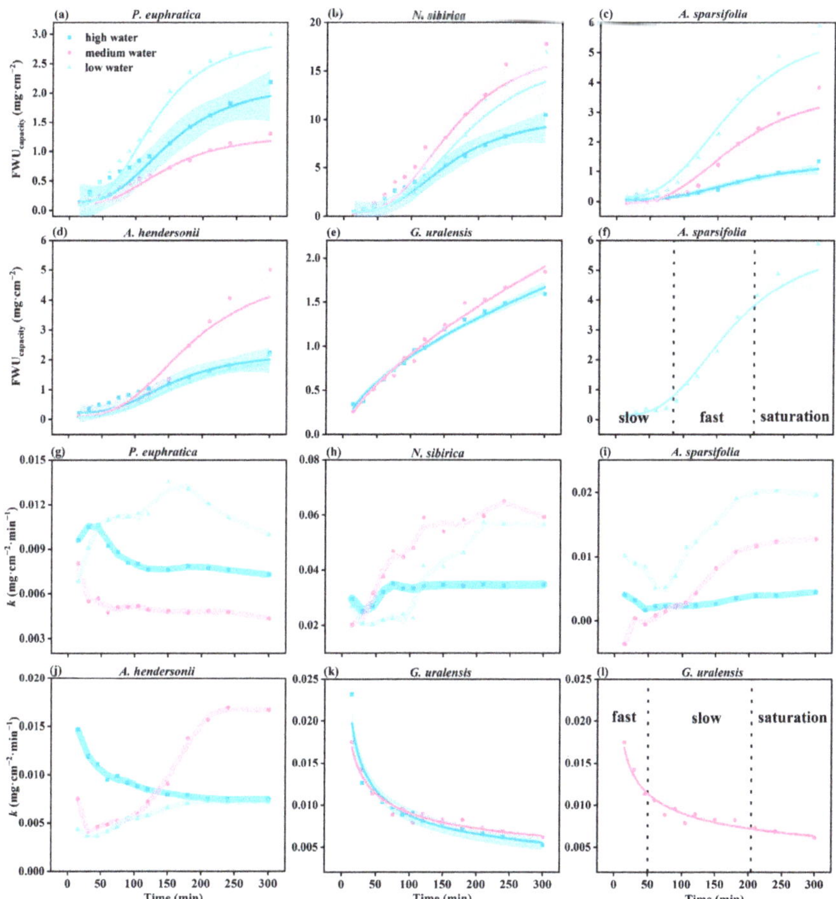

Figure 4. The variation in FWU$_{capacity}$ and k over time (**a**–**l**) under different water gradients. The parameters fit using each model are presented in Table S2 (N = 40).

3.2.3. FWU Characteristics under Different Water Gradients and Interspecific Differences

The water gradients and interspecific differences had significant effects on the FWU parameters (ΔM, FWUcapacity) ($p < 0.05$; Table S3).

Under high water, the FWU parameters of *N. sibirica* were significantly higher than those of the other plants ($p < 0.05$), followed by *H. halodendron*. Under medium water, *N. sibirica* had significantly higher parameters than the other plants ($p < 0.05$), followed by *G. uralensis* and *P. euphratica*. Under low water, *N. sibirica* had significantly higher parameters than the other plants ($p < 0.05$), while *A. hendersonii* had the lowest (Figure 5a).

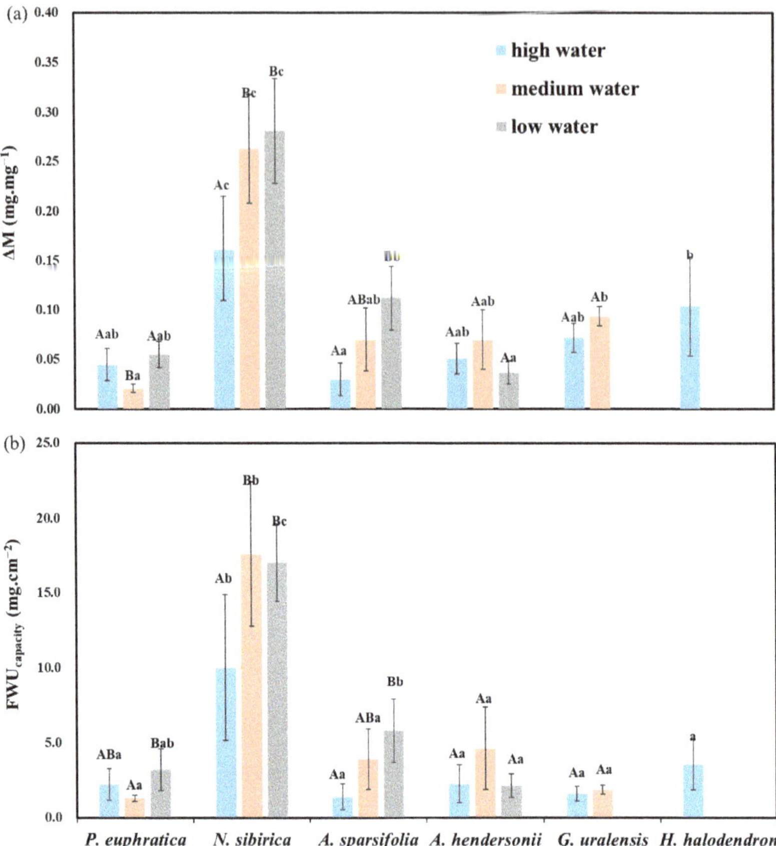

Figure 5. Differences in FWU parameters (ΔM and $\text{FWU}_{capacity}$) among water gradients and species. (**a**) The difference in foliar water uptake accumulation (ΔM) between species and water gradients; (**b**) $\text{FWU}_{capacity}$ among species and water gradients. Different lowercase letters indicate significant differences between species, and different uppercase letters indicate significant differences between water gradients (mean + SE; N = 4), α = 0.05.

Regarding the different water gradients, the FWU parameters of *N. sibirica* and *A. sparsifolia* under low water were significantly higher than those under high water ($p < 0.05$), and there was no clear rule among the other species (Figure 5).

3.3. Leaf Structural Characteristics

Among the six plants, *N. sibirica* had the thickest leaves, which were significantly thicker than the leaves of the other plants ($p < 0.05$), and the leaf dry matter content (LDMC) was significantly lower than that of the other plants ($p < 0.05$). *G. uralensis* had the thinnest leaves and the highest LDMC, while there was no significant difference between the LT and LDMC of *P. euphratica*, *A. hendersonii*, and *A. sparsifolia* (Table 3).

Table 3. Differences in leaf structure characteristics among three water gradients and species.

Gradient	Species	LT (mm)	LDMC (%)
High water	*P. euphratica*	0.463 ± 0.038 ABd	0.357 ± 0.037 Ac
	N. sibirica	0.652 ± 0.009 Ae	0.198 ± 0.026 Aa
	A. sparsifolia	0.382 ± 0.014 Ac	0.350 ± 0.017 Ac
	A. hendersonii	0.435 ± 0.078 Acd	0.276 ± 0.032 Ab
	G. uralensis	0.208 ± 0.008 Aa	0.400 ± 0.012 Ac
	H. halodendron	0.296 ± 0.037 b	0.389 ± 0.031 c
Medium water	*P. euphratica*	0.513 ± 0.080 Abc	0.253 ± 0.008 Bb
	N. sibirica	0.682 ± 0.115 Ad	0.215 ± 0.019 Aa
	A. sparsifolia	0.469 ± 0.016 Bb	0.312 ± 0.018 Bc
	A. hendersonii	0.635 ± 0.072 Bcd	0.234 ± 0.017 Aab
	G. uralensis	0.203 ± 0.004 Aa	0.393 ± 0.019 Ad
Low water	*P. euphratica*	0.410 ± 0.019 Ba	0.334 ± 0.026 Aa
	N. sibirica	0.728 ± 0.217 Ab	0.214 ± 0.038 Ab
	A. sparsifolia	0.489 ± 0.028 Ba	0.312 ± 0.019 Ba
	A. hendersonii	0.483 ± 0.046 Aa	0.282 ± 0.026 Aa

Note: LT represents leaf thickness; LDMC represents leaf dry matter content. Different lowercase letters indicate significant differences between species, and different uppercase letters indicate significant differences between water gradients (mean + SE; N = 4), $\alpha = 0.05$.

3.4. Hydraulic Parameters of Six Plants

The results of the two-way ANOVA showed that the water gradients and interspecific differences had significant effects on the turgor loss point (Ψ_{tlp}) ($p < 0.05$). The elastic modulus (ε) and osmotic adjustment ability (Ψ_{tlp}-π_0) were significantly different among species ($p < 0.05$). The saturated osmotic potential (π_0) was significantly different between species and under the interaction of species and water gradients ($p < 0.05$) (Table S3).

Ψ_{tlp} and π_0 showed differences among species. Under high water, the Ψ_{tlp} and π_0 of *P. euphratica*, *N. sibirica*, and *H. halodendron* were significantly higher than those of *G. uralensis*, *A. hendersonii*, and *A. sparsifolia* ($p < 0.05$). Under medium water, the Ψ_{tlp} and π_0 of *N. sibirica* were the highest, while those for *A. hendersonii* and *A. sparsifolia* were the lowest. Under low water, the Ψ_{tlp} and π_0 of *N. sibirica* were the highest, followed by those of *P. euphratica* (Figure 6a,b). Ψ_{tlp}-π_0 also showed differences among species, and the Ψ_{tlp}-π_0 values of *P. euphratica* and *N. sibirica* were significantly higher than those of the other plants ($p < 0.05$) (Figure 6d).

With the deepening of drought stress, the Ψ_{tlp} and π_0 of *G. uralensis* and *A. sparsifolia* increased or decreased, but there was no significant difference ($p > 0.05$). The Ψ_{tlp} and π_0 of *N. sibirica* under high water were significantly higher than those under low water ($p < 0.05$). The Ψ_{tlp} and π_0 of *P. euphratica* and *A. hendersonii* showed no apparent regularity (Figure 6).

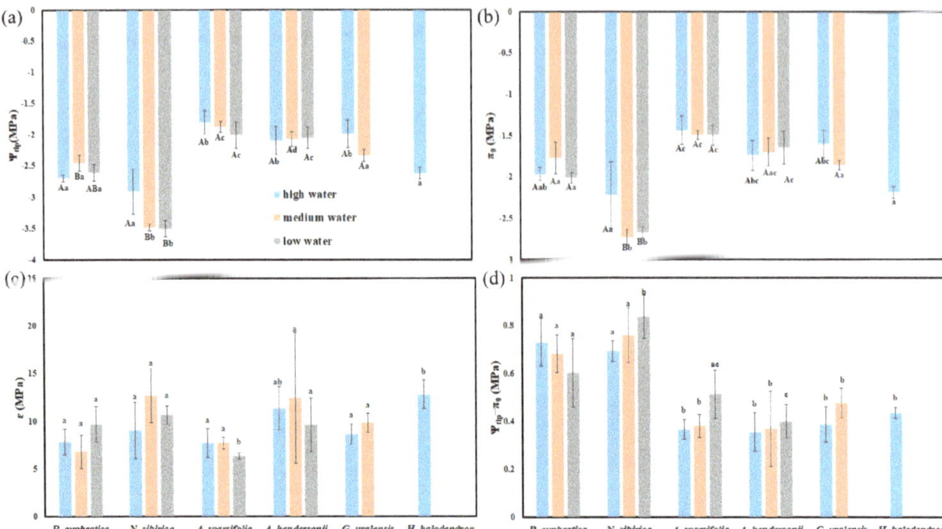

Figure 6. The differences in hydraulic parameters between water gradients and species. (**a**) The difference in the turgor loss point (Ψ_{tlp}); (**b**) the difference in the saturated osmotic potential (π_0); (**c**) the difference in the elastic mudulus (ε); (**d**) the difference in osmotic adjustment ability ($\Psi_{tlp}-\pi_0$). Different lowercase letters indicate significant differences between species, and different uppercase letters indicate significant differences between water gradients (mean + SE; N = 4), $\alpha = 0.05$.

3.5. Relationship between FWU, Leaf Structure, and Hydraulic Parameters

The MFA results (Figure 7) showed that the first and second axes accounted for 42.1% and 26.2% of the total variance, respectively. The hydraulic parameters had the most significant influence on the FWU parameters. The RV coefficients between the FWU parameters and hydraulic parameters, leaf wettability, and leaf structure were 0.38, 0.31, and 0.26, respectively (Figure 7). FWU$_{capacity}$ was positively correlated with LT and k_{45} ($R^2 = 0.66$; $R^2 = 0.3$) and negatively correlated with Ψ_{tlp}, π_0, and CA ($R^2 = -0.73$; $R^2 = -0.73$; $R^2 = -0.33$). The initial water uptake rate (k_{45}) was negatively correlated with CA ($R^2 = -0.59$). The turgor loss point (Ψ_{tlp}) was positively correlated with π_0 and CA ($R^2 = 0.96$; $R^2 = 0.48$), which was negatively correlated with ΔM, FWU$_{capacity}$, and LT ($R^2 = -0.71$; $R^2 = -0.73$; $R^2 = -0.48$) (Figure S1).

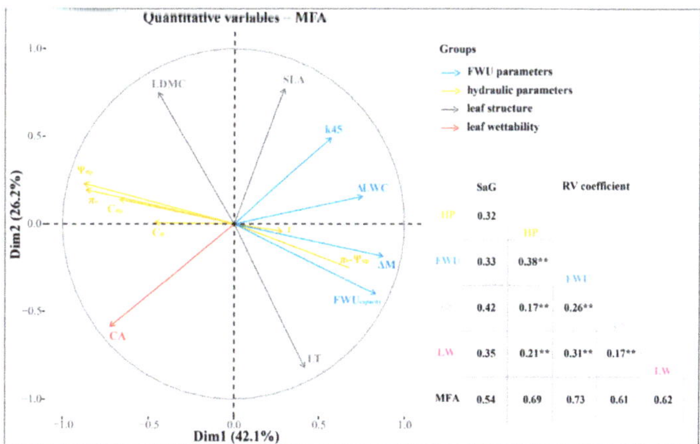

Figure 7. Multivariate factor analysis and correlation analysis of six plants' FWU parameters, leaf structure, and hydraulic parameters. Leaf structure (LS) includes LDMC and LT; FWU parameters (FWU) include ΔM, ΔLWC, FWU$_{capacity}$, and k_{45}; hydraulic parameters (HP) include Ψ_{tlp}, π_0, ε, and π_0-Ψ_{tlp}; leaf wettability (LW) represents CA; SaG denotes species and gradients; k_{45} represents the initial foliar water uptake rate; LDMC represents leaf dry matter content; SLA represents specific leaf area; C$_{ft}$ represents capacitance at full turgor; C$_{tlp}$ represents capacitance at turgor loss point; CA represents leaf contact angle. Blue, yellow, gray, and red represent FWU, HP, LS, and LW, respectively, ** $p < 0.01$.

4. Discussion

4.1. FWU Strategies of Six Plants

The leaves of all six plants showed water absorption phenomena after the soaking experiments, and different FWU strategies were adopted. *G. uralensis* showed a high k and low FWU storage. Consistent with the research results of Boanares et al. [5–7], there is a negative correlation between the k and FWU storage of plants. *N. sibirica* and *H. halodendron* showed a high k as well as high FWU storage. This result strongly proves that a high k and high FWU storage can also coexist in the same plant so that it has higher water use efficiency [5–7]. However, *P. euphratica*, *A. hendersonii*, and *A. sparsifolia* showed a low k and low FWU storage, which is different from the previous research results [5–7] and may have been caused by the differences in the climate and environment of the study area. The previous study area was mainly located in Brazil's tropical high-fog habitat. This study area is located in northwest China and belongs to a typical arid ecosystem [37]. Due to the difference in water input, plants living in arid areas have different water acquisition and maintenance mechanisms [48], resulting in different water adaptation strategies.

4.2. Effects of Leaf Wettability and Leaf Structure on FWU Strategies

The change in the foliar water uptake accumulation per unit area (FWU$_{capacity}$) of the six plants with time presents two patterns, which may be mainly related to the wettability of the leaves. Among them, the changes in the FWU$_{capacity}$ of *P. euphratica*, *N. sibirica*, *H. halodendron*, *A. sparsifolia*, and *A. hendersonii* showed a logistic curve growth pattern with time (Figure 3, Table S2). The existence of cuticles (hydrophobicity) in the early stage of FWU affects the leaf wettability, resulting in a low k (the first part of the logistic curve). With the extension of the water absorption time, FWU will promote the rehydration of the epidermis and accelerate the k (the second part of the logistic curve). With the increase in FWU, the k of leaves gradually decreases and finally approaches zero, and the leaves no longer absorb water (the third part of the logistic curve; Figure 4f), which is consistent with the research results of Guzmán-Delgado et al. [49]. However, *G. uralensis* leaves have high wettability (Table 2) [50]. The upper and lower cuticles are thin [51,52], which reduces the resistance of FWU [49,53–55] and then produces the maximum initial k (Figure 3), so the water absorption of *G. uralensis* leaves shows an exponential growth pattern with time (Figures 2 and 3e, Table S2), which can be divided into a fast water absorption period, slow water absorption period, and saturation period. This conclusion is consistent with the research results of Liang [45], Li [24], and Guzmán-Delgado [49].

The difference between leaf wettability and leaf structural characteristics shows that *N. sibirica* and *H. halodendron* adopt different adaptation methods to maintain high FWU strategies. *N. sibirica* leaves are thick (Table 3) and have high wettability (Table 2), which may be one of the reasons why *N. sibirica* has a high FWU capacity. Consistent with previous research results [17,56–58], plants with thicker leaves have higher FWU storage, while plants with hairy leaves or succulent leaves have higher leaf wettability, and their k is significantly higher than that of plants with lower leaf wettability. However, there was specificity between the leaf wettability, structural characteristics, and FWU strategies of *H. halodendron* leaves. The *H. halodendron* results showed that the leaves were thin (Table 3) and the leaf epidermis was hydrophobic (Table 2). According to the results of previous studies, the higher the leaf wettability, the higher the k [57,58], and the thicker the leaf, the higher the FWU storage [6–9], indicating that the FWU strategies of *H. halodendron* are theoretically a low k and low FWU storage. However, the actual research results are exactly the opposite. It may be that the stomata on the epidermis of the *H. halodendron* leaves are densely distributed, which leads to the hydrophobicity of the leaves [44]. At the same time, the stomata are also the primary way for the leaves to absorb water [59]. Densely distributed stomata will further promote FWU, thus producing a strong FWU capacity.

4.3. Effects of Stomatal Regulation on FWU Strategies

The turgor loss point (Ψ_{tlp}) is the osmotic potential at the initial plasmolysis, reflecting the plants' limited osmotic potential to maintain the lowest turgor. It is one of the best indicators to measure the drought tolerance of plants. The lower the Ψ_{tlp} value, the stronger the ability of plants to maintain turgor and the stronger their tolerance to drought [60–63]. In arid or semi-arid habitats with severe water shortage, the average value of the Ψ_{tlp} is about three times that of tropical rainforest species [64]. Secondly, the Ψ_{tlp} is one trait most closely related to the relative isohydric degree of species. The higher the Ψ_{tlp}, the more plants tend to be isohydric [64]. Meinzer et al. [65] proposed the use of the Ψ_{tlp} as an indicator to measure the degree of isohydric regulation, which has been proven in many plant species [66,67]. Therefore, based on the value of the Ψ_{tlp}, this paper judged the stomatal regulation behavior of plants. *N. sibirica* and *H. halodendron* are more inclined to adopt anisohydric regulation, while *P. euphratica. A. hendersonii, A. sparsifolia*, and *G. uralensis* are more inclined to adopt isohydric regulation.

This study found that the FWU strategies are related not only to leaf wettability and leaf structure but also to stomatal regulation behavior. The FWU capacity decreases with the increase in the Ψ_{tlp} of different plants (Figure 7). That is, anisohydric plants with faster water loss rates have a higher FWU capacity, and FWU can be used as a temporary water source to reduce their hydraulic risks. Species with a low FWU capacity will rely more on alternative strategies, such as stomatal regulation behavior that tends to be isohydric, to reduce water loss and maintain leaf turgor under drought stress, which is consistent with the results of Eller et al. [68]. For example, for *N. sibirica* and *H. halodendron*, as plants that are more inclined to adopt anisohydric regulation, their stomata are insensitive to environmental changes and always maintain a certain opening. Their leaves have a significantly higher FWU capacity than other plants [69–72]. *P. euphratica* is also more inclined to adopt anisohydric regulation, and its FWU capacity is significantly lower than that of *N. sibirica* and *H. halodendron* (Figure 5; Table S4). Compared with other works, the FWU capacity of trees in temperate arid areas is significantly lower than that of *P. euphratica* [66]. Therefore, when comparing trees, *P. euphratica* has a higher FWU capacity, which accords with the characteristics of plants with anisohydric regulation. However, for *A. sparsifolia, A. hendersonii*, and *G. uralensis*, as plants that are more inclined to adopt isobaric regulation, their stomata are sensitive to environmental changes. Due to the influence of plant internal factors, soil moisture status, and atmospheric vapor pressure deficit, plants will adopt a conservative strategy to reduce water loss by closing their stomata, so they have a relatively low FWU capacity [31,34].

With the deepening of drought stress, stomatal regulation behavior will also lead plants to adopt different FWU strategies. With the increase in drought stress, the FWU parameters of *N. sibirica* leaves increased significantly (Figure 5), and the Ψ_{tlp} decreased significantly (Figure 6), which is consistent with the results of Schreel et al. The FWU of anisohydric plants increases with the decrease in soil water potential [69,71], and the Ψ_{tlp} changes with the degree of soil drought [72]. However, the FWU parameters and Ψ_{tlp} of *P. euphratica* have no obvious regularity with the deepening of drought stress. The main reason may be that *P. euphratica* is an arbor, and its root system is very developed and has a high drought tolerance [73]. The water gradient selected in the plot is not sufficient to significantly impact its water physiology. In summary, anisohydric plants can reduce the effects of drought stress at the leaf level by increasing FWU [35].

4.4. Relationship between FWU and Leaf Wettability, Leaf Structure, and Stomatal Behavior

Comprehensive analysis showed that the effect of stomatal regulation on FWU capacity was higher than that of leaf wettability and structure (Figure 7b). For example, for *H. halodendron*, as a plant that is more biased towards anisohydric regulation, its stomata always maintain a certain degree of openness, and water loss is faster. Because the stomata are also the main route for FWU [58], although the epidermis of *H. halodendron* leaves is hydrophobic (Table 2) and the leaves are thin (Table 3), its FWU capacity is still significantly higher than that of other isohydric plants (plants that keep water by closing their stomata in stress environments). Therefore, stomatal regulation behavior is the main reason for high k as well as high FWU storage of anisohydric-regulating plants (*H. halodendron* and *N. sibirica*).

5. Conclusions

Our study found that the leaves of all six plants had absorbing water phenomena, and their water absorption ability was significantly different due to the influence of leaf wettability, leaf structure, and iso/anisohydric regulation behavior. Affected by the wettability of the leaves, the $FWU_{capacity}$ shows two changing laws with time: (i) when the wettability of the leaves is high, the $FWU_{capacity}$ shows an exponential growth pattern with time; (ii) when the leaf wettability is low, the change in $FWU_{capacity}$ with time shows a logistic curve growth mode. Moreover, the FWU capacity is most affected by iso/anisohydric regulation behavior compared with leaf wettability and structure. This study deepens our understanding of FWU strategies and provides a theoretical basis for understanding the mechanism of plant water use in arid regions.

Supplementary Materials: The following supporting information can be downloaded at: https://www.mdpi.com/article/10.3390/f14030551/s1. Figure S1: Pearson correlation analysis (heat map) between FWU parameters and leaf structural, leaf wettability, hydraulic parameters; Table S1: plant basic information table; Table S2: fitting model of FWUcapacity changing with time; Table S3: two-way ANOVA table; Table S4: analysis of leaf water relationship among species under water gradient.

Author Contributions: Conceptualization, H.W. and J.Y.; methodology, H.W.; software, H.W.; formal analysis, H.W.; investigation, H.W.; data curation, H.W.; writing—original draft preparation, H.W.; writing—review and editing, Z.L.; visualization, H.W.; supervision, J.Y.; project administration, J.Y.; funding acquisition, J.Y. All authors have read and agreed to the published version of the manuscript.

Funding: This research was financially supported by the National Natural Science Foundation of China (42171026), Xinjiang Uygur Autonomous Region innovation environment Construction special project & Science and technology innovation base construction project (PT2107) and Xinjiang Uygur Autonomous Region Graduate Research and Innovation Project (XJ2021G043).

Institutional Review Board Statement: Not applicable.

Informed Consent Statement: Not applicable.

Data Availability Statement: Data will be made available on request.

Conflicts of Interest: The authors declare no conflict of interest.

References

1. Burkhardt, J. Hygroscopic Particles on Leaves: Nutrients or Desiccants? *Ecol. Monogr.* **2010**, *80*, 369–399. [CrossRef]
2. Feng, T.J.; Zhang, Z.Q.; Zhang, L.X.; Xu, W.; He, J.S. Review on the influencing factors and functions of condensated water in arid and semi-arid ecosystems. *Acta Ecol. Sin.* **2021** *41*, 456–468.
3. Chin, A.R.O.; Guzmán-Delgado, P.; Sillett, S.C.; Kerhoulas, L.P.; Ambrose, A.R.; McElrone, A.R.; Zwieniecki, M.A. Tracheid Buckling Buys Time, Foliar Water Uptake Pays It Back: Coordination of Leaf Structure and Function in Tall Redwood Trees. *Plant Cell Env.* **2022**, *45*, 2607–2616. [CrossRef]
4. Darby, A.; Draguljić, D.; Glunk, A.; Gotsch, S.G. Habitat Moisture Is an Important Driver of Patterns of Sap Flow and Water Balance in Tropical Montane Cloud Forest Epiphytes. *Oecologia* **2016**, *182*, 357–371. [CrossRef] [PubMed]
5. Boanares, D.; Kozovits, A.R.; Lemos-Filho, J.P.; Isaias, R.M.S.; Solar, R.R.R.; Duarte, A.A.; Vilas-Boas, T.; França, M.G.C. Foliar Water-Uptake Strategies Are Related to Leaf Water Status and Gas Exchange in Plants from a Ferruginous Rupestrian Field. *Am. J. Bot.* **2019**, *106*, 935–942. [CrossRef]
6. Boanares, D.; Jovelina da-Silva, C.; Mary dos Santos Isaias, R.; Costa França, M.G. Oxidative Metabolism in Plants from Brazilian Rupestrian Fields and Its Relation with Foliar Water Uptake in Dry and Rainy Seasons. *Plant Physiol. Biochem.* **2020**, *146*, 457–462. [CrossRef] [PubMed]
7. Boanares, D.; Isaias, R.R.M.S.; de Sousa, H.C.; Kozovits, A.R. Strategies of Leaf Water Uptake Based on Anatomical Traits. *Plant Biol.* **2018**, *20*, 848–856. [CrossRef]
8. Lima, J.F.; Boanares, D.; Costa, V.E.; Moreira, A.S.F.P. Do Photosynthetic Metabolism and Habitat Influence Foliar Water Uptake in Orchids? *Plant Biol.* **2023**, *25*, 257–267. [CrossRef]
9. Gotsch, S.G.; Nadkarni, N.; Darby, A.; Glunk, A.; Dix, M.; Davidson, K.; Dawson, T.E. Life in the Treetops: Ecophysiological Strategies of Canopy Epiphytes in a Tropical Montane Cloud Forest. *Ecol. Monogr.* **2015**, *85*, 393–412. [CrossRef]
10. Pan, Z.L.; Guo, W.; Wang, T.; Li, Y.P.; Yang, S.J. Research progress on foliar water uptake. *Plant Physiol. J. China* **2021**, *57*, 19–32. [CrossRef]
11. Dawson, T.E.; Goldsmith, G.R. The Value of Wet Leaves. *New Phytol.* **2018**, *219*, 1156–1169. [CrossRef]
12. Li, Z.-K.; Gong, X.-W.; Wang, J.-L.; Chen, Y.-D.; Liu, F.-Y.; Li, H.-P.; Lü, G.-H. Foliar Water Uptake Improves Branch Water Potential and Photosynthetic Capacity in *Calligonum mongolicum*. *Ecol. Indic.* **2023**, *146*, 109825. [CrossRef]
13. Schreel, J.D.M.; Steppe, K. Foliar Water Uptake Changes the World of Tree Hydraulics. *NPJ Clim. Atmos. Sci.* **2019**, *2*, 1–2. [CrossRef]
14. Fan, X.; Hao, X.; Zhang, S.; Zhao, Z.; Zhang, J.; Li, Y. *Populus Euphratica* Counteracts Drought Stress through the Dew Coupling and Root Hydraulic Redistribution Processes. *Ann. Bot.* **2023**, mcac159. [CrossRef] [PubMed]
15. Akram, M.A.; Zhang, Y.; Wang, X.; Shrestha, N.; Malik, K.; Khan, I.; Ma, W.; Sun, Y.; Li, F.; Ran, J.; et al. Phylogenetic Independence in the Variations in Leaf Functional Traits among Different Plant Life Forms in an Arid Environment. *J. Plant Physiol.* **2022**, *272*, 153671. [CrossRef] [PubMed]
16. Dos Santos Garcia, J.; Boanares, D.; França, M.G.C.; Sershen; López-Portillo, J. Foliar Water Uptake in Eight Mangrove Species: Implications of Morpho-Anatomical Traits. *Flora* **2022**, *293*, 152100. [CrossRef]
17. Bryant, C.; Fuenzalida, T.I.; Zavafer, A.; Nguyen, H.T.; Brothers, N.; Harris, R.J.; Beckett, H.A.A.; Holmlund, H.I.; Binks, O.; Ball, M.C. Foliar Water Uptake via Cork Warts in Mangroves of the Sonneratia Genus. *Plant Cell Environ.* **2021**, *44*, 2925–2937. [CrossRef]
18. Martin, C.E.; von Willert, D.J. Leaf Epidermal Hydathodes and the Ecophysiological Consequences of Foliar Water Uptake in Species of Crassula from the Namib Desert in Southern Africa. *Plant Biol.* **2000**, *2*, 229–242. [CrossRef]
19. Neinhuis, C.; Barthlott, W. Seasonal Changes of Leaf Surface Contamination in Beech, Oak, and Ginkgo in Relation to Leaf Micromorphology and Wettability. *New Phytol.* **1998**, *138*, 91–98. [CrossRef]
20. Wagner, P.; Furstner, R.; Barthlott, W.; Neinhuis, C. Quantitative Assessment to the Structural Basis of Water Repellency in Natural and Technical Surfaces. *J. Exp. Bot.* **2003**, *54*, 1295–1303. [CrossRef]
21. Fernandez, V.; Brown, P.H. From Plant Surface to Plant Metabolism: The Uncertain Fate of Foliar-Applied Nutrients. *Front. Plant Sci.* **2013**, *4*, 289. [CrossRef] [PubMed]
22. Fernández, V.; Sancho-Knapik, D.; Guzmán, P.; Peguero-Pina, J.J.; Gil, L.; Karabourniotis, G.; Khayet, M.; Fasseas, C.; Heredia-Guerrero, J.A.; Heredia, A.; et al. Wettability, Polarity, and Water Absorption of Holm Oak Leaves: Effect of Leaf Side and Age. *Plant Physiol.* **2014**, *166*, 168–180. [CrossRef] [PubMed]
23. Pan, Z.L.; Guo, W.; Zhang, Y.J.; Schreel, J.D.M.; Gao, J.Y.; Li, Y.P.; Yang, S.J. Leaf Trichomes of Dendrobium Species (Epiphytic Orchids) in Relation to Foliar Water Uptake, Leaf Surface Wettability, and Water Balance. *Environ. Exp. Bot.* **2021**, *190*, 104568. [CrossRef]

24. Li, J.J.; Bai, G.S.; Zhang, R. Water absorption of common trees leaves in loess hilly and gully region of Northern Shaanxi. *Chin. Soil Water Conserv. Sci.* **2013**, *11*, 99–102. [CrossRef]
25. Guzmán-Delgado, P.; Laca, E.; Zwieniecki, M.A. Unravelling Foliar Water Uptake Pathways: The Contribution of Stomata and the Cuticle. *Plant Cell Environ.* **2021**, *44*, 1728–1740. [CrossRef]
26. Koch, K.; Hartmann, K.D.; Schreiber, L.; Barthlott, W.; Neinhuis, C. Influences of Air Humidity during the Cultivation of Plants on Wax Chemical Composition, Morphology and Leaf Surface Wettability. *Environ. Exp. Bot.* **2006**, *56*, 1–9. [CrossRef]
27. Chin, A.R.O.; Guzmán-Delgado, P.; Kerhoulas, L.P.; Zwieniecki, M.A. Acclimation of Interacting Leaf Surface Traits Affects Foliar Water Uptake. *Tree Physiol.* **2022**, tpac120. [CrossRef]
28. Roth-Nebelsick, A.; Hacke, U.G.; Voigt, D.; Schreiber, S.G.; Krause, M. Foliar Water Uptake in Pinus Species Depends on Needle Age and Stomatal Wax Structures. *Ann. Bot.* **2022**, mcac141. [CrossRef]
29. Goldsmith, G.R.; Bentley, L.P.; Shenkin, A.; Salinas, N.; Blonder, B.; Martin, R.E.; Castro-Ccossco, R.; Chambi-Porroa, P.; Diaz, S.; Enquist, B.J.; et al. Variation in Leaf Wettability Traits along a Tropical Montane Elevation Gradient. *New Phytol.* **2017**, *214*, 989–1001. [CrossRef]
30. Burkhardt, J.; Basi, S.; Pariyar, S.; Hunsche, M. Stomatal Penetration by Aqueous Solutions–an Update Involving Leaf Surface Particles. *New Phytol.* **2012**, *196*, 774–787. [CrossRef]
31. Luo, D.D.; Wang, C.K.; Jin, Y. Plant water-regulation strategies: Isohydric versus anisohydric behavior. *Chin. J. Plant Ecol.* **2017**, *41*, 1020–1032.
32. Tardieu, F.; Simonneau, T. Variability among Species of Stomatal Control under Fluctuating Soil Water Status and Evaporative Demand: Modelling Isohydric and Anisohydric Behaviours. *J. Exp. Bot.* **1998**, *49*, 419–432. [CrossRef]
33. Klein, T. The Variability of Stomatal Sensitivity to Leaf Water Potential across Tree Species Indicates a Continuum between Isohydric and Anisohydric Behaviours. *Funct. Ecol.* **2014**, *28*, 1313–1320. [CrossRef]
34. Eller, C.B.; Lima, A.L.; Oliveira, R.S. Cloud Forest Trees with Higher Foliar Water Uptake Capacity and Anisohydric Behavior Are More Vulnerable to Drought and Climate Change. *New Phytol.* **2016**, *211*, 489–501. [CrossRef]
35. Schreel, J.D.M.; von der Crone, J.S.; Kangur, O.; Steppe, K. Influence of Drought on Foliar Water Uptake Capacity of Temperate Tree Species. *Forests* **2019**, *10*, 562. [CrossRef]
36. Meng, X.Y.; Meng, B.C.; Wang, Y.J.; Liu, Z.H.; Ji, X.N.; Yu, D.L. Influence of Climate Change and Human Activities on Water Resources in Ebinur Lake in Recent 60 Years. *Hydrol. China* **2015**, *35*, 90–96.
37. Qin, W.H.; Meng, W.Q. Desert Salt Lake–Ebinur Lake National Nature Reserve. *Lifeworld China* **2020**, *7*, 16–27.
38. Wang, X.F. Development Countermeasures of Psammophytes in Northwest China. *Rural Technol.* **2022**, *13*, 115–117. [CrossRef]
39. Zhang, X.N.; Li, Y.; He, X.M.; Lv, G.H. Effects of soil water and salinity on relationships between desert plant functional diversity and species diversity. *Chin. J. Ecol.* **2019**, *38*, 2354–2360. [CrossRef]
40. Li, Y.; Sima, Y.Z.B.; Dong, Y.; Sheng, Y.C.; Tan, J. Analysis of Variation Rules and Abrupt Changes of Precipitation in Aibi Lake Oasis. *Water-Sav. Irrig. China* **2017**, *10*, 41–45.
41. Wang, S.Y.; Lv, G.H.; Jiang, L.M.; Wang, H.F.; Li, Y.; Wang, J.L. Multi-scale Analysis on Functional Diversity and Phylogenetic Diversity of Typical Plant Community in Ebinur Lake. *Chin. J. Ecol. Environ.* **2020**, *29*, 889–900. [CrossRef]
42. Rosado, B.H.P.; Holder, C.D. The Significance of Leaf Water Repellency in Ecohydrological Research: A Review. *Ecohydrology* **2013**, *6*, 150–161. [CrossRef]
43. Holder, C.D. Leaf Water Repellency of Species in Guatemala and Colorado (USA) and Its Significance to Forest Hydrology Studies. *J. Hydrol.* **2007**, *336*, 147–154. [CrossRef]
44. Shi, H.; Wang, H.X.; Li, Y.Y. Wettability on plant leaf surface and its ecological significance. *Acta Ecol. Sin.* **2011**, *31*, 4287–4298.
45. Liang, X.; Su, D.; Yin, S.; Wang, Z. Leaf Water Absorption and Desorption Functions for Three Turfgrasses. *J. Hydrol.* **2009**, *376*, 243–248. [CrossRef]
46. Tyree, M.T.; Hammel, H.T. The Measurement of the Turgor Pressure and the Water Relations of Plants by the Pressure-Bomb Technique. *J. Exp. Bot.* **1972**, *23*, 267–282. [CrossRef]
47. Leaf Pressure-Volume Curve Parameters. PROMETHEUS. Available online: https://prometheusprotocols.net/function/water-relations/pressure-volume-curves/leaf-pressure-volume-curve-parameters/ (accessed on 24 October 2022).
48. Reich, P.B.; Wright, I.J.; Cavender-Bares, J.; Craine, J.M.; Oleksyn, J.; Westoby, M.; Walters, M.B. The Evolution of Plant Functional Variation: Traits, Spectra, and Strategies. *Int. J. Plant Sci.* **2003**, *164*, 143–164. [CrossRef]
49. Guzmán-Delgado, P.; Mason Earles, J.; Zwieniecki, M.A. Insight into the Physiological Role of Water Absorption via the Leaf Surface from a Rehydration Kinetics Perspective. *Plant Cell Environ.* **2018**, *41*, 1886–1894. [CrossRef]
50. Ma, C.Y.; Wang, W.Q.; Zhao, Y.X.; Xiao, K. Study on leaf anatomical structure of Glycyrrhiza uralensis. *Chin. J. Tradit. Chin. Med.* **2009**, *34*, 1034–1037.
51. Yin, Q.L. Leaf Anatomical Structure of Main Plants and Its Environmental Adaptations in the Hilly-Gullied Platrau Region. Master's Thesis, Northwest University, Kirkland, WA, USA, 2015. Available online: http://kns.cnki.net/kcms/detail/frame/list.aspx?dbcode=CMFD&filename=1015333288.nh&dbname=CMFD201601&RefType=1&vl=p3BxlfZtEZV6ZlMXQ2T_6zj4w92WuEZpqvC4gPiHsafUhB_cklND0PZDZUpK1FZy (accessed on 26 November 2022).
52. Wang, S.J.; Ren, L.Q.; Han, Z.W.; Qiu, Z.M.; Zhou, C.H. non–smooth morphology of typical plant leaf surface and its anti–adhesion and hydrophobicity. *Chin. J. Agric. Eng.* **2005**, *9*, 16–19.

53. Liu, Y.X.; Ma, Y.L.; Lan, H.Y. Advances in morphology and function of plant non-glandular trichomes. *Plant Physiol. China* **2018**, *54*, 1527–1534. [CrossRef]

54. Schwerbrock, R.; Leuschner, C. Air Humidity as Key Determinant of Morphogenesis and Productivity of the Rare Temperate Woodland Fern Polystichum Braunii. *Plant Biol.* **2016**, *18*, 649–657. [CrossRef] [PubMed]

55. Yang, J.H.; Meng, N.; Zuo, F.Y.; Liu, Y.J.; Li, J.K. Leaf water potential, anatomical structure of epidermis and leaf salt-tolerant characteristics of wild *Nitraria tangutorum* B.and *Suaeda glauca* B. *J. Tianjin Agric. Univ. China* **2013**, *20*, 5–8.

56. Chen, L.; Yang, X.G.; Song, N.P.; Yang, M.X.; Xiao, X.P.; Wang, X. Leaf water uptake strategy of plant in the arid and semi–arid region of Ningxia. *J. Zhejiang Univ. China (Agric. Life Sci.)* **2013**, *39*, 565–574.

57. Boanares, D.; Ferreira, B.G.; Kozovits, A.R.; Sousa, H.C.; Isaias, R.M.S.; França, M.G.C. Pectin and Cellulose Cell Wall Composition Enables Different Strategies to Leaf Water Uptake in Plants from Tropical Fog Mountain. *Plant Physiol. Biochem.* **2018**, *122*, 57–64. [CrossRef]

58. Berry, Z.C.; Emery, N.C.; Gotsch, S.G.; Goldsmith, G.R. Foliar Water Uptake: Processes, Pathways, and Integration into Plant Water Budgets. *Plant Cell Environ.* **2019**, *42*, 410–423. [CrossRef]

59. Rascio, A.; Nicastro, G.; Carlino, E.; Di Fonzo, N. Differences for Bound Water Content as Estimated by Pressure-Volume and Adsorption Isotherm Curves. *Plant Sci.* **2005**, *169*, 395–401. [CrossRef]

60. Powell, T.L.; Wheeler, J.K.; de Oliveira, A.A.R.; Lola da Costa, A.C.; Saleska, S.R.; Meir, P.; Moorcroft, P.R. Differences in Xylem and Leaf Hydraulic Traits Explain Differences in Drought Tolerance among Mature Amazon Rainforest Trees. *Glob. Change Biol.* **2017**, *23*, 4280–4293. [CrossRef]

61. Zhu, S.-D.; Chen, Y.-J.; Ye, Q.; He, P.-C.; Liu, H.; Li, R.-H.; Fu, P.-L.; Jiang, G.-F.; Cao, K.-F. Leaf Turgor Loss Point Is Correlated with Drought Tolerance and Leaf Carbon Economics Traits. *Tree Physiol.* **2018**, *38*, 658–663. [CrossRef]

62. Huo, J.; Shi, Y.; Zhang, H.; Hu, R.; Huang, L.; Zhao, Y.; Zhang, Z. More Sensitive to Drought of Young Tissues with Weak Water Potential Adjustment Capacity in Two Desert Shrubs. *Sci. Total Environ.* **2021**, *790*, 148103. [CrossRef]

63. Bartlett, M.K.; Scoffoni, C.; Sack, L. The Determinants of Leaf Turgor Loss Point and Prediction of Drought Tolerance of Species and Biomes: A Global Meta-Analysis. *Ecol. Lett.* **2012**, *15*, 393–405. [CrossRef] [PubMed]

64. Meinzer, F.C.; Woodruff, D.R.; Marias, D.E.; Smith, D.D.; McCulloh, K.A.; Howard, A.R.; Magedman, A.L. Mapping 'Hydroscapes' along the Iso-to Anisohydric Continuum of Stomatal Regulation of Plant Water Status. *Ecol. Lett.* **2016**, *19*, 1343–1352. [CrossRef] [PubMed]

65. Proxies for Stringency of Regulation of Plant Water Status (Iso/Anisohydry): A Global Data Set Reveals Coordination and Trade-Offs among Water Transport Traits. *Tree Physiol.* **2019**, *39*, 122–134. [CrossRef] [PubMed]

66. Li, X.; Blackman, C.J.; Peters, J.M.R.; Choat, B.; Rymer, P.D.; Medlyn, B.E.; Tissue, D.T. More than Iso/Anisohydry: Hydroscapes Integrate Plant Water Use and Drought Tolerance Traits in 10 Eucalypt Species from Contrasting Climates. *Funct. Ecol.* **2019**, *33*, 1035–1049. [CrossRef]

67. Eller, C.B.; Burgess, S.S.O.; Oliveira, R.S. Environmental Controls in the Water Use Patterns of a Tropical Cloud Forest Tree Species, Drimys Brasiliensis (Winteraceae). *Tree Physiol.* **2015**, *35*, 387–399. [CrossRef]

68. Wu, Y.; Song, L.; Liu, W.; Liu, W.; Li, S.; Fu, P.; Shen, Y.; Wu, J.; Wang, P.; Chen, Q.; et al. Fog Water Is Important in Maintaining the Water Budgets of Vascular Epiphytes in an Asian Tropical Karst Forests during the Dry Season. *Forests* **2018**, *9*, 260. [CrossRef]

69. Schaepdryver, K.H.D.; Goossens, W.; Naseef, A.; Kalpuzha Ashtamoorthy, S.; Steppe, K. Foliar Water Uptake Capacity in Six Mangrove Species. *Forests* **2022**, *13*, 951. [CrossRef]

70. Carmichael, M.J.; White, J.C.; Cory, S.T.; Berry, Z.C.; Smith, W.K. Foliar Water Uptake of Fog Confers Ecophysiological Benefits to Four Common Tree Species of Southeastern Freshwater Forested Wetlands. *Ecohydrology* **2020**, *13*, 2240. [CrossRef]

71. Liu, Z.; Zhang, H.; Yu, X.; Jia, G.; Jiang, J. Evidence of Foliar Water Uptake in a Conifer Species. *Agric. Water Manag.* **2021**, *255*, 106993. [CrossRef]

72. Maréchaux, I.; Bartlett, M.K.; Iribar, A.; Sack, L.; Chave, J. Stronger Seasonal Adjustment in Leaf Turgor Loss Point in Lianas than Trees in an Amazonian Forest. *Biol. Lett.* **2017**, *13*, 20160819. [CrossRef]

73. Long, Y.X. Water Regulation Strategies of Five Dominant Woody Plants in Desert Forest of Ebinur Lake Basin. Master's Thesis, Xinjiang University, Urumqi, China, 2021. [CrossRef]

Disclaimer/Publisher's Note: The statements, opinions and data contained in all publications are solely those of the individual author(s) and contributor(s) and not of MDPI and/or the editor(s). MDPI and/or the editor(s) disclaim responsibility for any injury to people or property resulting from any ideas, methods, instructions or products referred to in the content.

Article

Interspecific Integration of Chemical Traits in Desert Plant Leaves with Variations in Soil Water and Salinity Habitats

Jifen Yang [1,2,3], Xueni Zhang [1,2,3,*], Danhong Song [1,2,3], Yongchang Wang [1,2,3] and Jingye Tian [1,2,3]

[1] College of Ecology and Environment, Xinjiang University, Urumqi 830017, China; yangjf005@163.com (J.Y.); sdh09290530@163.com (D.S.); wangyc526@163.com (Y.W.); jytian859@163.com (J.T.)
[2] Key Laboratory of Oasis Ecology of Education Ministry, Urumqi 830017, China
[3] Xinjiang Jinghe Observation and Research Station of Temperate Desert Ecosystem Ministry of Education, Urumqi 830017, China
* Correspondence: xnzhang@xju.edu.cn

Abstract: Understanding the relationship between soil environmental conditions and the interspecific integration of plant traits might shed light on how plants adapt to their environment. In order to clarify the adaptation strategies of desert plants in the various habitats, this study calculated interspecific trait integration (ITI) and plant trait networks (PTN) by selecting plants from high water-salinity habitat (HSM) with salt stress and low water-salinity habitat (LSM) with drought stress in the Ebinur Lake region. Eight different phytochemical traits were taken into consideration, including carbon (C), nitrogen (N), phosphorus (P), sulfur (S), potassium (K), calcium (Ca), sodium (Na), and magnesium (Mg). Six soil factors were chosen, including soil pH, water content (SVWC), electrical conductivity (EC), soil nitrogen (N), phosphorus (P), and potassium (K). The results obtained are shown below: (1) the relationship between plant leaf chemical traits was closer in HSM than in LSM, and the correlation between C and other leaf chemical traits was significant in HSM and insignificant in LSM; (2) the correlations between soil factors and ITI were not statistically significant; however, in both soil water-salinity habitats, the strength of fit between SVWC and ITI was the greatest, while the strength of fit between EC and ITI was the smallest; and (3) according to the PTN, C and Ca are the two most central traits for the growth of desert leaf chemical plants in Ebinur Lake, which is consistent with the results of the PCA. Coordination of plant leaf traits along water-salinity gradients involves many different combinations of traits, and the use of ITI and PTN can quantify the complex relationships between multiple traits to a greater extent, highlighting the multivariate mechanisms of plant response and adaptation to soil habitats. This information will help expand and optimize our ability to observe and predict desert plant responses to habitat change, providing powerful insights for assessing desert plant survival strategies.

Keywords: plant leaf chemical traits; interspecific trait integration; soil water and salinity habitats; plant trait network; desert plant

Citation: Yang, J.; Zhang, X.; Song, D.; Wang, Y.; Tian, J. Interspecific Integration of Chemical Traits in Desert Plant Leaves with Variations in Soil Water and Salinity Habitats. *Forests* **2023**, *14*, 1963. https://doi.org/10.3390/f14101963

Academic Editor: Marta Pardos

Received: 4 August 2023
Revised: 17 September 2023
Accepted: 21 September 2023
Published: 28 September 2023

Copyright: © 2023 by the authors. Licensee MDPI, Basel, Switzerland. This article is an open access article distributed under the terms and conditions of the Creative Commons Attribution (CC BY) license (https://creativecommons.org/licenses/by/4.0/).

1. Introduction

Functional traits are measurable attributes of individual plants that determine how they acquire and compete for resources and tolerate stressful conditions [1,2]. Ecology has long faced challenges in comprehending the functional connections between plant communities and their environments [3]. To investigate the relationship between traits and the environment, community ecologists have mostly concentrated on the mean and variance of traits at the species level [4]. Interspecific trait integration has not received as much attention along local environmental gradients as trait mean values and other metrics of functional diversity have [5]. The idea of trait integration stems from the discovery that an individual's or a species' traits can vary in a coordinated way, favoring some functions over others [6,7]. For instance, species adapted to severe conditions may prioritize resource acquisition functions over resource conservation activities, whereas species adapted to softer habitats may follow the reverse strategy [8,9]. Understanding how species' multidimensional functional niches along environmental gradients, limitations, and changes in resource supply affect plant adaptation patterns and community structure requires research on trait integration. It also reveals ecological processes or key drivers of community assembly [2,10].

Trait integration indicates how multiple traits differ from one another [11], providing useful insights into the functional tradeoffs underlying biodiversity patterns [8]. On a local or small scale, topographic and soil factors determine the distribution of traits [12]. Delhaye et al. proposed that the increase in trait integration with soil metal toxicity in plant communities supports the idea that highly constraining environments select increasingly coordinated sets of functional traits, possibly driving the decrease in species richness [2]. According to this, abiotic gradients may have a direct or indirect impact on species richness [13], and trait integration and how it responds to the environment may not only reflect a plant's strategy for adaptation but also have an impact on changes in the variety of a community. Gianoli and Palacio-Lopez came to the conclusion that, in some situations, flexibility and integration could be additional coping mechanisms for stress [14]. Some academics think that a trait's plasticity may vary depending on how well it integrates in various situations [15]. In other words, integration qualities have more similar plasticity than non-integration traits, and their similarity in plastic responses predicts their integration in the environment [16]. The leaves are the primary site of photosynthesis and a vital organ for plants to keep the hydrological system in balance. They are quite plastic and sensitive to environmental changes [12]. Multiple components of plant fitness help mechanically determine the environmental distribution of the species. They are therefore often used to explore plant adaptation strategies to the environment. The correlation and integration between attributes may indicate the adaptation methods of plants in arid settings, according to Yang et al.'s study of leaf traits in typical small tree and shrub plants in arid regions of northwest China [17]. Over the course of their long-term evolutionary history, these dominant species have gradually evolved a number of unique morphological and physiological adaptation traits in response to droughts [18]. Through the trait integration approach, we can understand how the interrelationships between multiple traits change [11]. Trait integration can be greatly changed by changing the external environment [6]. In order to better understand the community trait space and the ecological processes that shape it, trait integration analysis is a complementary method [2].

It is still challenging to effectively describe and integrate diverse social structures and predict ecosystem-level responses to environmental disturbance using trait assessments at the individual level [19]. Recently, some scholars have studied the interdependence between network analyses and traits by constructing plant trait networks to quantify their parameters. Using network analyses, they can visualize and quantify trait integration [7,20,21]. Burton et al. found a small clustering of a priori trait characteristics when employing network analyses to explore trait integration and functional differences within coexisting species of plants [22]. Biogeochemical cycles in desert ecosystems are often slower than those in forest and aquatic ecosystems, which leads to arid soils and poor net primary productivity [23]. Gao et al. improved our comprehension of how vulnerable arid ecosystems are to environmental change by building functional trait networks that included leaf, root, and component biomass in desert ecosystems in China. These networks revealed the complex relationships between the three and the key traits of the ecosystem [24]. Desert plants are a type of zonal vegetation that thrives in arid regions and has special functional characteristics. The building of desert plants' communities is significantly influenced by the trade-off relationship between their functional traits and adversity strategies [25]. In order to compare phenotype-based tactics among different species, Westoby recommended the use of trait-based dimensions. A collection of associated traits known as the trait dimension reflects the limitations and trade-offs that make up the plant phenotype [26].

The "leaf economic spectrum (Leaf economics spectrum, LES)"—a well-known relationship among an established group of leaf economic features such as specific leaf area, leaf nitrogen concentration, leaf longevity, and net photosynthesis—is a prime example [27]. The leaf economic spectrum arranges plants on a specific axis, and one end represents a fast investment-return strategy larger than the leaf surface, higher leaf nitrogen content, faster photosynthetic and respiratory rate, and short leaf life, while the other end represents a slow investment-return strategy opposite to the above characteristics [28]. However, individual plant traits are usually combined into multiple functional systems for growth and adaptation to stressful environments. Recent studies have proposed how the leaf economic spectrum, which simplifies the relationship between traits, blurs the overall pattern of plant adaptation [29,30]. Therefore, shifting the trait-based ecology perspective from axis to network view provides a better understanding of the interdependence of multiple physiological functions, which is essential to understanding the ecology and evolution of plant morphological and functional diversity [21].

Due to its role in offering mechanical stability and the supply of nutrients for plants [12], soil is thought to play a significant role in affecting leaf functional traits on an isolated regional scale [18]. The ecological environment in arid desert areas is extremely fragile, and desert plants are usually under stressful conditions of drought, salinization, and nutrient depletion [31]. Environmental elements such as soil pH, moisture, salinity, and nutrients, which change plant stoichiometric ratios through plant-soil feedback, typically have a major impact on a plant's functional traits [32]. Phylogenetically more stable than forms that are frequently tested, phytochemical elements might possibly be more directly related to ecosystem function [33]. For plant growth and the regulation of several physiological processes, the elements C, N, and P within leaves are essential [34]. By blocking sodium uptake, encouraging the absorption of K, Ca, and nitrate ions, boosting the exocytosis of Cl^-, and causing the synthesis of anti-salt compounds, Ca can improve plant salt resistance [35]. As crucial osmoregulatory components, potassium (K) and sodium (Na) aid desert plants in their tolerance to both drought and salinity stress [36]. The grouping and structure of plant groups can be revealed by analyzing the variance of plant leaf features in desert settings. This research also aids in understanding how plants react to local changes in the environment [12]. Numerous studies have shown that higher interdependence between traits allows plants to effectively acquire and mobilize resources [37], and resource-poor plants may face stronger choices and therefore tend to have tighter trait correlations and trade-offs [38]. Although it is commonly established that gradients in the environment affect average trait values, it is less clear how these gradients affect local populations' levels of trait integration [13]. In general, higher-stress circumstances are likely to result in an increase in trait integration [39]. It is possible to quantify the strategies that plants employ using a relatively limited number of traits, according to the integration of traits and trait spectra across resources (such as light, water, carbon, and nutrients) and organs [40]. In this study, we hope to study the relationship between plant leaf chemical traits and soil habitats by constructing ITI and PTN, to understand the mechanism of promotion or inhibition of different habitats on desert plants, and to comprehensively elaborate the adaptive responses of plants to their environments from a systematic perspective. Based on this, we are focusing on the following questions: (1) Which soil-water-salinity habitat has a stronger positive correlation between ITI and soil factors? We hypothesized that ITI is more closely correlated positively with soil factors in low-soil water-salinity habitat than in high-soil water-salinity habitat because interspecific trait integration is facilitated by environmental stress. (2) Is there a significant variation in PTN between the two habitats in terms of soil, water, and salinity? Because interplant trait connections might exhibit considerable changes as impacted by environmental conditions, we hypothesized that the differences in PTN complexity and central traits between the two-soil water-salinity habitats were significant. By testing the above hypotheses, this study hopes to reveal the adaptation strategies of desert plants to stressful environments from the perspective of interspecific trait integration and network formation.

2. Materials and Methods

2.1. Study Area

The Ebinur Lake Wetland National Nature Reserve (44°30′–45°09′ N, 82°36′–83°50′ E) is located in northwest Jinghe County, Bortala Mongolian Autonomous Prefecture, Xinjiang, China, in the lowest depression in the southwest margin of Junggar Basin and the water and salt collection center [41]. The climate is hot and dry throughout the year and is influenced by the normal temperate continent's dry climate. The highest and lowest recorded temperatures are 41.3 °C and −36.4 °C, respectively, with an annual average temperature of 7.8 °C. The yearly sunshine hours are 2699.87, the annual evaporation is 2221.3 mm, and the annual precipitation is 105.17 mm. One of the water sources for Ebinur Lake is the Aqikesu River, which flows through the study area and is situated on the east side of the lake. The soil around riverbanks had a high water and salt content, which gradually decreased as one moved further away from the river. Depending on how far away from the river you are, the area's vegetation mix differs. Major plant species close to the river channel are shrub plants *Lycium ruthenicum, Halostachys capsica*, and herbaceous plants *Salsola aperta*, licorice, *Glycyrrhiza uralensis*, etc. Away from the river, the main distribution of small trees is *Tamarix ramosissima*, shrubby *Kalidium foliatum*, herbaceous *Salsola collina, Halocnemum strobilaceum, Salsola ruthenica*, etc. The common species in both habitats are *Arbor Populus euphratica, Haloxylon ammodendron, Nitraria sibirica*, and *Apocynum venetum*. The herbaceous plants are *Phragmites australis, Suaeda microphylla, Halimodendron halodendron, Suaeda salsa, Reaumuria soongorica, Alhagi sparsifolia, Suaeda glauca*, and *Karelinia capsica* [41].

2.2. Plot Setting

In the Ebinur Lake Wetland National Nature Reserve, the experimental area is situated on the north bank of the Aqikesu River. The vegetation in this location is typical of the desert; the climate and surroundings are the same. The slope and appearance are the same at the same moment. There is no influence from the landscape because the height difference is so modest (between 290 and 310 m). In addition, the study area is not a human activity area and is not disturbed by human beings. According to previous studies, there is a significant gradient change of soil water and salt in the direction perpendicular to the Aqikesu River in the experimental area. Therefore, the transect of water and salt change is selected for investigation and experimentation.

The precise transect and plot layout approach is as follows: The Aqikesu River is divided into three transects, each with a 5 km interval. The soil water and salinity along the transects gradually decreased with distance from the river, and the transects were not susceptible to human disturbance. Ten to twelve 10 m × 10 m samples ($n = 32$) were created across each transect, with 0.5 km between each sample. The number of cultivated species in two sample plots is too small, which affects the subsequent analysis; therefore, this study selected 30 sample plots for later analysis.

2.3. Sample Collection

Soil sampling: In accordance with the idea of random and uniform sampling, three sites were chosen in each sample plot, and soil samples were collected from these three locations for each sample plot. This was undertaken in order to accurately reflect the soil water, salt, and nutrient status of the sample plots. Following the removal of the litter layer, soil samples ranging from 0 to 15 cm were collected. Soil samples were quickly mixed and homogenized before being placed in self-sealing bags and given a number. In the area of sampling points, soil volumetric water content (SVWC) was measured using time-domain reflectometry (TDR, Field Scout TM TDR 300, Spectrum Technologies, Inc., Plainfield, IL, USA). For different indices, other samples were naturally dried in the air.

Plant sampling: Using GPS, the latitude and longitude of the centers of each square were recorded, along with the names of each species that was present and the number of individuals that were there. Choose the primary plant species in the location of the samples, choose three individuals at random from each species, and note the plant height, crown width, and DBH (base) values of each individual; three identically sized mature leaves that were healthy, unharmed, and from three distinct individuals were flattened with clear plastic plates before being photographed. Twenty to thirty leaves, weighing about twenty grams, were collected from each of the above three individual plants using twig clippers, placed in an envelope bag, and returned to the laboratory for subsequent testing.

2.4. Sample Processing and Testing

The soil was dried, and other soil factors were determined indoors. Soil pH was determined using the acidimeter method (PHS-3C, Shanghai Yidian Scientific Instrument Co., Ltd., Shanghai, China), and soil electrical conductivity (EC) was determined using a conductivity meter (DDS-307, Shanghai Yidian Scientific Instrument Co., Ltd., Shanghai, China), with a soil-to-water ratio of 1:5. Kjeldahl nitrogen determination, $HClO_4$–H_2SO_4 molybdenum–antimony colorimetry, and atomic absorption spectrometry (HITACHI Z-2000) were each used to measure soil nitrogen, phosphorus, and potassium.

Leaf samples were pulverized and screened for leaf traits after being dried in a 75 °C oven for 48 h. Leaf C content by potassium dichromate-sulfate oxidation method; leaf N by Kjeldahl method (with H_2SO_4 used to speed up digestion); leaf P by first absorbing nitrogenous and perchloric acids and then using the molybdenum antimony colorimetric (Z−2000, Hitachi High-Technologies Trading (Shanghai) Co., Ltd., Shanghai, China) method; leaf S content by $BaSO_4$ turbidimetry; and leaf Ka, Na, Mg, and Ca content by atomic absorption spectroscopy. Specific experimental methods are described in the literature [42].

2.5. Data Processing and Analysis

Prior to completing numerical and statistical analysis, all trait indicators for each species were averaged. The Shapiro–Wilk test and Levene's test were then used to analyze all variables for normality and the chi-square hypothesis, respectively. By using cluster analysis (class average method), the habitat for soil water and salinity was identified. The 30 samples were then divided into two habitats: habitats with more water and salinity are more strongly stressed by salinity and are defined as high water-salinity habitat (HSM), and habitats with less water and salinity are defined as low water-salinity habitat (LSM). We utilized one-way analysis of variance (one-way ANOVA) to assess the variation in plant characteristics and soil physicochemical components according to water and salinity gradients. Both Pearson correlation analysis and principal component analysis (PCA) were used to investigate the association between plant leaf chemical characteristics in habitats with high and low soil water and salinity. The variance in interspecific trait integration (ITI) along the gradient was then explored using the range of eigenvalues (i.e., the difference between the eigenvalues of the first and last principal component axes) for each PCA. The variation in ITI was analyzed in relation to soil factors using a one-dimensional linear regression method. Finally, the topology of trait relationships was visualized using the IGRAPH package in R to form a plant trait network [28]. Leaf chemical traits are shown as nodes, while the correlations between them are shown as edges. The adjacency matrix $A = [a_{i,j}]$, where $a_{i,j} \in [0, 1]$, was generated by assigning associations above the significance threshold as 1 and relationships below the significance threshold as 0. We defined significance thresholds of $|r| > 0.2$ and $p < 0.05$ [21,43]. For each characteristic, two measures of network centrality were computed: degree (D), which is the number of edges on a node, and weighted degree (Dw), which is the total of a node's significant correlation coefficients [44]. Excel 2016, Origin 2023, and R 4.1.0 were used for data collection and statistical analysis.

3. Results

3.1. Leaf Chemical Traits and Their Relationships in Response to High and Low Soil Water-Salinity Habitats

With the exception of soil phosphorus (P) and potassium (K), soil factors were significantly higher in high soil water and salinity habitat than in low soil water and salinity habitat ($p < 0.05$) (Table 1). In the high and low soil water-salinity habitats, respectively, there were 16 and 17 plant species present. Since many of the plants in the two habitats shared characteristics, a total of 23 plant species were identified in the two habitats. Plants in the low soil water salinity habitat had significantly greater K and Na contents than those in the high soil water-salinity habitat, whereas the high soil water-salinity habitat had significantly higher Ca contents than the low soil water-salinity habitat.

Table 1. Soil properties and plant leaf chemical traits (mean ± standard deviation).

Features	Habitat HSM	LSM
Number of plant species	16	17
Electrical conductivity/(mS/cm)	12.02 [a] ± 4.93	1.96 [b] ± 1.27
Soil volume water concentration/%	16.64 [a] ± 2.89	2.86 [b] ± 2.35
pH	8.64 [a] ± 0.28	8.06 [b] ± 0.24
Soil N concentration/(g/kg)	0.46 [a] ± 0.18	0.16 [b] ± 0.10
Soil P concentration/(g/kg)	0.46 [a] ± 0.26	0.39 [a] ± 0.17
Soil K concentration/(g/kg)	17.37 [a] ± 5.90	16.34 [a] ± 5.61
Leaf C content	432.32 [a] ± 87.76	439.03 [a] ± 109.01
Leaf N content	19.66 [a] ± 5.28	19.25 [a] ± 4.66
Leaf P content	0.98 [a] ± 0.58	0.88 [a] ± 0.28
Leaf S content	18.52 [a] ± 13.15	16.28 [a] ± 12.49
Leaf K content	10.80 [b] ± 5.65	16.51 [a] ± 7.45
Leaf Ca content	21.48 [a] ± 11.52	14.55 [b] ± 6.85
Leaf Na content	20.13 [b] ± 25.83	39.79 [a] ± 40.73
Leaf Mg content	6.35 [a] ± 4.63	6.97 [a] ± 4.69

Note: a, b—different letters indicate significant differences in trait indicators among different soil, water, and salinity habitats ($p < 0.05$).

3.2. Relationship between Plant Leaf Chemical Traits under Different Soil Water and Salinity Habitats

Overall, trait correlations were slightly higher in high soil water-salinity habitats than in low soil water-salinity habitats (Figure 1). In high soil water and salinity habitat, C showed significant correlations with other chemical traits except S and P, significant positive correlations with Ca, and significant negative correlations with Mg, Na, N, and K. In low soil water and salinity habitat, C has a weak correlation with other chemical traits except Ca. Na showed significant positives with S and N, and it was significantly negative with Ca.

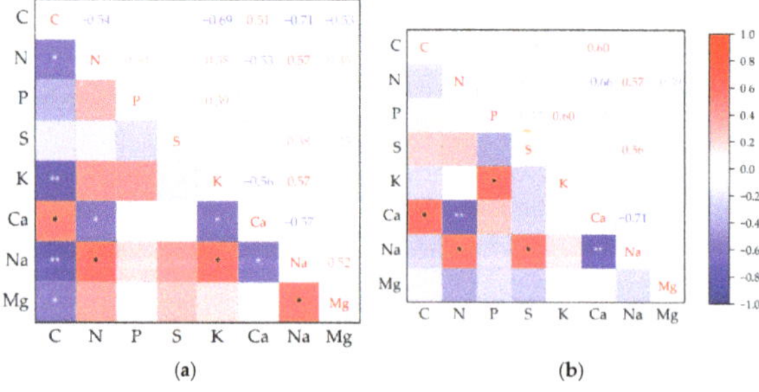

Figure 1. Pearson correlation analysis among plant leaf chemical traits in high and low soil water and salinity habitat. (**a**) high soil water and salinity habitat; (**b**) low soil water and salinity habitat. Note: $^*p \leq 0.05$; $^{**}p \leq 0.01$.

The first two principal components (PC) of the high soil water-salinity habitat explained 44.88% and 19.29% of the total variance, respectively (Table 2). PC1 mainly represented Na, N, K, C, and Ca. S explained the least for PC1, and the first principal component axis showed a large negative correlation between C, Ca, and other chemical traits. PC2 mainly represents P, S, and Mg, with C explaining the least for PC2 (Figure 2).

Table 2. Eigenvalues and explained variance (%) of high and low soil water and salinity habitat.

	Habitat	PC1	PC2	PC3	PC4
Eigenvalues	HSM	3.59	1.54	1	0.78
	LSM	2.84	1.83	1.48	0.81
Explained	HSM	44.88	19.29	12.53	9.79
variance (%)	LSM	35.52	22.83	18.5	10.07

Note: HSM-high soil water and salinity habitat; LSM-low soil water and salinity habitat.

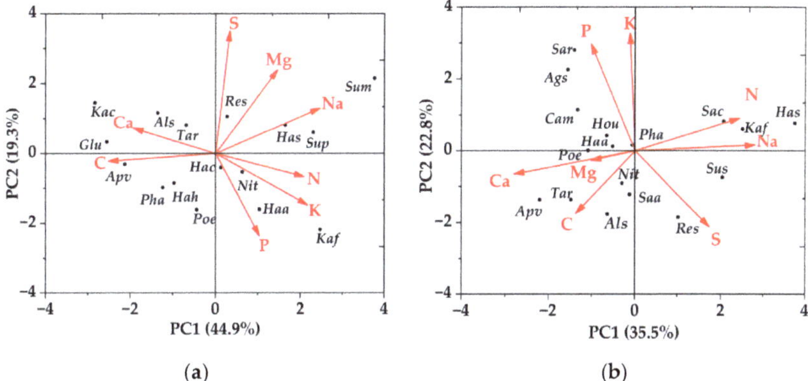

Figure 2. Principal component analysis of plant leaf chemical traits in high and low soil water and salinity habitat. (**a**) high soil water and salinity habitat; (**b**) low soil water and salinity habitat. *Als—Alhagi sparsifolia; Ags—Agriophyllum squarrosum; Apv—Apocynum venetum; Cam—Calligonum mongolicum; Hou—Horaninowia ulicina; Glu—Glycyrrhiza uralensis; Hac—Halostachys caspica; Hah—Halimodendron halodendron; Has—Halocnemum strobilaceum; Kac—Karelinia caspica; Kaf—Kalidium foliatum; Nit—Nitraria tangutorum; Pha—Phragmites australis; Res—Reaumuria songarica; Saa—Salsola aperta; Sac—Salsola collina; Sum—Suaeda microphylla; Sup—Suaeda prostrata; Sar—Salsola ruthenica; Tar—Tamarix ramosissima; Sus—Suaeda salsa; Haa—Haloxylon ammodendron; Sup—Suaeda prostrata.*

The first two principal components (PC) of the low soil water-salinity habitat explained 35.52% and 22.83% of the total variance, respectively. PC1 mainly represented Na, N, C, and Ca; K explained the least for PC1 (Table 2). All chemical traits were positively correlated with PC1, except Na, N, and S, which were negatively correlated with PC1. PC2 mainly represents K and P. Na and Mg have the least explanation for PC2. Shrub plants, *Haloxylon ammodendron* and *Populus euphratica*, were distributed around P in both high and low soil water and salinity habitat, and the distribution of trees and herbs is not irregular (Figure 2).

3.3. Interspecific Trait Integration and Trait Network Changes along Soil Water and Salinity Habitats

Under low soil water and salinity habitat, SVWC, N, and P were positively correlated with ITI, with N having the highest strength of linear fit for ITI and K being negatively correlated with ITI with a higher strength of linear fit. Under high soil water and salinity levels, all soil factors except P were positively correlated with ITI, with SVWC having the highest linear fit strength for ITI. Overall, SVWC had the strongest linear fit to ITI, and EC had the weakest linear fit to ITI in both soil water-salinity habitats (Figure 3).

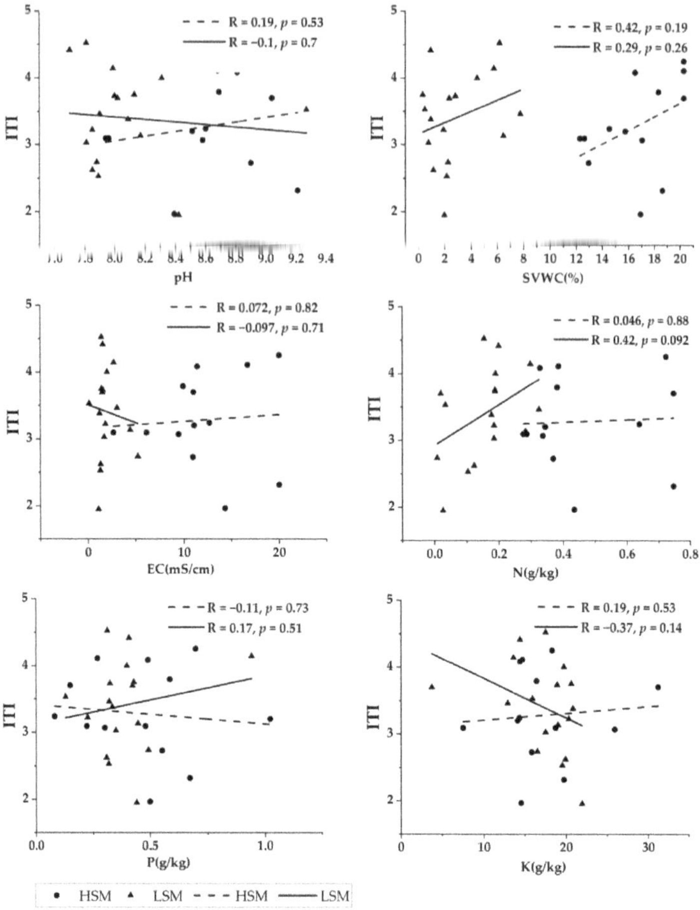

Figure 3. The relationship between pH, SVWC (soil volume and water content), EC (electrical conductivity), N (nitrogen), P (phosphorus), K (potassium), and ITI (interspecific trait integration) under high and low soil water and salinity habitat.

The plant trait network showed a stronger association of plant leaf chemical traits in high soil water-salinity habitats than in low soil water-salinity habitats, and not all chemical traits were directly connected in the trait network (Figure 4). In high soil water-salinity habitats, the greater correlations between K, Ca, Na, C, Mg, and N connected into a trait network, with C serving as the central trait that connected all other chemical traits with the exception of S and P. The central trait in the low soil water-salinity habitat was Ca; Na also had the same degree as the central trait (D = 3), with a lower weighting than the central trait (Table 3); and Mg was not connected to any other chemical traits in the trait network.

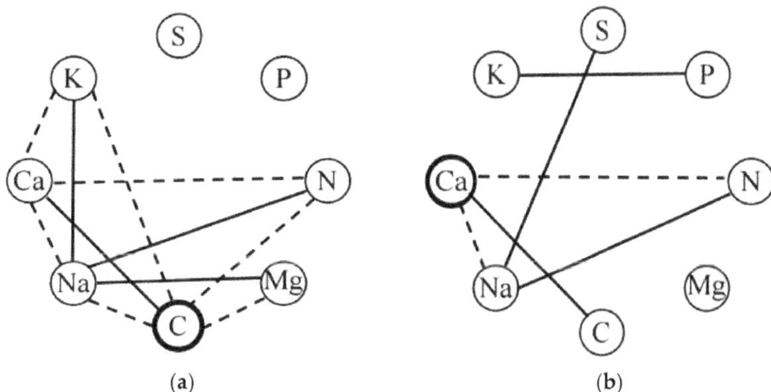

Figure 4. Plant trait networks in high and low soil water and salinity habitat. (**a**) high soil water and salinity habitat; (**b**) low soil water and salinity habitat. Black solid dashed line edges indicate positive and negative correlations, respectively. Only significant relationships ($p < 0.05$) are displayed. The traits denoted by black circles have the highest weighted degree (the total of all significant coefficients of correlation for a node) centrality value.

Table 3. Parameters of plant trait networks in high and low soil water-salinity habitats.

Trait	HSM		LSM	
	Dw	**D**	**Dw**	**D**
N	1.64	3	1.23	2
P	0.00	0	0.60	1
S	0.00	0	0.56	1
K	1.82	3	0.60	1
Ca	2.17	4	1.97	3
Na	2.94	5	1.84	3
C	2.98	5	0.60	1
Mg	1.05	2	0.00	0
Overall network	12.60	22	7.39	12

Note: HSM-high soil water and salinity habitat; LSM-low soil water and salinity habitat. Dw, or weighted degree, is the total of all significant correlation coefficients for a node; D, or degree, is the total number of edges for a node.

4. Discussion

4.1. Leaf Chemical Traits and Their Relationships in Response to High and Low Soil Water-Salinity Habitats

During long-term natural selection, plants growing in different habitats can optimize their resource allocation to cope with environmental stresses, and the differences in chemical elements contained in plants are one of the manifestations [45]. The current investigation demonstrated that the K, Ca, and Na contents of plants varied significantly in different water and salinity habitats (Table 1). The Ca content of plants in the high soil water and salinity habitat was significantly higher than in the low soil water and salinity habitat, which is consistent with the findings of Zhang et al. In contrast to the habitat of Ebinur Lake with high soil water and salinity, which has a higher content of soil powder and clay particles, the low soil water and salinity habitat is primarily made up of desert sand with a sandy texture [31]. At the level of Ebinur Lake plant arid plant groups, this led to a large decrease in leaf Ca content with decreasing soil water salinity because soil conditions have a strong impact on the variance of Ca content [46,47]. This suggests that plants adapt to environmental change by changing their leaf chemical traits. C, N, and P are vital nutrients that enable plant growth and are closely associated with crucial metabolic processes [48]. The ratio of C:N:P in terrestrial plants can indicate how well-adapted the plant is to the particular growth conditions in a given area [49]. In this study, there was a significant negative correlation between C and N (P), and both N and P showed a significant positive correlation in high soil water and salinity habitat. This is in line with the findings of the Sterner et al. study, which claims that one of the shared leaf chemical traits of higher terrestrials is the substantial negative correlation between leaf C and N (P) and the positive correlation between N and P [49,50]. This is one of the most basic characteristics of plants and is also a strong guarantee for stable community growth and development [51]. However, compared to the above characteristics, the relationship between C and P in plants with low soil water and salinity habitat is different. This may be because the plants get their N and P directly from the soil [52]. Environmental changes will also lead to significant differences in trait relationships, which are typically exhibited by a loss of trait correlation in stressful environments [16]. This theory was supported by the relationship between K and N, which showed that these two variables were significantly positively associated in high soil water-salinity habitat but not in low soil water-salinity habitat. Furthermore, different water-saline ecosystems support diverse plant species. Different trait correlations within species and between species can result from the degree of individual access to resources [53]. Ecological strategies that have evolved through evolutionary processes include synergies and trade-offs between various functioning traits [54].

The PCA revealed synergies or trade-offs between desert plant leaf chemical traits in both soil water and salinity habitats. Na, N, C, and Ca had the largest explanations of PC1 in the two-soil water and salinity habitats, and C and Ca showed a negative correlation with PC1. This suggests that Na, N, C, and Ca are relatively significant leaf chemical traits of desert plants in Ebinur Lake. Ca is a crucial nutrient element and messenger substance that plays a crucial role in both nutrient and signal transduction as an important biogenic element [33]. Osmoregulation is a crucial physiological defense mechanism for plant drought resistance in arid and desolate desert regions, and Na is a crucial osmoregulation ingredient for plants to adapt to an arid environment [55]. These chemical traits have special biochemical functions in plant growth, development, and reproduction [56]. Due to water scarcity and low nutrient levels, perennial herbs, shrubs, and trees of various kinds dominate the structure and functionality of desert ecosystems in the Ebinur Lake area. It happens frequently that various species respond differently to environmental stress [57]. The importance of elemental P for shrubs was demonstrated by PCA, which revealed that the shrub species *Haloxylon ammodendron* and *Populus euphratica* had high P content in both high and low soil water and salinity habitat. The amount of P in a leaf indicates how well a plant can absorb nutrients from the soil and is a crucial component of many proteins and genetic materials [58]. Desert plants have undergone long-term deductive evolution that has produced distinctive adaptive traits and functional countermeasures [59]. However, PCA failed to identify any distinct patterns in the distribution of trees and herbs. This is likely because changes in plant functional traits are impacted by a mix of environmental screening and biological interactions, which results in a convergence of traits that adopt the same growth strategy.

4.2. Relationship between ITI and Various Soil Physical and Chemical Factors

Interactions between different functional traits produce complex relationships, which are reflected in the synergistic response of different functional traits to environmental changes [60]. In our investigation, soil SVWC and N had the strongest relationships with ITI, which also tended to rise in both soil water-salinity habitats. The effect of pH on ITI was not significant in both soil water-salinity habitats. This may be due to the fact that in the Ebinur Lake area, plants have adapted to salinity habitats through physiological adaptations, and therefore leaf chemical traits are not significantly affected by pH. Nitrogen (N) and phosphorus (P) are the two nutrients that most often limit plant growth in terrestrial ecosystems [52]. According to this study, N is the main factor limiting plant growth and yield formation in habitats with low soil water and salinity, probably because saline plants have higher N uptake efficiency under low N conditions [61]. N supply and drought stress had similar interactions on plant growth [62], which could explain the positive correlation and high fit of soil SVWC and N with ITI in both soil water-salinity habitats. P was positively correlated with ITI in low soil water-salinity habitats, but ITI tended to decrease in high soil water-salinity habitats. One of the causes of the decline in ITI in high soil water-salinity habitats may be the accumulation of soil carbonate and a rise in soil alkalinity, which might result in the underutilization of P [24]. Additionally, P becomes insoluble in alkaline soils when it mixes with Ca, Mg, and other alkaline earth metals [63]. This can also result in a decrease in ITI since it makes it harder for plants to absorb and use P from the soil. The effect of potassium (K) on ITI was different in the two-soil water-salinity habitats, with a stronger negative correlation between K and ITI in the low soil water-salinity habitat. This may be due to the reduced mobility of K^+ under drought and saline habitats in low soil water-salinity habitats, where the efficiency of K^+ diminishes with lowering soil water content [62]. Soil water content and salinity have been established in earlier studies to be important environmental variables affecting desert plants [41]. In our study, ITI was positively connected with SVWC in both soil water and salinity habitats, and the association between ITI and SVWC was more significant in desert plants than EC. Studies have indicated that the main factor restricting the proper growth and development of desert plants in arid locations is the soil water content [12,64]. By limiting nutrient availability through mineralization as well as by reducing nutrient diffusion and mass movement in the soil, drought stress and the associated reduction in soil moisture can lower plant nutrient uptake [65]. Small variations in water will show up in the soil and vegetation, especially in arid and semi-arid areas. In order to adapt to drought stress in various water and salinity ecosystems, plants therefore chose combinations of diverse traits, or drought adaptation strategies [66]. According to the findings of PTN, where the network structure is more complex in high soil water and salinity habitat than in low soil water and salinity habitat, indicating more combinations between leaf chemical traits, the difference in the strength of fit between ITI and SVWC in high and low soil water salinity habitats may be caused by differences in the selected chemical trait combinations. The ITI and SVWC fits were higher in high soil water and salinity habitat because trait integration is the combination of individual qualities at progressively more sophisticated levels to determine the fitness and ultimate ecological processes of an organism [67]. The association between EC and ITI varied between the two soil water and salinity habitats, and neither habitat had a strong linear fit. In high soil water and salinity habitat, ITI was positively connected with EC, whereas in low soil water and salinity habitat, it was negatively correlated. This may be because soil salinity stress can limit plant growth by altering processes like photosynthesis, oxidative stress, osmotic stress, and ion imbalance [68]. For instance, too much Na^+ in the soil might affect how plants use and process K^+, changing plant growth, mineral distribution, and photosynthetic rates [69]. Light stress and salt stress are the main environmental factors limiting photosynthetic efficiency, and salt stress has short- or long-term effects on photosynthesis [70]. Although salt stress has been reported to inhibit photosynthesis [71], it has also been reported that photosynthesis is not slowed down by salinity and is even stimulated by low salt concentrations [72]. Therefore, salt stress limits photosynthesis, which results in less combination and integration of leaf chemical traits, which may explain why ITI declines with rising EC in low soil water and salinity habitat. Additionally, there are several plant species in two soil water-salinity habitats, and plant species vary widely in how salinity affects their ability to flourish [73,74], which also influences the ITI among plants.

Reduced density, dwarf plants, changes in species mix, simplistic structures, and poor water quality will eventually result in decreased production and the deterioration of ecosystem function. Therefore, HSM is more suited to the survival of desert plants than LSM. The data presented above contradict hypothesis (1), which states that ITI under LSM should be higher than LSM and more closely tied to the soil habitat. This could be due to two factors. The first is because there is not a strong enough environmental gradient for selection. Delhaye et al. additionally stated that an extreme environmental gradient is necessary for ITI research [2]. The second reason may be that this study selected only chemical traits and not traits like morphology and physiology. Because the environmental filter chooses particular combinations of traits rather than individual trait values, as shown by a greater ITI between some traits at the community level, for instance, a decrease in soil water may have a negative impact on the levels of phosphate and nitrogen in the leaves [44]. As a result, the relationship between leaf N and P in the LSM is inconsistent with the study by Sterner et al. [50]. Reich et al. also argued that which function is optimized, or the extent to which traits are coordinated, is usually determined by the biophysical, evolutionary and/or prevailing environmental constraints operating under a particular selection regime [8,60,75]. The ITI of leaf chemical traits along a weaker environmental gradient was investigated in this work. The drawback of this study stems from the conclusion that there was little association between soil factors. Future investigations should use plant morphological, physiological, and chemical traits to evaluate ITI on a stronger environmental gradient. To analyze changes in plant functional strategy, ITI can potentially be used with other functional diversity indices. In addition, selection for plant leaf chemical traits affects the complexity of plant trait networks and central traits [28]. The central leaf chemical traits of high and low soil water and salinity habitat are C and Ca, respectively, and Na also had the same degree as Ca, further confirming the results of the PCA. C, Ca, and Na are relatively important chemical traits of desert plant leaves in the Ebinur Lake area. This is in line with hypothesis (2) that distinct water and salinity habitats have different central leaf chemical traits. Highly connected traits are expected as a result of biophysical and/or selection processes that favor the efficient use and acquisition of resources within and between plant tissues [40]. It is important to take into account plant community composition, growth type (or functional group), and strategic variations brought on by certain landforms (such as deserts) when considering plant growth trade-off strategies at the regional scale [12]. The results indicate a more complex network structure and more leaf chemical trait connections in high soil water and salinity habitat than in low soil water and salinity habitat. The studies suggest that communities with less species diversity in relatively cold and arid climates have less trait connectivity, simpler trait network topology, and lower trait correlations [56]. Resources become more scarce, and abiotic circumstances become more harsh or varied as habitats change from benign to stressful or from high to low productivity ecosystems, reducing the size of the community's possible niche space [13]. A complex network of trait correlations reflecting numerous concurrent selective processes in adaptation or community construction will develop from adaptation across resource gradients [76]. To ascertain which traits are required to represent ecological strategies, it is also vital to comprehend the link between traits and functional differentiation among species within local groups.

5. Conclusions

In this paper, eight plant leaf chemical traits were selected based on the analysis of two habitats in the Ebinur Lake area, but the plant species were not identical under different soil, water, and salinity habitats. The relationship between plant leaf chemical traits was closer in HSM than in LSM, and the relationship between C and other leaf chemical traits was significant in HSM but not in LSM. Indicating that stressful habitats may lead to the loss of correlations among chemical traits, in both soil water-salinity habitats, the strength of fit between SVWC and ITI was the greatest, while the strength of fit between EC and ITI was the smallest. Contrary to hypothesis (1) of this study, the relationship between soil factors and ITI in both habitats was not statistically significant. This could be the result of selecting only chemical traits and not traits like morphology and physiology, as well as insufficient gradients in the habitats. The plant trait network determined that C and Ca were the two central leaf chemical traits for desert plant growth in Ebinur Lake. This agrees with both hypothesis (2) and the PCA findings.

Author Contributions: Conceptualization, J.Y. and X.Z.; methodology, J.Y. and X.Z.; software, J.Y. and D.S.; validation, J.Y.; investigation, J.Y., D.S., Y.W. and J.T.; formal analysis, J.Y.; data curation, J.Y. and J.T.; writing—original draft preparation, J.Y. and Y.W.; writing—review and editing, J.Y. and X.Z.; supervision, X.Z.; funding acquisition, X.Z. All authors have read and agreed to the published version of the manuscript.

Funding: This research was supported by the Natural Science Foundation of Xinjiang Uygur Autonomous Region (2023D01C12) and National Natural Science Foundation of China (31700354).

Institutional Review Board Statement: Not applicable.

Informed Consent Statement: Not applicable.

Data Availability Statement: Not available.

Conflicts of Interest: The authors declare no conflict of interest.

References

1. Violle, C.; Navas, M.L.; Vile, D.; Kazakou, E.; Fortunel, C.; Hummel, I.; Garnier, E. Let the concept of trait be functional! *Oikos* **2007**, *116*, 882–892. [CrossRef]
2. Delhaye, G.; Bauman, D.; Séleck, M.; Ilunga Wa Ilunga, E.; Mahy, G.; Meerts, P.; Spasojevic, M. Interspecific trait integration increases with environmental harshness: A case study along a metal toxicity gradient. *Funct. Ecol.* **2020**, *34*, 1428–1437. [CrossRef]
3. Zhang, S.H.; Zhang, Y.; Xiong, K.N.; Yu, Y.H.; Min, X.Y. Changes of leaf functional traits in karst rocky desertification ecological environment and the driving factors. *Glob. Ecol. Conserv.* **2020**, *24*, e01381. [CrossRef]
4. Akram, M.A.; Zhang, Y.; Wang, X.; Shrestha, N.; Malik, K.; Khan, I.; Ma, W.; Sun, Y.; Li, F.; Ran, J.; et al. Phylogenetic independence in the variations in leaf functional traits among different plant life forms in an arid environment. *J. Plant Physiol.* **2022**, *272*, 153671. [CrossRef]
5. Mouchet, M.A.; Villéger, S.; Mason, N.W.H.; Mouillot, D. Functional diversity measures: An overview of their redundancy and their ability to discriminate community assembly rules. *Funct. Ecol.* **2010**, *24*, 867–876. [CrossRef]
6. Murren, C.J. Phenotypic integration in plants. *Plant. Spec. Biol.* **2002**, *17*, 89–99. [CrossRef]
7. Messier, J.; Mcgill, B.J.; Enquist, B.J.; Lechowicz, M.J. Trait variation and integration across scales: Is the leaf economic spectrum present at local scales? *Ecography* **2017**, *40*, 685–697. [CrossRef]
8. Reich, P.B.; Walters, M.B.; Ellsworth, D.S. From tropics to tundra: Global convergence in plant functioning. *Proc. Natl. Acad. Sci. USA* **1997**, *94*, 13730–13734. [CrossRef]
9. Laughlin, D.C. The intrinsic dimensionality of plant traits and its relevance to community assembly. *J. Ecol.* **2014**, *102*, 186–193. [CrossRef]
10. Tonsor, S.J.; Scheiner, S.M. Plastic trait integration across a CO_2 gradient in Arabidopsis thaliana. *Am. Nat.* **2007**, *169*, E119–E140. [CrossRef]
11. Pigliucci, M. Phenotypic integration: Studying the ecology and evolution of complex phenotypes. *Ecol. Lett.* **2003**, *6*, 265–272. [CrossRef]
12. Li, S.; Wang, H.; Gou, W.; White, J.F.; Kingsley, K.L.; Wu, G.; Su, P. Leaf functional traits of dominant desert plants in the Hexi Corridor, Northwestern China: Trade-off relationships and adversity strategies. *Glob. Ecol. Conserv.* **2021**, *28*, e01666. [CrossRef]
13. Silva, J.L.A.; Souza, A.F.; Vitória, A.P. Leaf trait integration mediates species richness variation in a species-rich neotropical forest domain. *Plant. Ecol.* **2021**, *222*, 1183–1195. [CrossRef]
14. Gianoli, E.; Palacio-López, K. Phenotypic integration may constrain phenotypic plasticity in plants. *Oikos* **2009**, *118*, 1924–1928. [CrossRef]
15. Parsons, K.J.; Mcwhinnie, K.; Pilakouta, N., Walker, L. Does phenotypic plasticity initiate developmental bias? *Evol. Dev.* **2020**, *22*, 56–70. [CrossRef] [PubMed]
16. Matesanz, S.; Blanco-Sanchez, M.; Ramos-Munoz, M.; De La Cruz, M.; Benavides, R.; Escudero, A. Phenotypic integration does not constrain phenotypic plasticity: Differential plasticity of traits is associated to their integration across environments. *New Phytol.* **2021**, *231*, 2359–2370. [CrossRef]
17. Yang, X.D.; Anwar, E.; Zhou, J.; He, D.; Gao, Y.C.; Lv, G.H.; Cao, Y.E. Higher association and integration among functional traits in small tree than shrub in resisting drought stress in an arid desert. *Environ. Exp. Bot.* **2022**, *201*, 104993. [CrossRef]
18. Kadioglu, A.; Terzi, R. A dehydration avoidance mechanism: Leaf rolling. *Bot. Rev.* **2007**, *73*, 290–302. [CrossRef]
19. Funk, J.L.; Larson, J.E.; Ames, G.M.; Butterfield, B.J.; Cavender-Bares, J.; Firn, J.; Laughlin, D.C.; Sutton-Grier, A.E.; Williams, L.; Wright, J. Revisiting the Holy Grail: Using plant functional traits to understand ecological processes. *Biol. Rev.* **2017**, *92*, 1156–1173. [CrossRef]
20. Wright, I.J.; Falster, D.S.; Pickup, M.; Westoby, M. Cross-species patterns in the coordination between leaf and stem traits, and their implications for plant hydraulics. *Physiol. Plant.* **2006**, *127*, 445–456. [CrossRef]
21. Poorter, H.; Lambers, H.; Evans, J.R. Trait correlation networks: A whole-plant perspective on the recently criticized leaf economic spectrum. *New Phytol.* **2014**, *201*, 378–382. [CrossRef] [PubMed]

22. Burton, J.L.; Perakis, S.S.; Brooks, J.R.; Puettmann, K.J. Trait integration and functional differentiation among co-existing plant species. *Am. J. Bot.* **2020**, *107*, 628–638. [CrossRef]
23. Yang, X.D.; Zhang, X.N.; Lv, G.H.; Ali, A. Linking Populus euphratica Hydraulic Redistribution to Diversity Assembly in the Arid Desert Zone of Xinjiang, China. *PLoS ONE* **2014**, *9*, e109071. [CrossRef] [PubMed]
24. Gao, D.X.; Wang, S.; Wei, F.L.; Wu, X.T.; Zhou, S.; Wang, L.X.; Li, Z.D.; Chen, P.; Fu, B.J. The vulnerability of ecosystem structure in the semi-arid area revealed by the functional trait networks. *Ecol. Indic.* **2022**, *139*, 108894. [CrossRef]
25. Luo, Y.; Chen, Y.; Peng, Q.W.; Li, K.H.; Mohammat, A.; Han, W.X. Nitrogen and phosphorus resorption of desert plants with various degree of propensity to salt in response to drought and saline stress. *Ecol. Indic.* **2021**, *125*, 107488. [CrossRef]
26. Stuart Chapin, F.; Kellar, A.; Francisco, P. Evolution of Suites of Traits in Response to Environmental Stress. *Am. Nat.* **1993**, *142*, S78–S92. [CrossRef]
27. Wright, I.J.; Reich, P.B.; Westoby, M.; Ackerly, D.D.; Baruch, Z.; Bongers, F.; Cavender-Bares, J.M.; Chapin, T.; Cornelissen, J.H.C.; Diemer, M.W.; et al. The worldwide leaf economics spectrum. *Nature* **2004**, *428*, 821–827. [CrossRef]
28. Li, Y.; Liu, C.C.; Xu, L.; Li, M.X.; Zhang, J.H.; He, N.P. Leaf Trait Networks Based on Global Data: Representing Variation and Adaptation in Plants. *Front. Plant Sci.* **2021**, *12*, 710530. [CrossRef]
29. Bruelheide, H.; Dengler, J.; Purschke, O.; Lenoir, J.; Jiménez-Alfaro, B.; Hennekens, S.M.; Botta-Dukát, Z.; Chytrý, M.; Field, R.; Jansen, F.; et al. Global trait–environment relationships of plant communities. *Nat. Ecol. Evol.* **2018**, *2*, 1906–1917. [CrossRef]
30. Sack, L.; Buckley, T.N. Trait multi-functionality in plant stress response. *Integr. Comp. Biol.* **2020**, *60*, 98–112. [CrossRef]
31. Zhang, X.N.; Li, Y.; Yang, X.D.; He, X.M.; Lv, G.H. Responses of leaf chemical trait and economics spectrum in desert plants to varied soil water and salinity. *Chin. J. Ecol.* **2018**, *37*, 1299–1306. [CrossRef]
32. Furey, G.N.; Tilman, D. Plant biodiversity and the regeneration of soil fertility. *Proc. Natl. Acad. Sci. USA* **2021**, *118*, e2111321118. [CrossRef] [PubMed]
33. Furey, G.N.; Tilman, D. Plant chemical traits define functional and phylogenetic axes of plant biodiversity. *Ecol. Lett.* **2023**, *26*, 1394–1406. [CrossRef] [PubMed]
34. Zhang, J.H.; Zhao, N.; Liu, C.C.; Yang, H.; Li, M.L.; Yu, G.R.; Wilcox, K.; Yu, Q.; He, N.P. C:N:P stoichiometry in China's forests: From organs to ecosystems. *Funct. Ecol.* **2018**, *32*, 50–60. [CrossRef]
35. Jiang, T.H.; Zhan, X.H.; Xu, Y.C.; Zhou, L.X.; Zong, L.G. Roles of calcium in stress-tolerance of plants and its ecological significance. *Chin. J. Appl. Ecol.* **2005**, *16*, 971–976.
36. Wang, L.L.; Zhang, X.F.; Xu, S.J. Is salinity the main ecological factor that influences foliar nutrient resorption of desert plants in a hyper-arid environment? *BMC Plant Biol.* **2020**, *20*, 461. [CrossRef]
37. Flores-Moreno, H.; Fazayeli, F.; Banerjee, A.; Datta, A.; Kattge, J.; Butler, E.E.; Atkin, O.K.; Wythers, K.; Chen, M.; Anand, M.; et al. Robustness of trait connections across environmental gradients and growth forms. *Glob. Ecol. Biogeogr.* **2019**, *28*, 1806–1826. [CrossRef]
38. Liu, C.C.; Li, Y.; Xu, L.; Chen, Z.; He, N.P. Variation in leaf morphological, stomatal, and anatomical traits and their relationships in temperate and subtropical forests. *Sci. Rep.* **2019**, *9*, 5803. [CrossRef]
39. Dwyer, J.M.; Laughlin, D.C. Selection on trait combinations along environmental gradients. *J. Veg. Sci.* **2017**, *28*, 672–673. [CrossRef]
40. Reich, P.B. The world-wide 'fast–slow' plant economics spectrum: A traits manifesto. *J. Ecol.* **2014**, *102*, 275–301. [CrossRef]
41. Tusifujiang, Y.L.K.; Zhang, X.N.; Gong, L. The relative contribution of intraspecific variation and species turnover to the community-level foliar stoichiometric characteristics in different soil moisture and salinity habitats. *PLoS ONE* **2021**, *16*, e0246672. [CrossRef] [PubMed]
42. Dong, M. *Observation and Analysis of Terrestrial Biocommunities*; Standards Press of China: Beijing, China, 1997.
43. Kleyer, M.; Trinogga, J.; Cebrián-Piqueras, M.A.; Trenkamp, A.; Fløjgaard, C.; Ejrnaes, R.; Bouma, T.J.; Minden, V.; Maier, M.; Mantilla-Contreras, J.; et al. Trait correlation network analysis identifies biomass allocation traits and stem specific length as hub traits in herbaceous perennial plants. *J. Ecol.* **2019**, *107*, 829–842. [CrossRef]
44. Rosas, T.; Mencuccini, M.; Barba, J.; Cochard, H.; Saura-Mas, S.; Martinez-Vilalta, J. Adjustments and coordination of hydraulic, leaf and stem traits along a water availability gradient. *New Phytol.* **2019**, *223*, 632–646. [CrossRef]
45. Cao, Y.B.; Wang, B.T.; Wei, T.T.; Ma, H. Ecological stoichiometric characteristics and element reserves of three stands in a closed forest on the Chinese loess plateau. *Environ. Monit. Assess.* **2016**, *188*, 80. [CrossRef]
46. Asner, G.P.; Martin, R.E.; Tupayachi, R.; Anderson, C.B.; Sinca, F.; Carranza-Jiménez, L.; Martinez, P. Amazonian functional diversity from forest canopy chemical assembly. *Proc. Natl. Acad. Sci. USA* **2014**, *111*, 5604–5609. [CrossRef]
47. He, M.Z.; Song, X.; Tian, F.P.; Zhang, K.; Zhang, Z.S.; Chen, N.; Li, X.R. Divergent variations in concentrations of chemical elements among shrub organs in a temperate desert. *Sci. Rep.* **2016**, *6*, 20124. [CrossRef]
48. Sardans, J.; Rivas-Ubach, A.; Peñuelas, J. The C:N:P stoichiometry of organisms and ecosystems in a changing world: A review and perspectives. *Perspect. Plant Ecol. Evol. Syst.* **2012**, *14*, 33–47. [CrossRef]
49. Li, Z.; Han, L.; Liu, Y.H.; An, S.Q.; Leng, X. C, N and P stoichiometric characteristics in leaves of Suaeda salsa during different growth phase in coastal wetlands of China. *Chin. J. Plant Ecol.* **2013**, *36*, 1054–1061. [CrossRef]
50. Sterner, R.W.; Elser, J.J. Ecological Stoichiometry: Biology of Elements from Molecules to the Biosphere. *J. Plankton Res.* **2003**, *25*, 1183. [CrossRef]

51. Wu, T.G.; Wu, M.; Liu, L.; Xiao, J.H. Seasonal variations of leaf nitrogen and phosphorus stoichiometry of three herbaceous species in Hangzhou Bay coastal wetlands, China. *Chin. J. Plant Ecol.* **2010**, *34*, 23–28. [CrossRef]

52. Güsewell, S. N:P ratios in terrestrial plants: Variation and functional significance. *New Phytol.* **2004**, *164*, 243–266. [CrossRef]

53. Vannoordwijk, A.J.D.; De Jong, G. Acquisition and Allocation of Resources: Their Influence on Variation in Life History Tactics. *Am. Nat.* **1986**, *128*, 137–142. [CrossRef]

54. Eviner, V.T.; Iii, F.S.C. Functional Matrix: A Conceptual Framework for Predicting Multiple Plant Effects on Ecosystem Processes. *Annu. Rev. Ecol. Evol. Syst.* **2003**, *34*, 455–485. [CrossRef]

55. Li, S.X.; Zhou, X.R.; Wang, S.M. Positive Functions of Sodium in Plants. *J. Desert Res.* **2008**, *28*, 485–490.

56. Li, Y.; Liu, C.C.; Sack, L.; Xu, L.; Li, M.X.; Zhang, J.H.; He, N.P. Leaf trait network architecture shifts with species-richness and climate across forests at continental scale. *Ecol. Lett.* **2022**, *25*, 1442–1457. [CrossRef]

57. Powell, T.L.; Wheeler, J.K.; De Oliveira, A.a.R.; Da Costa, A.C.L.; Saleska, S.R.; Meir, P.; Moorcroft, P.R. Differences in xylem and leaf hydraulic traits explain differences in drought tolerance among mature Amazon rainforest trees. *Glob. Chang. Biol.* **2017**, *23*, 4280–4293. [CrossRef]

58. Zhang, K.; He, M.Z.; Li, X.R.; Tan, H.J.; Gao, Y.H.; Li, G.; Han, G.J.; Wu, Y.Y. Foliar carbon, nitrogen and phosphorus stoichiometry of typical desert plants across the Alashan Desert. *Acta Ecol. Sin.* **2014**, *34*, 6538–6654. [CrossRef]

59. Lanta, V.; Lepš, J. Effect of functional group richness and species richness in manipulated productivity–diversity studies: A glasshouse pot experiment. *Acta Oecol.* **2006**, *29*, 85–96. [CrossRef]

60. He, D.; Biswas, S.R.; Xu, M.S.; Yang, T.H.; You, W.H.; Yan, E.R. The importance of intraspecific trait variability in promoting functional niche dimensionality. *Ecography* **2020**, *44*, 380–390. [CrossRef]

61. Abliz, A.; Lv, G.H.; Zhang, X.N.; Gong, Y.M. Carbon, nitrogen and phosphorus stoichiometry of photosynthetic organs across Ebinur Lake Wetland Natural Reserve of Xinjiang, Northwest China. *Chin. J. Ecol.* **2015**, *34*, 2123–2130. [CrossRef]

62. Hu, Y.; Schmidhalter, U. Drought and salinity: A comparison of their effects on mineral nutrition of plants. *J. Plant Nutr. Soil Sci.* **2005**, *168*, 541–549. [CrossRef]

63. Huang, C.Y.; Xu, J.M. *Pedology Soil Study*; China Agricultural Publishing House: Beijing, China, 1983.

64. Li, W.J.; Lv, G.H.; Zhang, L.; Wang, H.F.; Li, Z.K.; Wang, J.L.; Ma, H.Y.; Liu, Z.D. Analysis of potential water source differences and utilization strategies of desert plants in arid regions. *Ecol. Environ.* **2019**, *28*, 1557–1566. [CrossRef]

65. He, M.; Dijkstra, F.A. Drought effect on plant nitrogen and phosphorus: A meta-analysis. *New Phytol.* **2014**, *204*, 924–931. [CrossRef]

66. Pivovaroff, A.L.; Pasquini, S.C.; De Guzman, M.E.; Alstad, K.P.; Stemke, J.S.; Santiago, L.S. Multiple strategies for drought survival among woody plant species. *Funct. Ecol.* **2016**, *30*, 517–526. [CrossRef]

67. Marks, C.O. The causes of variation in tree seedling traits: The roles of environmental selection versus chance. *Evolution* **2007**, *61*, 455–469. [CrossRef]

68. Meng, X.Q.; Zhou, J.; Sui, N. Mechanisms of salt tolerance in halophytes: Current understanding and recent advances. *Open Life Sci.* **2018**, *13*, 149–154. [CrossRef]

69. Bejaouia, F.; Salasb, J.J.; Nouairid, I.; Smaouia, A.; Abdellya, C.; Martínez Forceb, E.; Youssefc, N.B. Changes in chloroplast lipid contents and chloroplast ultrastructure in Sulla carnosa and Sulla coronaria leaves under salt stress. *J Plant. Physiol.* **2016**, *198*, 32–38. [CrossRef]

70. Parida, A.K.; Das, A.B. Salt tolerance and salinity effects on plants: A review. *Ecotoxicol. Environ. Saf.* **2005**, *60*, 324–349. [CrossRef]

71. Chaudhuri, K.; Choudhuri, M.A. Effects of short-term NaCl stress on water relations and gas exchange of two jute species. *Biol. Plant.* **1997**, *40*, 373–380. [CrossRef]

72. Kurban, H.; Saneoka, H.; Nehira, K.; Adilla, R.; Premachandra, G.S.; Fujita, K. Effect of salinity on growth, photosynthesis and mineral composition in leguminous plant *Alhagi pseudoalhagi* (Bieb.). *Soil. Sci. Plant. Nutr.* **1999**, *45*, 851–862. [CrossRef]

73. Batterton, J.C., Jr.; Baalen, C.V. Growth Responses of Blue-green Algae to Sodium Chloride Concentration. *Arch. Mikrobiol.* **1971**, *76*, 151–165. [CrossRef] [PubMed]

74. Moisander, P.H.; Mcclinton, E.; Paerl, H.W. Salinity Effects on Growth, Photosynthetic Parameters, and Nitrogenase Activity in Estuarine Planktonic Cyanobacteria. *Microb. Ecol.* **2014**, *43*, 432–442. [CrossRef] [PubMed]

75. Diaz, S.; Kattge, J.; Cornelissen, J.H.; Wright, I.J.; Lavorel, S.; Dray, S.; Reu, B.; Kleyer, M.; Wirth, C.; Prentice, I.C.; et al. The global spectrum of plant form and function. *Nature* **2016**, *529*, 167–171. [CrossRef]

76. Freschet, G.T.; Kichenin, E.; Wardle, D.A. Explaining within-community variation in plant biomass allocation: A balance between organ biomass and morphology above vs below ground? *J. Veg. Sci.* **2015**, *26*, 431–440. [CrossRef]

Disclaimer/Publisher's Note: The statements, opinions and data contained in all publications are solely those of the individual author(s) and contributor(s) and not of MDPI and/or the editor(s). MDPI and/or the editor(s) disclaim responsibility for any injury to people or property resulting from any ideas, methods, instructions or products referred to in the content.

 *forests*

Article

Variations in Physiological and Biochemical Characteristics of *Kalidium foliatum* Leaves and Roots in Two Saline Habitats in Desert Region

Lamei Jiang [1,2], Deyan Wu [1,2], Wenjing Li [1,2], Yuehan Liu [1,2], Eryang Li [1,2], Xiaotong Li [1,2], Guang Yang [3] and Xuemin He [1,2,*]

[1] College of Ecology and Environment, Xinjiang University, Urumqi 830017, China; jianglam0108@126.com (L.J.)
[2] Key Laboratory of Oasis Ecology of Education Ministry, Xinjiang University, Urumqi 830017, China
[3] Farmland Irrigation Research Institute, Xinxiang 453000, China
* Correspondence: hxm@xju.edu.cn

Abstract: Salt stress is a key environmental factor that has adverse effects on plant growth and development. High salinity induces a series of structural and functional changes in the morphological and anatomical features. The physiological and biochemical changes in *K. foliatum* in response to salt stress in natural environments are still unclear. Based on this, this study compared and analyzed the differences in the physiological and biochemical indicators between the leaf and root tissues in high-salt and low-salt habitats, selecting *K. foliatum* as the research object. The results showed that the chlorophyll contents in the leaves of *K. foliatum* decreased in the high-salt habitat, while the thicknesses of the upper and lower epidermises, as well as the thicknesses of the palisade tissue, significantly increased. The high-salt environment led to decreases in the N and P contents in the leaves and root tissues of *K. foliatum*, resulting in changes in the stoichiometric ratio of elements. The concentrations of C, N, and P in the roots of *K. foliatum* were lower than those in the leaves. The accumulation of Na^+ in the *K. foliatum* roots was greater than that in the leaves, and the roots could promote the transport of sodium ions to the leaves. The contents of starch and soluble sugar in the leaves showed higher proportions in the high-salt habitat than in the low-salt habitat, while the changes in the roots and leaves were the opposite. As the salt content increased, the proline contents in the leaves and roots of *K. foliatum* significantly increased, and the proline contents in the roots of *K. foliatum* were lower than those in the leaves. The leaves and roots exhibited higher levels of peroxidase and superoxide enzymes in the high-salinity habitat than in the low-salinity habitat. The superoxide dismutase (SOD) activity of the *K. foliatum* leaves and catalase (CAT) activity of the roots were the "central traits" in the high-salt habitat. In the low-salt habitat, the leaf malondialdehyde (MDA) and root C/N were the central traits of the leaves and roots, indicating that *K. foliatum* adapts to changes in salt environments in different ways.

Keywords: *Kalidium foliatum*; salt stress; desert; biochemical characteristics

Citation: Jiang, L.; Wu, D.; Li, W.; Liu, Y.; Li, E.; Li, X.; Yang, G.; He, X. Variations in Physiological and Biochemical Characteristics of *Kalidium foliatum* Leaves and Roots in Two Saline Habitats in Desert Region. *Forests* **2024**, *15*, 148. https://doi.org/10.3390/f15010148

Academic Editor: Angelo Rita

Received: 11 November 2023
Revised: 31 December 2023
Accepted: 4 January 2024
Published: 11 January 2024

Copyright: © 2024 by the authors. Licensee MDPI, Basel, Switzerland. This article is an open access article distributed under the terms and conditions of the Creative Commons Attribution (CC BY) license (https://creativecommons.org/licenses/by/4.0/).

1. Introduction

Salinization is one of the most serious environmental problems in the world, affecting the growth, development, and productivity of many plants. More than 6% of the world's land is affected by salinization, with 20% of the arable land and 33% of the agricultural irrigation area affected by high salinity. By 2050, it is expected that over 50% of the arable land will be salinized [1,2]. The serious impact of salt stress on plants and soil has prompted scholars to conduct extensive research on the response and adaptation mechanisms of plants to salt stress. According to the responses of plants to salt, they can be classified into halophytes and sweet-soil plants [3]. Halophytes can grow in soil environments with salinities greater than 200 mol/L, while sweet-soil plants cannot survive at similar levels of salinity. Halophytes have unique structures, including salt gland structures (vesicular head cells, hairy head cells, stem cells, and basal cells), fleshy tissues, and thick cork layers, to withstand salt stress [3].

Plants exhibit various physiological adaptation mechanisms to cope with salt stress, including the regulation of photosynthesis and energy metabolism, selective ion absorption/elimination, nutrient balance, osmotic regulation, and the accumulation of antioxidant enzymes. Photosynthetic pigments play a crucial role in photosynthesis by converting light energy into chemical energy, but salt stress can lead to severe pigment damage [4]. As the salt concentration increases, both chlorophyll a and chlorophyll b in kidney beans significantly decrease. The decrease in the chlorophyll levels in salt-stressed plants is considered a typical characteristic of oxidative stress [5]. Salt stress significantly increases the thicknesses of the palisade tissue and stratum corneum, while reducing the thickness of the sponge tissue [6]. Salt can affect the absorption of nutrients by roots, such as by increasing the Na^+ uptake and reducing the Ca^{2+} uptake, leading to nutrient imbalance [7]. The limitation of Na^+ accumulation in the aboveground parts of rice under salt stress may be related to its own salt stress tolerance [8]. Under high-salt conditions, the Na^+/Ca^{2+} and Na^+/K^+ ratios in plant cells increase sharply, causing damage to the cell membrane and cell leakage, further exacerbating the passive accumulation of Na^+ in the plant body [9]. Rong et al. (2015) found that there is a significant negative correlation between the phosphorus concentration in *Tamarix chinensis* leaves and soil salinity in the coastal wetland of Laizhou Bay, China [10]. Salt reduced the N contents in mint leaves and the P contents in the roots [11]. Salt stress also limits other biosynthetic functions, including amino acid and protein synthesis [12]. Salt stress leads to starch accumulation in rice leaves and an increase in the soluble sugar content [13]. The increase in starch or sugars in leaves may be an adaptive response to salt stress [14] or a sign of growth arrest (insufficient carbon use) [15,16]. Proline is an amino acid and osmotic protector, and an important signaling molecule. The accumulation of proline under salt stress enhances the plant water absorption and antioxidant capacity, reducing the accumulation of toxic ions [17,18]. Plants under salt stress can produce a large amount of reactive oxygen species (ROS), which cause membrane lipid peroxidation [19]. In order to prevent or reduce the damage caused by reactive oxygen species and enable them to continue their beneficial functions, the plant's antioxidant defense system plays a role in controlling the content of reactive oxygen species. Superoxide dismutase (SOD), peroxidase (POD), and catalase (CAT) are the main protective enzymes in the enzymatic defense system [18]. Some studies have found that the activities of POD and CAT do not change significantly at low salt concentrations but increase sharply at high salt concentrations. High salinity may stimulate the synthesis of POD and CAT [20], but excessive salt stress can lead to impaired enzyme activity [21]. Based on the concept of green development, cultivating salt-tolerant varieties is an effective way to develop, utilize, and improve saline soil, which contributes to sustainable economic development. Conducting research on the salt tolerance mechanism of plants is of great significance for elucidating the impact of soil salinization on plant growth, as well as for the ecological protection and sustainable development of saline soil.

Kalidium foliatum belongs to the family Chenopodiaceae and the genus *Kalidium*, and it is a true halophyte. *Kalidium foliatum* is widely distributed in the arid areas of northwest China. Within a certain range of salt concentration, salt significantly changes the structure of *K. foliatum* in chloroplasts and nuclei [22], and its rhizosphere exhibits a significant salt island effect, which can significantly enrich multiple ions [23]. Currently, research on the salt tolerance characteristics of *K. foliatum* mainly focuses on the leaf anatomical structure and physiological characteristics, while there is less research on the physiological adaptation characteristics of *K. foliatum* in natural salt habitats. Based on this, this study took the typical halophyte *K. foliatum* in the desert ecosystem as the research object, selected high-salt and low-salt habitats as the sampling sites, and analyzed the salt tolerance adaptation characteristics of *K. foliatum* in detail with an emphasis on the roots and leaves. It clarifies the internal morphological structures, element and stoichiometric ratios, ion distribution characteristics, and changes in antioxidant substances in the two salt habitats to provide a theoretical basis for cultivating excellent salt-tolerant varieties and for the comprehensive evaluation of the plant salt tolerance.

2. Materials and Methods

2.1. Overview of the Research Area

The Ebinur Lake Wetland National Nature Reserve (ELWNNR) in Xinjiang (44°43′ N–45°12′ N, 82°35′ E–83°40′ E) is the lowest depression and water salt concentration center in the southwest margin of the Junggar Basin. The ELWNNR is centered around the water body of Ebinur Lake, with a total area of 2670.85 km^2, of which the desert area accounts for more than 50% of the protected area. The climate in this area belongs to the temperate continental arid climate, with about 170 strong wind days throughout the year and a maximum wind speed of 55.0 m/s. The average annual temperature is 7.8 °C, and the extreme lowest and highest temperatures can reach −36.4 °C and 41.3 °C. The precipitation is scarce and unevenly distributed, with an average annual precipitation of 90.9 mm and an average annual evaporation of 1662 mm [24]. The soil in the watershed is rich in minerals, and the soil types are gray-brown desert soil, gray desert soil, and windblown sand soil. The hidden soil types are meadow soil, saline (saline) soil, and swamp soil, with a high degree of soil salinization and severe changes in the soil water and salt [25]. There are various types of sandy vegetation, mesophytic vegetation, and aquatic vegetation distributed in the area. The main plants in the area are *Populus euphratica*, *Haloxylon ammodendron*, *Halimoderon halodendron*, *Alhagi sparsifolia*, *Reaumuria soongarica*, *Nitraria roborowskii*, *K. foliatum*, *Apocynum venetum*, and *Phragmites australis* [26].

2.2. Sample Site Layout and Sample Collection

2.2.1. Layout of Sample Plots

According to the research results of the previous team, starting from the East Bridge Management and Protection Station of the Ebinur Lake Wetland National Nature Reserve, two 100 m × 100 m plots were set perpendicular to the banks of the Aqikesu River, south of the banks and north of the banks, and meter sample plots were used to determine the salt contents of the two plots. The soil salt contents of plots A and B were 10.15 ± 0.07 g/kg and 3.62 ± 0.08 g/kg, respectively (Figure 1). According to the soil salinity grading index [27] and the results of the soil salinity content measurements, the soil salinity contents in the desert areas are mild salinization at 2.0–4.0 g/kg, moderate salinization at 4.0–6.0 g/kg, and severe salinization at 6.0–20.0 g/kg. Therefore, there is a significant difference in the salt contents between sample plots A and B in this study, which can be divided into two habitats: a high-salt habitat and low-salt habitat. The soil physicochemical properties of the two saline habitats were shown in Table 1. We simultaneously investigated the number of species of *K. foliatum* in various fields for the random sampling of *K. foliatum* individuals. The main accompanying species in the high-salt habitat are *Tamarix ramosissima*, *Nitraria tangutorum*, *Halostachys caspica*, *Halonemum strobilaceum*, and *P. australis*. In the low-salt habitat, the main accompanying species are *H. ammodendron*, *H. halodendron*, *R. soongarica*, *P. australis*, and *Nitraria roborowskii*.

2.2.2. Sample Collection

Plant sample collection: 18 healthy *K. foliatum* plants with consistent growth and size were selected from each salt habitat. First, the leaf and root tissues of the *K. foliatum* were collected, and every 3 leaves and roots of the 18 *K. foliatum* plants from each salt habitat were mixed. A total of 24 plant samples were obtained from each salt habitat, which were quickly placed in liquid nitrogen for the subsequent sequencing of the physiological indicators. We selected 3–5 healthy leaves from each plant and immediately fixed them with FAA fixation solution. We screened the roots of each plant and randomly selected 3 fine roots from them. After washing the fine roots, we fixed them with electron-microscope fixing solution and immediately brought them back to the laboratory after sampling.

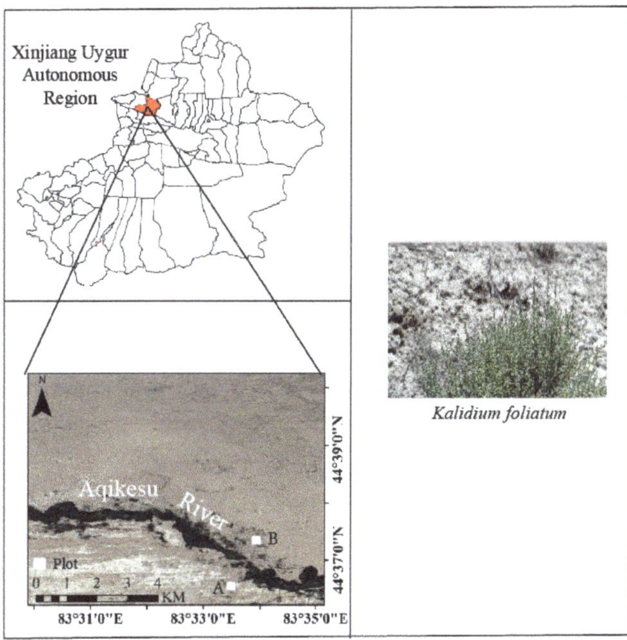

Figure 1. Schematic diagram of plot layout and plant morphology (A and B represent high-salt and low-salt sample plots).

Table 1. The physical and chemical characteristics of soil in two habits.

	H	L
Soil water content (%)	15.25 ± 0.61	12.05 ± 0.18
Soil Organic carbon content (g/kg)	3.93 ± 0.19	2.84 ± 0.13
Soil total nitrogen content (g/kg)	0.64 ± 0.16	0.48 ± 0.02
Soil total phosphorus content (g/kg)	0.53 ± 0.01	0.51 ± 0.01
pH	8.33 ± 0.04	8.00 ± 0.03
Soil ammonium nitrogen content (mg/kg)	4.47 ± 0.14	5.03 ± 0.17
Soil nitrate nitrogen content (mg/kg)	28.33 ± 0.80	34.24 ± 0.63
Soil available phosphorus content (mg/kg)	23.96 ± 0.86	22.57 ± 1.04

2.2.3. Sample Determination

We cut the leaves into thin slices, observed and measured them using an optical microscope, and analyzed the following parameters using an image analysis system: the leaf thickness, upper-epidermis thickness, lower-epidermis thickness, upper-epidermis length, lower-epidermis length, fence tissue thickness, and other indicators. We sliced the leaves and roots into slices, quickly vacuumed them, placed them on an observation platform, opened the scanning electron microscope software (SU 8100), and connected it to the scanning electron microscope device. We set the acceleration voltage and magnification of the scanning electron microscope on the software (SU 8100) interface, and we adjusted them according to the actual needs. We focused the specimen in the electron microscope and adjusted the lens to make the image clear and visible. We performed the necessary calibration, such as adjusting the scanning speed and scanning mode of the electron beam. We clicked the "Start Scanning" button on the software (SU 8100) interface to start observing and taking samples. The details and position of the image were adjusted through the zoom and navigation functions on the software (SU 8100) interface to obtain the desired observation results.

The concentrations of chlorophyll a, chlorophyll b, and carotenoids were determined using the ethanol acetone method, according to the Harmut (1987) method [28]. The content of soluble protein was determined according to Bradford's (1976) method [29]. According to Khelil et al. (2007), the method was slightly modified to measure the soluble sugar and starch contents [30]. The sample was extracted with ethanol in an 80 °C water bath for 30 min, and the soluble sugar content in the supernatant was determined according to Lepesant et al. (1972) [31]. The residue was extracted with 52% $HClO_4$ and the supernatant was obtained for the starch content [32]. The activity of superoxide dismutase (SOD) was measured using the nitrogen blue tetrazolium photochemical reduction method [33]. The peroxidase (POD) activity was measured using the guaiacol colorimetric method [34]. The activity of catalase (CAT) was determined via the UV absorption method [35]. The content of malondialdehyde (MDA) was determined using the phenobarbitone acid method [36]. The acidic ninhydrin staining method was used to determine the proline (PRO) content [36]. The contents of carbon, nitrogen, and phosphorus in the leaves and roots were determined using potassium dichromate titration, Kjeldahl nitrogen determination, and molybdenum antimony resistance colorimetry [27]. An Inductively Coupled Plasma Emission Spectrometer (ICP-OES) (Thermo, Waltham, MA, USA, iCAP7400) was used to determine the concentrations of Na^+, K^+, Ca^{2+}, and Mg^{2+}. Cl^- and sulfate anions were titrated using silver nitrate titration and EDTA complexometric titration, respectively [27].

2.3. Data Analysis Methods

Based on the *stats* package in R software (R4.1.2), Shapiro–Wilk's and Levene's methods were used to test the normality and homogeneity of the variance in the data. An independent-sample *t*-test ($p < 0.05$) was used to evaluate the differences in the photosynthetic pigments, leaf anatomical structures, nutrient elements, inorganic ions and their stoichiometric ratios, osmoregulation substances, antioxidant enzyme activities, and malondialdehyde contents between the two saline habitats. The characteristics of plant trait networks can reflect the interaction between plants and the environment from a holistic perspective. The nodes represent the plant traits, and the lines between the nodes represent the correlations between the plant traits [37]. The correlations between the plant traits indicate the interactions between the traits [38]. The *igraph* package in R software (R4.1.2) was used to calculate the correlations between the root and leaf traits and the characteristics of the network (degree and weighting): the degree (D), which refers to the number of edges on nodes, and the degree weight (Dw), which refers to the sum of significant correlation coefficients of nodes [39]. A correlation coefficient greater than 0.7 and $p < 0.05$ was considered to be a significant correlation between traits. The trait with the highest degree represented the "central trait" in the network [37]. Finally, Cytoscape software (Cytoscape 3.9.1) was used to visualize the network of plant functional traits. R 4.0.5 was used for statistical analysis and image rendering.

3. Results and Analysis

3.1. Differences in Chlorophyll, Leaf, and Root Anatomical Structures of K. foliatum under Two Saline Habitats

There were significant differences in the chlorophyll a, chlorophyll b, and total chlorophyll contents between the two saline habitats, both of which were higher in low-saline habitats than in high-saline habitats. The carotenoid content in the high-salt habitat was slightly lower than that in the low-salt habitat, and there was no significant difference between the carotenoids in the low-salt habitat (Figure 2).

The thicknesses of the *K. foliatum* leaves were greater in the high-salinity habitat (1325.82 ± 101.53 μm) than those in the low-salt habitat (1262.14 ± 23.87 μm). There were significant differences in the thicknesses of the upper and lower epidermises, the lengths of the lower epidermises, the thicknesses of the palisade tissue, the vascular bundle apertures, and the thickness ratios of the palisade tissue observed under an optical microscope (Table 2).

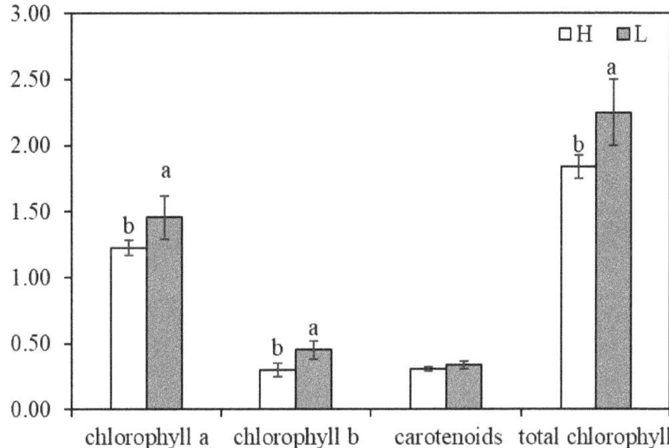

Figure 2. Differences in chlorophyll content of *K. foliatum* under two saline habitats (difference lowercase indicates significant difference between the high salt and lower salt).

Table 2. Differences in anatomical indicators of *K. foliatum* leaves under two saline habitats.

Index	High-Salt Habitat	Low-Salt Habitat
Leaf thickness (μm)	1325.82 ± 101.53 a	1262.14 ± 23.87 a
Upper-epidermis thickness (μm)	44.53 ± 2.38 a	33.48 ± 1.69 b
Lower-epidermis thickness (μm)	46.42 ± 1.59 a	35.44 ± 2.1 b
Length of upper epidermis (μm)	44.34 ± 3.79 a	38.31 ± 3.29 a
Length of lower epidermis (μm)	53.87 ± 4.90 a	41.36 ± 3.17 b
Palisade tissue thickness (μm)	23.40 ± 1.24 a	17.11 ± 1.20 b
Vascular bundle aperture (μm)	7.71 ± 0.40 a	5.97 ± 0.31 b
Thickness ratio of palisade (%)	3.60 ± 0.15 a	2.83 ± 0.24 b

Note: Different lowercase letters indicate significant differences in indicators, while the same letter indicates insignificant differences.

The scanning electron microscopy results of the *K. foliatum* leaves indicate that the stomatal structure was clearly visible on the upper epidermises of the *K. foliatum* leaves in both saline habitats (Figure 3B,F). In the low-salt habitat, the surfaces of the stomatal guard cells in *K. foliatum* were smooth with only a few folds (Figure 3F,G), while, in the high-salt habitat, the surfaces of the stomatal guard cells were convex, forming a clear wrinkled structure (Figure 3B,C). A large number of white salt crystals were clearly visible on the upper epidermises of the *K. foliatum* leaves (Figure 3B), while, in the low-salt habitat, only a few crystals were visible on the upper epidermises of the *K. foliatum* leaves (Figure 3F).

The epidermal cells, four–five layers of cortical parenchyma cells, a large number of cavities, and central vascular bundle tissue can be seen in the roots of *K. foliatum* under a scanning electron microscopic (Figure 4). The vascular tissue consists of the primary xylem, primary phloem, and parenchyma cells, with the xylem located at the center of the root (Figure 4). No salt crystals were found on the transverse section of the *K. foliatum* roots in the high-salt habitat, but a small amount of salt crystals was found in the fine roots of the *K. foliatum* roots in the low-salt habitat (Figure 4G).

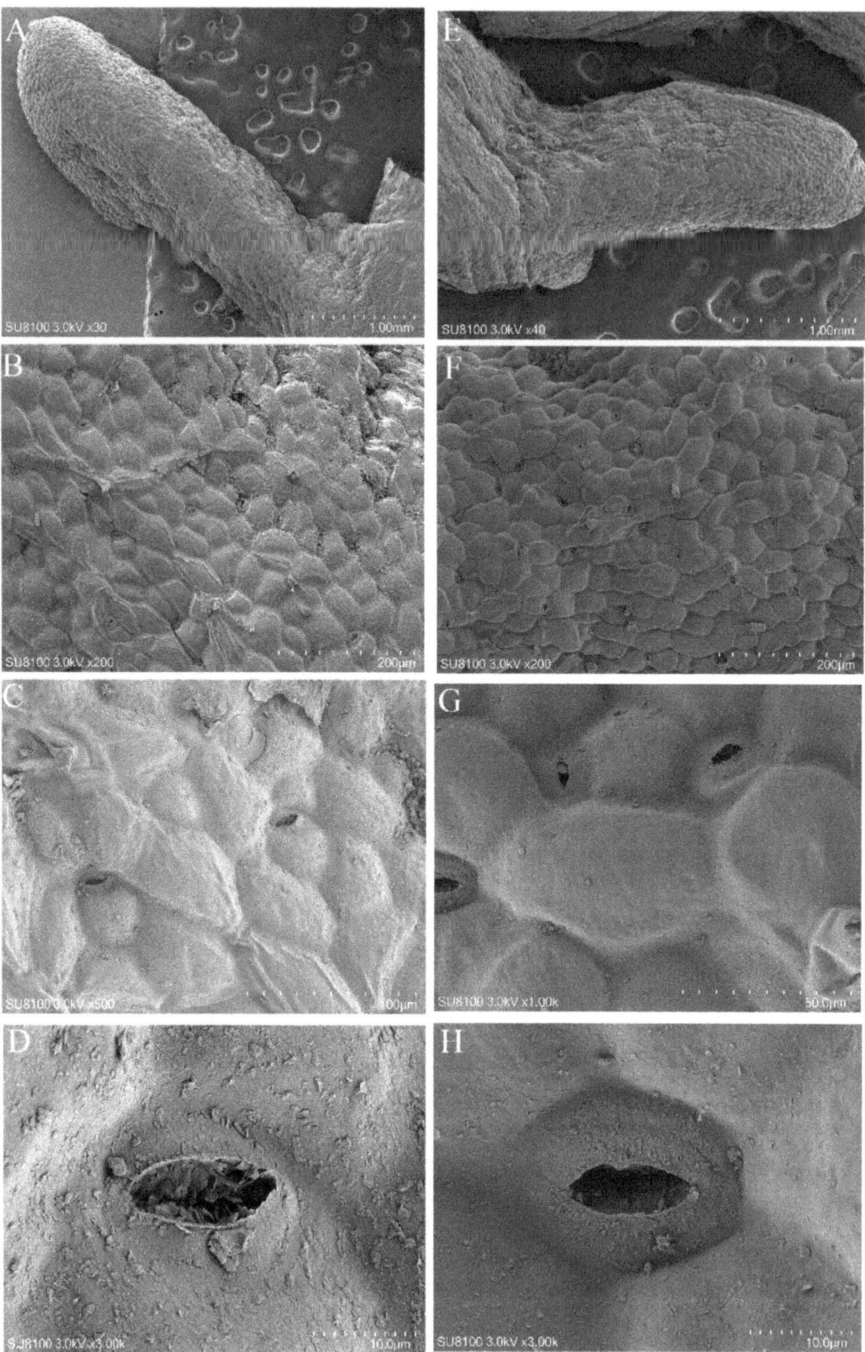

Figure 3. Scanning electron microscopic images of *K. foliatum* leaf in high-salt habitat showing (**A–D**) overall appearance structure, guard cell and stomata morphology of the upper epidermis of the leaf, and (**E–H**) overall appearance structure, guard cell, and stomata morphology of the lower epidermis of the leaf.

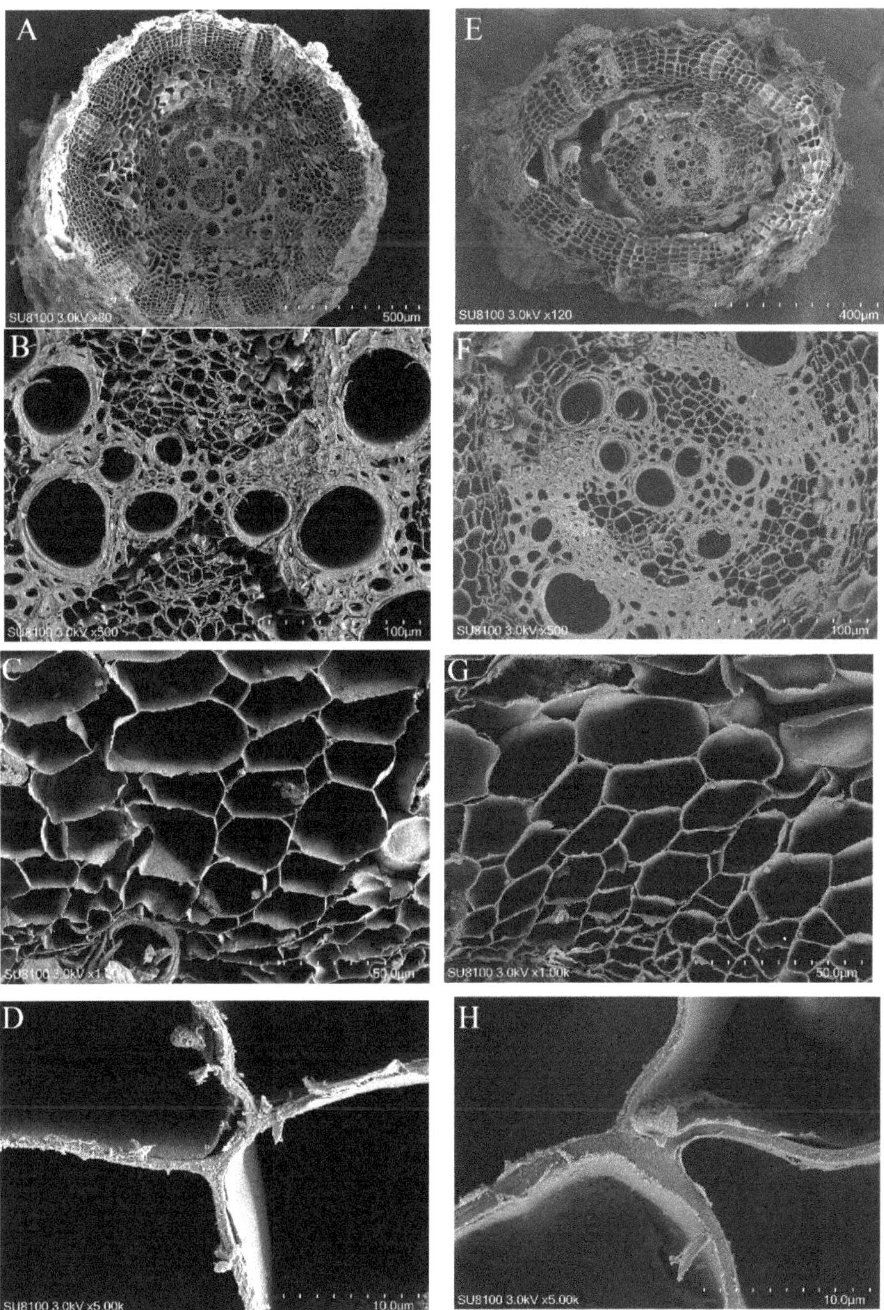

Figure 4. Scanning electron microscopic structure of *K. foliatum* leaf in low-salt habitat ((**A–D**) represent the overall appearance of the structure, vessel, and thin-wall cell morphology of fine roots (<0.2 mm) in high-salt habitats, while (**E–H**) represent the overall appearance of the structure, vessel, and thin-wall cell morphology of fine roots (<0.2 mm) in low-salt habitats).

3.2. Differences in Nutrient Contents and Stoichiometric Ratios of K. foliatum Leaves and Roots in Two Saline Habitats

The total carbon content of the *K. foliatum* leaves in the low-salt habitat was significantly higher than that in the high-salt habitat. The results of the leaf element stoichiometry showed that there were significant differences in the carbon–nitrogen-ratio and carbon–phosphorus-ratio contents of the *K. foliatum* leaves under the two different salt habitats, which were both greater in the low-salt habitat than in the high-salt habitat (Figure 5).

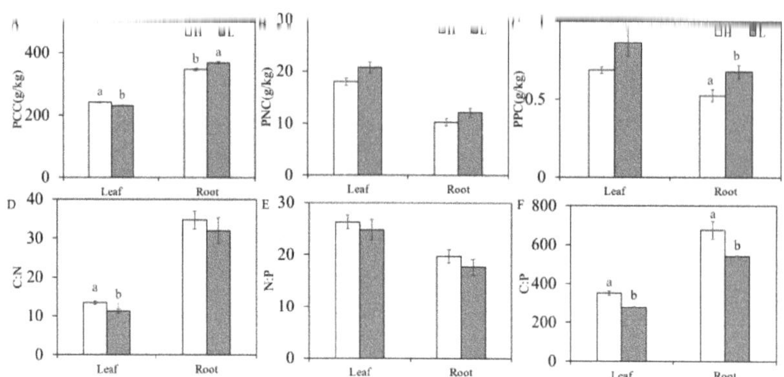

Figure 5. Differences in nutrient contents and stoichiometric ratios of *K. foliatum* leaves and roots under two saline habitats (H and L represent high-salt and low-salt habitats. (**A–F**) Plant carbon contents (PCCs), nitrogen contents (PNCs), phosphorus contents (PPCs), and stochiometric ratios of C-N (C:N), N-P (N:P), and C-P (C:P). Different lowercase letters represent significant differences between leaves and roots in different saline habitats, while unlabeled letters represent insignificant differences. The number of repeats was six).

There was a significant difference in the carbon and phosphorus contents of the *K. foliatum* roots between the two saline habitats, both showing higher carbon and phosphorus contents in the low-saline habitat than in the high-saline habitat. The results of the elemental stoichiometry of the roots indicated that the carbon–nitrogen ratio and carbon–phosphorus ratio of the *K. foliatum* roots in both saline habitats were higher in the low-salinity habitat than in the high-salinity habitat, and there was a significant difference in the nitrogen–phosphorus ratios between the two saline habitats (Figure 5).

3.3. Differences in Inorganic Ion Contents and Stoichiometric Ratios between K. foliatum Leaves and Roots under Two Different Salt Habitats

There were significant differences in the potassium and magnesium ion contents in the leaves of *K. foliatum* under the two different salt habitats, both of which were lower in the low-salt habitat than in the high-salt habitat (Figure 6). The results of the ion content stoichiometry of the leaves showed that there were significant differences in the K^+/Na^+ and Mg^{2+}/Na^+ ratios of the *K. foliatum* leaves under the two salt habitats, which were both lower in the low-salt habitat than in the high-salt habitat (Figure 7).

There were significant differences in the potassium ion contents, calcium ion contents, magnesium ion contents, and sodium ion contents in the roots of *K. foliatum* under the two different salt habitats, all of which were lower in the low-salt habitat than in the high-salt habitat (Figure 6). The results of the ion content stoichiometry of the roots indicate that there was a significant difference in the Ca^{2+}/Na^+ ratios between the *K. foliatum* roots in the two saline habitats, with a lower ratio in the lower-salinity habitat than in the higher-salinity habitat (Figure 7).

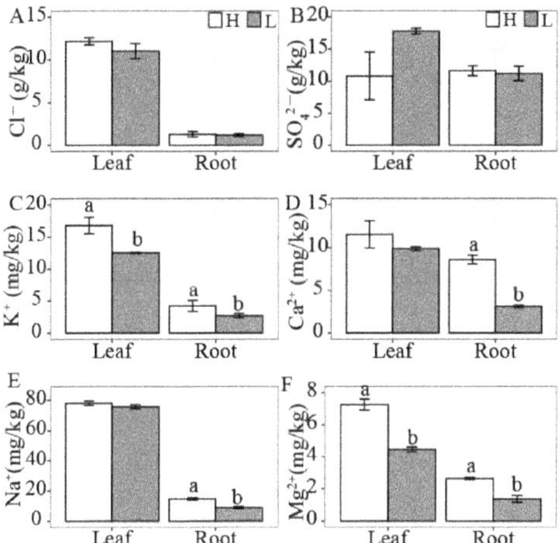

Figure 6. Differences in inorganic ion contents of *K. foliatum* leaves and roots under two saline habitats (H and L represent high-salt and low-salt habitats. Different lowercase letters represent significant differences in inorganic ion contents of *K. foliatum* leaves and roots in different saline habitats, while unlabeled letters represent insignificant differences. (**A–F**) Concentrations of Na^+, K^+, Ca^{2+}, Mg^{2+}, Cl^-, and sulfate anions. The number of repeats was six).

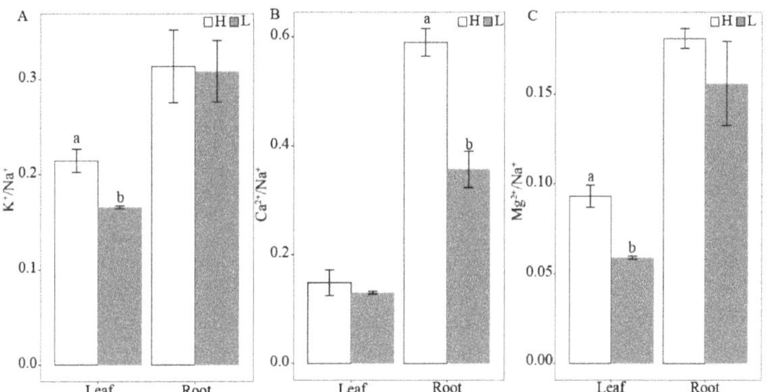

Figure 7. Differences in the stoichiometries of inorganic ion contents of *K. foliatum* leaves and roots under two saline habitats (H and L represent high-salt and low-salt habitats. Different lowercase letters represent significant differences in inorganic ion contents of *K. foliatum* leaves and roots in different saline habitats, while unlabeled letters represent insignificant differences. (**A–C**) The ratios of K^+/Na^+, Ca^{2+}/Na^+, and Mg^{2+}/Na^+. The number of repeats was six).

3.4. Differences in Osmoregulation Substances in K. foliatum Leaves and Roots under Two Different Salt Habitats

The soluble protein expression of the *K. foliatum* leaves in the two different salt habitats was higher in the low-salt habitat than in the high-salt habitat. The contents of starch, soluble sugar, and proline in the high-salt habitat were higher than those in the low-salt habitat. The soluble sugar and starch contents of the *K. foliatum* roots in the two different salt habitats showed significantly higher levels in the low-salt habitat than those in the high-salt habitat. The contents of soluble protein and proline in the high-salt habitat were significantly higher than those in the low-salt habitat (Figure 8).

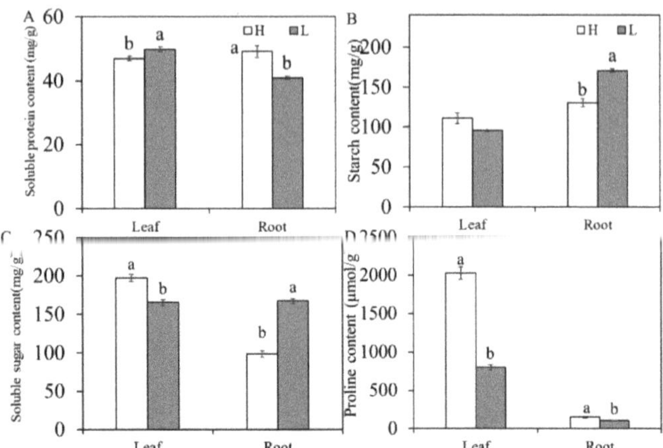

Figure 8. Differences in osmoregulation substance of *K. foliatum* leaves and roots under two saline habitats (H and L represent high-salt and low-salt habitats. Different lowercase letters represent significant differences in inorganic ion contents of *K. foliatum* leaves and roots in different saline habitats, while unlabeled letters represent insignificant differences. (**A–D**) Soluble protein content, starch content, soluble sugar content, and proline content. The number of repeats was six).

3.5. Differences in Antioxidant and Membrane Lipid Peroxidation Substances in K. foliatum Leaves and Roots under Two Different Salt Habitats

The contents of superoxide dismutase (SOD), peroxidase (POD), and catalase (CAT) in the *K. foliatum* leaves under the two different salt habitats were higher in the high-salt habitat than in the low-salt habitat. The contents of superoxide dismutase, peroxidase, and malondialdehyde (MDA) in the *K. foliatum* roots of both saline habitats showed higher levels in the high-saline habitat than in the low-saline habitat. The contents of superoxide dismutase, catalase, and malondialdehyde in the roots showed significant differences between the two habitats (Figure 9).

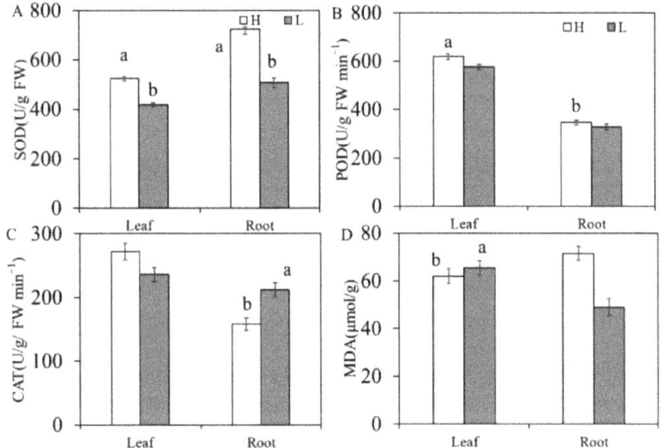

Figure 9. Differences in antioxidant and membrane lipid peroxidation substances under two saline habitats (H and L represent high-salt and low-salt habitats. Different lowercase letters represent significant differences in inorganic ion contents of *K. foliatum* leaves and roots in different saline habitats, while unlabeled letters represent insignificant differences. (**A–D**) Contents of SOD, POD, CAT, and MDA. The number of repeats was six).

3.6. Network Association of Physiological Traits of K. foliatum Leaves and Roots in Two Salt Habitats

In the physiological trait network of the *K. foliatum* leaves and roots in the high-salt habitat, the SOD of the *K. foliatum* leaves had the highest degree and was the "central trait" of the leaves in the high-salt habitat, while the root CAT activity was the "central trait" of the roots in the high-salt habitat (Figure 10, Tables 3 and 4). In the physiological trait network of the *K. foliatum* leaves and roots in the low-salt habitat, the leaf MDA and root C/N were the central traits of the leaves and roots (Figure 10, Tables 3 and 4). Moreover, there was a greater number of positive correlations between the leaves and roots in the high-salt habitat (Figure 10).

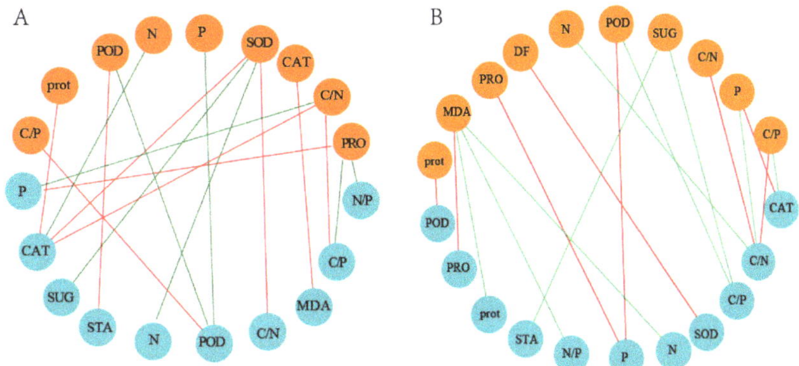

Figure 10. Physiological character networks of *K. foliatum* leaves and roots in two salt habitats ((**A**,**B**) represent high-salt habitat and low-salt habitat, respectively. The orange circles represent leaf traits, the blue circles represent root traits, the red lines represent significant positive correlations, and the green lines represent significant negative correlations. SUG, prot, and STA represent soluble sugar, soluble protein, and starch contents).

Table 3. Network parameters of leaf and root physiological traits in high-salt habitat.

Name	Type	Degree	Weight Degree
C/N	leaf	3	2.600
C/P	leaf	1	0.829
CAT	leaf	1	0.886
prot	leaf	1	1.000
N	leaf	1	0.829
P	leaf	1	0.829
POD	leaf	2	1.771
PRO	leaf	3	2.657
SOD	leaf	4	3.543
CAT	root	4	3.600
C/N	root	1	0.886
C/P	root	2	1.714
STA	root	1	0.943
MDA	root	1	0.886
N	root	1	0.886
N/P	root	1	0.886
P	root	2	1.710
POD	root	3	2.486
SUG	root	1	0.943

Note: SUG, prot, and STA represent soluble sugar, soluble protein, and starch content.

Table 4. Network parameters of leaf and root physiological traits in low-salt habitat.

Name	Type	Degree	Weight Degree
C/N	leaf	1	0.829
C/P	leaf	2	1.714
prot	leaf	1	0.829
STA	leaf	1	0.829
MDA	leaf	4	3.599
N	leaf	1	0.829
P	leaf	2	1.714
POD	leaf	2	1.771
PRO	leaf	1	0.886
SUG	leaf	2	1.829
CAT	root	2	1.771
C/N	root	4	3.314
CP	root	2	1.771
protein	root	1	0.942
STA	root	1	0.886
N	root	1	0.886
N/P	root	1	0.829
P	root	2	1.829
POD	root	1	0.829
PRO	root	1	0.943
SOD	root	1	0.829

Note: SUG, prot, and STA represent soluble sugar, soluble protein, and starch contents.

4. Discussion

Salt reduces photosynthetic pigments, such as chlorophyll and carotenoids, by disrupting biosynthetic pathways [40]. Photosynthesis is a prerequisite for maintaining normal plant growth, and its intensity can serve as an indicator of plant growth and resilience, while photosynthetic pigments directly affect the photosynthetic capacity of plants. Chlorophyll is located in the thylakoid membrane and acts as an "antenna", absorbing red and blue wavelengths of light energy and transferring it to the reaction center of the photosystem [41]. Carotenoids are auxiliary pigments that can absorb light from the blue–green region of the solar spectrum and transfer the absorbed energy to chlorophyll molecules [42]. This study showed that *K. foliatum* chlorophyll a and chlorophyll b were significantly reduced in the high-salt soil environment (Figure 2), consistent with the research results of other scholars on tomatoes [43]. In indoor control experiments, the contents of chlorophylls and carotenoids were significantly reduced via a higher concentration of $NaCl + Na_2SO_4$ under a single salt stress, and their contents increased compared to the control group [44]. The *K. foliatum* in this study came from natural habitats, and the presence of multiple salts in the soil simultaneously affected its growth. Therefore, the results of this study were consistent with those of indoor mixed salt stress. This study found that, although there was a significant difference in the chlorophyll contents of *K. foliatum* under the two different salt habitats, the difference in the change was small (Table 2). This may be due to various factors affecting the degree of chlorophyll change, such as the plant species, salt type, and salt concentration. Moreover, the chlorophyll contents of salt-tolerant plants change less than those of salt-sensitive plants [45], so the amplitude of the change was relatively small. Carotenoids are effective antioxidants, so an increase in the carotenoid contents in plants helps protect chloroplasts and maintain higher chlorophyll contents [46]. This study found that, compared to the low-salt habitat, the palisade tissue of the *K. foliatum* leaves in the high-salt habitat significantly thickened, indicating that *K. foliatum* can increase the proportion of palisade tissue by regulating the development of mesophyll cells, thereby alleviating the photoinhibition caused by salt stress. The increase in the leaf epidermal thickness in the high-salt habitat may be an adaptation strategy of *K. foliatum* to salt stress in order to better prevent water loss on the leaf surface and improve the water retention efficiency [47]. In high-salt habitats, a large amount of salt crystals precipitate from *K. foliatum* leaves, while, in low-salt environments, salt crystals precipitate less, indicating that salt stress promotes an increase in plant salt secretion (Figure 3).

In this study, the concentrations of C, N, and P in the *K. foliatum* roots were lower than those in the leaves (Figure 5). This is because, compared to leaves, roots typically have relatively lower concentrations of C, N, and P, as their main function is to transport absorbed water and nutrients to the leaves [48]. Secondly, leaves are the assimilation organs for plants to obtain energy and synthesize photosynthesis; therefore, carbohydrates are effectively accumulated in the leaves. The N, P contents and N:P ratio of the *K. foliatum* leaves were higher than those of the roots (Figure 5), indicating that the two organs allocate available nutrients in different ways to cope with soil environmental changes [49]. For example, under soil salt stress, plants can regulate limited nutrients between organs, especially with a higher proportion of N and P allocated to photosynthetic organs (leaves) than non-photosynthetic organs (roots) to maintain normal carbon assimilation or plant growth [50]. Salt stress causes physiological limitations on plants, including osmotic stress, nutrient imbalance, and interference with photosynthesis, which can affect the plant growth and alter the C:N:P stoichiometric ratio between the plant organs [51]. In this study, in addition to the root carbon contents, the concentrations of N and P in the *K. foliatum* leaves were lower under high-salinity conditions (Figure 5), indicating that salt stress affects plant growth and development by limiting nutrient acquisition [10]. Under salt stress, plants typically exhibit nutrient metabolism disorders due to the presence of a large number of anions (such as chloride and sulfate ions) in the soil that may compete for nutrients, thereby reducing nutrient absorption and organic matter accumulation [52]. Generally speaking, P is important for the composition of ATPase, which plays a crucial role in plant photosynthesis and energy metabolism [10]. The decrease in the P concentration in *K. foliatum* leaves in high-salt soil is mainly due to the presence of more anions in the soil, which can affect the P concentration, leading to a decrease in the plant photosynthesis rate and thereby affecting the plant absorption of P [53]. The decrease in the nitrogen content in *K. foliatum* roots may be due to the reduced nitrogen accumulation in the plants due to increased chloride absorption, resulting in a decrease in the ability of *K. foliatum* to absorb and utilize nitrogen [54]. Some energy and nutrients are allocated to the osmotic regulation of the roots to maintain an osmotic balance, thereby reducing the absorption of N and P by the roots [51,55]. In the high-salt habitat, the N and P contents of the *K. foliatum* roots decreased, while the N:P ratio increased. Affected by root osmotic regulation, the nutrient transport from root to leaf is limited, resulting in a decrease in the available nutrients for leaf photosynthesis and a decrease in the photosynthetic capacity [55].

The increases in the Na^+ content and K^+ exosmosis inhibit the absorption of other mineral elements by plant roots; that is, osmotic pressure and ion stress caused by high concentrations of ions disrupt the normal physiological metabolism of plants, and even cause death. Potassium is an essential nutrient element that plays an important role in enzyme activation, osmotic regulation, expansion generation, membrane potential regulation, and cytoplasmic pH homeostasis [56]. Due to the similarity in the physical and chemical properties between Na^+ and K^+ (i.e., similar ion radii and ion hydration energies), Na^+ and K^+ compete in the metabolic process, inhibiting the activities of many enzymes that require K^+ to function [57,58]. This study indicated that the accumulation of Na^+ in the roots was smaller than in the leaves (Figure 6), which is consistent with the research results of Wang et al. [59], indicating that the roots of this species can promote the transport of sodium ions to the leaves. This can protect the root system with relatively vigorous metabolism, ensuring the absorption of water and other nutrients by the root system, and it can reduce the leaf osmotic potential, promote upward water transport, and reduce the salt concentration in the plant body. At the same time, it can take salt ions out of the body through leaf aging and withering [60,61]. This study indicates that, despite the accumulation of a large amount of Na^+ in the leaves, they had no obvious symptoms of salt damage. Previous studies have shown that a large amount of Na^+ in plant leaves enters the vacuoles during salt stress, leading to regionalization. This reduces the salt concentration in the cytoplasm to maintain normal metabolic activity, and it increases the ion concentration in the vacuoles to reduce the cell's water potential and ensure normal water absorption. Some people believe that the survival of plants under salt stress requires a high K^+/Na^+ ratio in the cytoplasm. This study found that the K^+/Na^+ ratio of the roots and leaves significantly increased in the high-salt habitat (Figure 7). In addition, this study also found that the K^+ contents in the leaves increased in the high-salt habitat, indicating that the roots can specifically transport K^+ from the roots to the leaves, thereby maintaining a low K^+/Na^+ ratio in the leaves to reduce ion toxicity caused by salt stress. Although Cl^- has an auxiliary effect on promoting chlorophyll synthesis, high concentrations of Cl^- can affect photosynthesis and disrupt the cell expansion balance [62]. In this study, the Cl^- concentrations in the leaves in the high-salt habitat were higher than those in the low-salt habitat, which is consistent with Ehsan et al. (2010)'s finding that the Cl^- content in broad bean leaves increases with the increasing salt content [63]. Moreover, a high chloride ion concentration can reduce the photosynthetic capacity and photon yield, damaging the PSII structure.

In plants, sugar, as a metabolic resource and structural component of cells, participates in cell osmotic regulation in salt stress environments [64,65]. In this study, the soluble sugar content of the *K. foliatum* leaves increased in the high-salt habitat (Figure 8), which is consistent with the research results of Wang et al. [66]. Salt stress will first affect the plant roots, and the excessive ROS induced by salt stress seriously affect the root growth [67]. Proline is an important osmoregulation substance that plays an important role in plant responses to osmotic and salt stress by protecting the plant cell membranes and proteins, and it serves as a reactive oxygen species scavenger [68]. Proline synthesis is considered the main pathway for plant metabolite accumulation under abiotic stress [69]. The changes in the proline contents of different species under salt stress may vary. Proline rapidly accumulates in mangroves and Australian wild rice under salt stress [70,71], but it decreases in both pearl millet and wheat [72]. This study found that, with the increase in the salt content, the proline contents in the leaves and roots of the *K. foliatum* significantly increased (Figure 8), and the proline contents in the *K. foliatum* roots were lower than those in the leaves, which is consistent with the research results on *Populus euphratica*. Similarly, the proline contents of the *K. foliatum* leaves grown in the desert saline alkali land of the Hexi Corridor in Gansu Province significantly increased with the increase in the soil salinity [73]. Under NaCl stress, the content of proline in *P. euphratica* roots is low, but it is high in the branches and leaves [74]. This may be because plants can adjust the proportion of proline in their roots and leaves appropriately through low levels of proline accumulation. Under salt stress, the balance between ROS production and clearance is disrupted, leading to the accumulation of ROS in plant cells and inducing oxidative damage [75]. Plants have antioxidant enzyme systems to alleviate oxidative damage caused by ROS under salt stress. This study found that the activity of antioxidant enzymes (SOD and POD) in the leaves and roots of *K. foliatum* increased under the high-salt condition (Figure 9), indicating that *K. foliatum* can increase the activity of antioxidant enzyme systems to alleviate oxidative damage caused by ROS under salt stress. The enhancement of the antioxidant enzyme system activity also alleviates the ion toxicity caused by high Na^+ contents in roots under salt stress. The SOD content decreased with the increase in the $NaCl + Na_2SO_4$ concentration, and it gradually increased after reaching the minimum at a salt concentration of 250 mM. Under a single salt stress, the SOD activity reached its maximum at a salt concentration of 250 mM. This indicated that the response of *K. foliatum* in natural habitats to salt stress was different from that of *K. foliatum* cultured indoors [44]. This study found that the activities of SOD and CAT in the leaves were higher than those in the roots, which is consistent with the results obtained by Daccord et al. for legumes [76]. As a product of lipid peroxidation, MDA is a good indicator for measuring oxidative membrane damage. MDA is an indicator of plant oxidative stress [77], and it accumulates continuously to form oxidative stress [20], commonly used to evaluate the degree of plant oxidative damage [78]. This study found that, under salt stress, the salt content increased and the MDA content in the roots significantly increased, promoting the production of reactive oxygen species in the roots and enhancing lipid peroxidation. Some studies have also found that excessive salt content can significantly reduce the MDA content, which may be attributed to the role of antioxidants in plants [79], triggering antioxidant reactions.

Research on the correlation between the leaf traits and root traits has been ongoing, and it has been found that there is indeed a certain correlation between the two within a certain region [80]. The number of positive correlations between the roots and leaves in the high-salt habitat is greater than that in the low-salt habitat (Figure 10), indicating that salt can alter the interaction between the plant roots and leaves. This is consistent with Yang's finding that the correlation between the leaf chemical traits in high-salt habitats is higher, and the network is more complex [81]. Numerous studies have shown that higher correlations between the traits enable plants to effectively acquire resources [81], while plants in desert areas may face stronger environmental pressures, often resulting in closer trait correlations and trade-offs [82]. The SOD activity in the leaves and CAT activity in roots of medium *K. foliatum* are the "central traits" in high-salt habitats. In low-salt habitats, the leaf MDA and root C/N are the central traits of the leaves and roots (Tables 3 and 4), indicating that *K. foliatum* adapts to changes in salt environments in different ways. There are many insignificant correlations in the parts not shown in this study, which may be due to greater variability in and uncertainty about the plant root traits or lead to insignificant correlations. In summary, the correlation between the leaf and root traits is relatively complex [83], so the specific correlation between the two needs further exploration.

5. Conclusions

The results of this study indicate that *K. foliatum* can adapt to high-salinity habitats by regulating the development of the mesophyll cells in the leaves and increasing the proportion of palisade tissue in the mesophyll. In high-salt habitats, large amounts of salt crystals precipitate from *K. foliatum* leaves, while, in low-salt environments, salt crystals precipitate less, indicating that salt stress promotes an increase in plant salt secretion. The salt content leads to changes in the distribution ratio of the carbon, nitrogen, and phosphorus elements in the leaves and roots of *K. foliatum*. The soluble protein of the *K. foliatum* leaves of the two different salt habitats showed higher levels in the low-salt habitat than in the high-salt habitat, while the protein changes in the roots were the opposite to those in the leaves. The leaves respond to salt stress by reducing soluble proteins, while the roots exhibit the opposite trend. The leaves respond to salt stress by increasing starch and soluble sugars, while the changes in the roots are the opposite to those in the leaves. *K. foliatum* leaves and roots adapt to high-salt stress by increasing their proline contents and SOD and POD activities. The plant trait network indicates that the key quantitative traits of *K. foliatum* leaves and roots in high-salt and low-salt habitats are significantly different.

Author Contributions: Investigation, L.J., D.W., W.L., Y.L., E.L., X.L. and G.Y.; conceptualization, L.J. and X.H.; methodology, L.J.; software, L.J., D.W. and W.L.; writing—original draft preparation, L.J.; writing—review and editing, L.J.; supervision, X.H.; funding acquisition, X.H. All authors have read and agreed to the published version of the manuscript.

Funding: This research was financially supported by the National Natural Science Foundation of China (32260266), Special project of central government guiding local science and technology development (ZYYD2023A03) and the National Natural Science Foundation of China (32101360).

Institutional Review Board Statement: Not applicable.

Informed Consent Statement: Not applicable.

Data Availability Statement: Data are contained within the article.

Conflicts of Interest: The authors declare that they have no conflicts of interest.

References

1. Jamil, A.; Riaz, S.; Ashraf, M.; Foolad, M.R. Gene expression profiling of plants under salt stress. *Crit. Rev. Plant Sci.* **2011**, *30*, 435–458. [CrossRef]
2. Kumar, P.; Sharma, P.K. Soil salinity and food security in India. *Front. Sustain. Food Syst.* **2020**, *4*, 533781. [CrossRef]
3. Flowers, T.J.; Colmer, T.D. Salinity tolerance in halophytes. *New Phytol.* **2008**, *179*, 945–963. [CrossRef]
4. Rao, Y.; Peng, T.; Xue, S. Mechanisms of plant saline-alkaline tolerance. *J. Plant Physiol.* **2023**, *281*, 153916. [CrossRef]
5. Taïbi, K.; Taïbi, F.; Abderrahim, L.A.; Ennajah, A.; Belkhodja, M.; Mulet, J.M. Effect of salt stress on growth, chlorophyll content, lipid peroxidation and antioxidant defence systems in *Phaseolus vulgaris* L. *S. Afr. J. Bot.* **2016**, *105*, 306–312. [CrossRef]
6. Ghazali, G.E. *Suaeda vermiculata* Forssk. ex J.F. Gmel.: Structural characteristics and adaptations to salinity and drought: A review. *Int. J. Bot.* **2020**, *1*, 1V 22.
7. Neel, J.P.S.; Alloush, G.A.; Belesky, D.P.; Clapham, W.M. Influence of rhizosphere ionic strength on mineral composition, dry matter yield and nutritive value of forage chicory. *J. Agron. Crop Sci.* **2002**, *188*, 398–407. [CrossRef]
8. Teng, X.X.; Cao, W.L.; Lan, H.X.; Tang, H.J.; Bao, Y.M.; Zhang, H.S. OsNHX2, an Na$^+$/H$^+$ antiporter gene, can enhance salt tolerance in rice plants through more effective accumulation of toxic Na in leaf mesophyll and bundle sheath cells. *Acta Physiol. Plant.* **2017**, *39*, 113. [CrossRef]
9. Liu, J.; Gao, H.; Wang, X.; Zheng, Q.; Wang, C.; Wang, X.; Wang, Q. Effects of 24-epibrassinolide on plant growth, osmotic regulation and ion homeostasis of salt-stressed canola. *Plant Biol.* **2014**, *16*, 440–450. [CrossRef]
10. Rong, Q.; Liu, J.; Cai, Y.; Lu, Z.; Zhao, Z.; Yue, W.; Xia, J. Leaf carbon, nitrogen and phosphorus stoichiometry of *Tamarix chinensis* Lour. in the Laizhou Bay coastal wetland, China. *Ecol. Eng. J. Ecotechnol.* **2015**, *76*, 57–65. [CrossRef]
11. Chrysargyris, A.; Papakyriakou, E.; Petropoulos, S.A.; Tzortzakis, N. The combined and single effect of salinity and copper stress on growth and quality of Mentha spicata plants. *J. Hazard. Mater.* **2019**, *368*, 584–593. [CrossRef]
12. Seki, M.; Ishida, J.; Narusaka, M.; Fujita, M.; Nanjo, T.; Umezawa, T.; Kamiya, A.; Nakajima, M.; Enju, A.; Sakurai, T. Monitoring the expression pattern of around 7000 Arabidopsis genes under aba treatments using a full-length cDNA microarray. *Funct. Integr. Genom.* **2002**, *2*, 282–291. [CrossRef]
13. Theerawitaya, C.; Boriboonkaset, T.; Cha-Um, S.; Supaibulwatana, K.; Kirdmanee, C. Transcriptional regulations of the genes of starch metabolism and physiological changes in response to salt stress rice (*Oryza sativa* L.) seedlings. *Physiol. Mol. Biol. Plants* **2012**, *18*, 197–208. [CrossRef]
14. Thalmann, M.; Santelia, D. Starch as a determinant of plant fitness under abiotic stress. *New Phytol.* **2017**, *214*, 943–951. [CrossRef]
15. Henry, C.; Bledsoe, S.W.; Griffiths, C.A.; Paul, M.J.; Kollman, A.; SAKR, S.; Lagrimini, M. Differential role for trehalose metabolism in salt-stressed maize. *Plant Physiol.* **2015**, *169*, 1072–1089. [CrossRef]
16. Oury, V.; Caldeira, C.F.; Prodhomme, D.; Pichon, J.-P.; Gibon, Y.; Tardieu, F.; Turc, O. Is change in ovary carbon status a cause or a consequence of maize ovary abortion in water deficit during flowering? *Plant Physiol.* **2016**, *171*, 997–1008. [CrossRef]
17. Tang, X.; Mu, X.; Shao, H.; Wang, H.; Brestic, M. Global plant-responding mechanisms to salt stress: Physiological and molecular levels and implications in biotechnology. *Crit. Rev. Biotechnol.* **2015**, *35*, 425–437. [CrossRef]
18. Zhang, Q.; Dai, W. Plant Response to Salinity Stress. In *Stress Physiology of Woody Plants*; Dai, W., Ed.; CRC Press: Boca Raton, FL, USA, 2019.
19. Demidchik, V. Mechanisms of oxidative stress in plants: From classical chemistry to cell biology. *Environ. Exp. Bot.* **2015**, *109*, 212–228. [CrossRef]
20. Kaya, C.; Akram, N.A.; Ashraf, M.; Sonmez, O. Exogenous application of humic acid mitigates salinity stress in maize (*Zea mays* L.) plants by improving some key physico-biochemical attributes. *Cereal Res. Commun.* **2018**, *46*, 67–78. [CrossRef]
21. Grebosz, J.; Badowiec, A.; Weidner, S. Changes in the root proteome of *Triticosecale* grains germinating under osmotic stress. *Acta Physiol. Plant.* **2014**, *36*, 825–835. [CrossRef]
22. Wang, G.; Jin, J. Effects of NaCl on physiology and leaf ultrastructure in the halophyte *Kalidium foliatum*. *Nord. J. Bot.* **2015**, *33*, 232–238. [CrossRef]
23. Zhang, Z.D.; Gu, M.Y.; Tang, Q.Y.; Chu, M.; Zhu, J.; Sun, J.; Yang, R.; Xu, W.L. Screening of Salt-tolerant and Growth-promoting Bacteria in the rhizosphere of *Kalidium foliatum* and the Functional Identification in Pot Experiments. *J. Agric. Sci. Technol.* **2021**, *23*, 186–192.
24. Yang, X.-D.; Anwar, E.; Zhou, J.; He, D.; Gao, Y.-C.; Lv, G.-H.; Cao, Y.-E. Higher association and integration among functional traits in small tree than shrub in resisting drought stress in an arid desert. *Environ. Exp. Bot.* **2022**, *201*, 104993. [CrossRef]
25. Gong, Y.; Ling, H.; Lv, G.; Chen, Y.; Guo, Z.; Cao, J. Disentangling the influence of aridity and salinity on community functional and phylogenetic diversity in local dryland vegetation. *Sci. Total. Environ.* **2019**, *653*, 409–422. [CrossRef]
26. Jiang, L.; Hu, D.; Wang, H.F.; Lv, G. Discriminating ecological processes affecting different dimensions of α- and β-diversity in desert plant communities. *Ecol. Evol.* **2022**, *12*, e8710. [CrossRef]
27. Lu, R.K. *Soil Agrochemical Analysis Methods*; China Agricultural Science and Technology Press: Beijing, China, 2000.
28. Harmut, A.J.M.E. Chlorophylls and carotenoids: Pigments of photosynthetic membranes. *Methods Enzymol.* **1987**, *148*, 350–383.
29. Bradford, M.M. A rapid and sensitive method for the quantitation of microgram quantities of protein utilizing the principle of protein-dye binding. *Anal. Biochem.* **1976**, *72*, 248–254. [CrossRef]
30. Khelil, A.; Menu, T.; Ricard, B. Adaptive response to salt involving carbohydrate metabolism in leaves of a salt-sensitive tomato cultivar. *Plant Physiol. Biochem.* **2007**, *45*, 551–559. [CrossRef]

31. Lepesant, J.A.; Kunst, F.; Lepesant-Kejzlarová, J.; Dedonder, R. Chromosomal location of mutations affecting sucrose metabolism in *Bacillus subtilis* Marburg. *Mol. Gen. Genet.* **1972**, *118*, 135–160. [CrossRef]

32. Gomes, H.T.; Bartos, P.M.C.; Silva, C.O.; Amaral, L.I.V.D.; Scherwinskipereira, J.E. Comparative biochemical profiling during the stages of acquisition and development of somatic embryogenesis in African oil palm (*Elaeis guineensis* Jacq.). *Plant Growth Regul.* **2014**, *74*, 199–208. [CrossRef]

33. Stewart, R.R.; Bewley, J.D. Lipid peroxidation associated with accelerated aging of soybean axes. *Plant Physiol.* **1980**, *65*, 245–248. [CrossRef]

34. Hammerschmidt, R.; Nuckles, E.M.; Kuć, J. Association of enhanced peroxidase activity with induced systemic resistance of cucumber to *Colletotrichum lagenarium*. *Physiol. Plant Pathol.* **1982**, *20*, 73–82. [CrossRef]

35. Panda, A.; Rangani, J.; Kumari, A.; Parida, A.K. Efficient regulation of arsenic translocation to shoot tissue and modulation of phytochelatin levels and antioxidative defense system confers salinity and arsenic tolerance in the Halophyte *Suaeda maritima*. *Environ. Exp. Bot.* **2017**, *143*, 49–171. [CrossRef]

36. Hodges, D.M.; DeLong, J.M.; Forney, C.F.; Prange, R.K. Improving the thiobarbituric acid-reactive-substances assay for estimating lipid peroxidation in plant tissues containing anthocyanin and other interfering compounds. *Planta* **1999**, *207*, 604–611. [CrossRef]

37. Yang, J.F.; Zhang, X.N.; Song, D.H.; Tian, J.Y. Interspecific Integration of Chemical Traits in Desert Plant Leaves with Variations in Soil Water and Salinity Habitats. *Forest* **2023**, *14*, 1963. [CrossRef]

38. Li, Y.; Liu, C.; Sack, L.; Xu, L.I.; Li, M.; Zhang, J.; He, N.; Penuelas, J. Leaf trait network architecture shifts with species-richness and climate across forests at continental scale. *Ecol. Lett.* **2022**, *25*, 1442–1457. [CrossRef]

39. Rosas, T.; Mencuccini, M.; Barba, J.; Cochard, H.; Saura-Mas, S.; Martinez-Vilalta, J. Adjustments and coordination of hydraulic, leaf and stem traits along a water availability gradient. *New Phytol.* **2019**, *223*, 632–646. [CrossRef]

40. Kiani-Pouya, A.; Rasouli, F. The potential of leaf chlorophyll content to screen bread-wheat genotypes in saline condition. *Photosynthetica* **2014**, *52*, 288–300. [CrossRef]

41. Wu, S.; Wang, J.; Yan, Z.; Wu, J. Monitoring tree-crown scale autumn leaf phenology in a temperate forest with an integration of Planet Scope and drone remote sensing observations. *ISPRS J. Photogramm. Remote Sens.* **2021**, *171*, 36–48. [CrossRef]

42. Larcher, W. Photosynthesis as a tool for indicating temperature stress events. In *Ecophysiology of Photosynthesis*; Schulze, E.D., Caldwell, M.M., Eds.; Springer: Berlin/Heidelberg, Germany, 1995.

43. Parvin, K.; Hasanuzzaman, M.; Bhuyan, M.B.; Nahar, K.; Mohsin, S.M.; Fujita, M. Comparative physiological and biochemical changes in tomato (*Solanum lycopersicum* L.) under salt stress and recovery: Role of antioxidant defense and glyoxalase systems. *Antioxidants* **2019**, *8*, 350. [CrossRef]

44. Gong, D.H.; Wang, G.Z.; Si, W.T.; Zhou, Y.; Liu, Y.; Jia, J. Effects of salt stress on photosynthetic pigments and activity of ribulose-1,5-bisphosphate carboxylase/oxygenase in *Kalidium foliatum*. *Russ. J. Plant Physiol.* **2018**, *65*, 98–103. [CrossRef]

45. Rao, G.C.; Rao, G.R. Pigment composition and chlorophyllase activity in pigeon pea (*Cajanus indicus* Spreng) and Gingelley (*Sesamum indicum* L.) under NaCl salinity. *Indian J. Exp. Biol.* **1981**, *19*, 768–770.

46. Kacharava, N.; Chanishvili, S.; Badridze, G.; Chkhubianishvili, E.; Janukashvili, N. Effect of seed irradiation on the content of antioxidants in leaves of Kidney bean, Cabbage and Beet cultivars. *Aust. J. Crop Sci.* **2009**, *3*, 137.

47. Ashraf, M.; Ozturk, M.; Ahmad, M.S.A. *Plant Adaptation and Phytoremediation: Structural and Functional Adaptations in Plants for Salinity Tolerance*; Springer: Berlin/Heidelberg, Germany, 2010.

48. Luo, Y.; Peng, Q.; Li, K.; Gong, Y.; Han, W. Patterns of nitrogen and phosphorus stoichiometry among leaf, stem and root of desert plants and responses to climate and soil factors in Xinjiang, China. *Catena* **2021**, *199*, 105100. [CrossRef]

49. He, M.Z.; Zhang, K.; Tan, H.J.; Hu, R.; Su, J.Q.; Wang, J.; Huang, L.; Zhang, Y.F.; Li, X.R. Nutrient levels within leaves, stems, and roots of the xeric species *Reaumuria soongoricain* relation to geographical, climatic, and soil conditions. *Ecol. Evol.* **2015**, *5*, 1494–1503. [CrossRef]

50. Thompson, K.E.N.; Parkinson, J.A.; Band, S.R.; Spencer, R.E. A comparative study of leaf nutrient concentrations in a regional herbaceous flora. *New Phytol.* **1997**, *36*, 679–689. [CrossRef]

51. Sun, X.; Gao, Y.; Wang, D.; Chen, J.; Zhang, F.; Zhou, J.; Yan, X.; Li, Y.; Luo, Z.B. Stoichiometric variation of halophytes in response to changes in soil salinity. *Plant Biol.* **2017**, *19*, 360–367. [CrossRef]

52. Balba, M.A. *Management of Problem Soils in Arid Ecosystems*; CRC Press: Boca Raton, FL, USA, 1995.

53. Joshi, S.S. Effect of salinity stress on organic and mineral constituents in the leaves of pigeonpea (*Cajanus cajan* L. var C-11). *Plant Soil* **1984**, *82*, 69–76. [CrossRef]

54. Ramoliya, P.J.; Patel, H.M.; Joshi, J.B.; Pandey, A.N. Effect of salinization of soil on growth and nutrient accumulation in seedlings of Prosopis cineraria. *J. Plant Nutr.* **2006**, *29*, 283–303. [CrossRef]

55. Azzeme, A.M.; Abdullah, S.; Aziz, M.A.; Wahab, P.E.M. Oil palm leaves and roots differ in physiological response, antioxidant enzyme activities and expression of stress-responsive genes upon exposure to drought stress. *Acta Physiol. Plant.* **2016**, *38*, 52. [CrossRef]

56. Barragan, V.; Leidi, E.O.; Andrés, Z.; LRubio Luca, A.D.; Fernandez, J.A.; Cubero, B.; Pardo, J.M. Ion exchangers NHX1 and NHX2 mediate active potassium uptake into vacuoles to regulate cell turgor and stomatal function in Arabidopsis. *Plant Cell* **2012**, *24*, 1127–1142. [CrossRef] [PubMed]

57. Mansour, M.M.F. Plasma membrane permeability as an indicator of salt tolerance in plants. *Biol. Plant.* **2012**, *57*, 1–10. [CrossRef]

58. Patade, V.Y.; Bhargava, S.; Suprasanna, P. Effects of NaCl and iso-osmotic peg stress on growth, osmolytes accumulation and antioxidant defense in cultured sugarcane cells. *Plant Cell Tissue Organ. Cult.* **2012**, *108*, 279–286. [CrossRef]
59. Wang, X.J.; Yang, D.; Yi, F.Y.; Zhang, Y.Y.; Saixiyala, B.S.; Sun, H.L. Research Progress on the Halophyte *Kalidium foliatum* and Its Resources Characteristics. *Anim. Husb. Feed. Sci.* **2015**, *36*, 64–67.
60. Luo, D.; Wu, Z.B.; Shi, Y.J. Effects of salt stress on leaf analomical structure and ion absorption, transportation and distribution of three Ping' ou hybrid hazelnut seedlings. *Acta Ecol. Sin.* **2022**, *42*, 1876–1888.
61. Tang, X.Q.; Li, H.Y.; Yang, X.Y.; Liu, Z.X.; Zhang, H.X. Effects of short-time salt stress on distribution and balance of Na^+ and K^+ in *Nitraria sibirica* Pall. *seedlings. For. Res.* **2017**, *30*, 1022–1027.
62. Lekklar, C.; Chadchawan, S.; Boon-Long, P.; Pfeiffer, W.; Chaidee, A. Salt stress in rice: Multivariate analysis separates four components of beneficial silicon action. *Protoplasma* **2019**, *166*, 771–787. [CrossRef]
63. Ehsan, T.; Pichu, R.; Mcdonald, G.K. High concentrations of Na^+ and Cl^- ions in soil solution have simultaneous detrimental effects on growth of faba bean under salinity stress. *J. Exp. Bot.* **2010**, *61*, 4449–4459.
64. Kaplan, F.; Kopka, J.; Haskell, D.W.; Zhao, W.; Schiller, K.C.; Gatzke, N.; Sung, D.Y.; Guy, C.L. Exploring the temperature-stress metabolome of Arabidopsis. *Plant Physiol.* **2004**, *1364*, 4159–4168. [CrossRef]
65. Rosa, M.; Prado, C.; Podazza, G.; Interdonato, R.; Prado, F.E. Soluble sugars-metabolism, sensing and abiotic stress: A complex network in the life of plants. *Plant Signal. Behav.* **2009**, *4*, 388. [CrossRef]
66. Wang, Y.; Stevanato, P.; Yu, L.; Zhao, H.J.; Sun, X.W.; Sun, F.; Li, J.; Geng, G. The physiological and metabolic changes in sugar beet seedlings under different levels of salt stress. *J. Plant Res.* **2017**, *1306*, 1079–1093. [CrossRef] [PubMed]
67. Wang, C.; Bi, S.T.; Liu, C.Y.; Li, M.; He, M. Arabidopsis ADF7 inhibits VLN1 to regulate actin filament dynamics and ROS accumulation in root hair development responses to osmotic stress. *Authorea* **2021**, *2*. [CrossRef]
68. Guo, R.; Yang, Z.Z.; Li, F.; Yan, C.; Zhong, X.; Qi, L.; Xu, X.; Li, H.; Long, Z. Comparative metabolic responses and adaptive strategies of wheat (*Triticum aestivum*) to salt and alkali stress. *BMC Plant Biol.* **2015**, *15*, 170. [CrossRef]
69. Kishor, P.B.K.; Sreenivasulu, N. Is proline accumulation per se correlated with stress tolerance or is proline homeostasis a more critical issue? *Plant Cell Environ.* **2014**, *37*, 300–311. [CrossRef] [PubMed]
70. Afefe, A.A.; Khedr, A.; Abbas, M.S.; Soliman, A. Responses and tolerance mechanisms of mangrove trees to the ambient salinity along the Egyptian red sea coast. *Limnol. Rev.* **2021**, *21*, 3–13. [CrossRef]
71. Nguyen, H.T.; Bhowmik, S.D.; Long, H.; Cheng, Y.; Mundree, S.; Mundree, S.; Hoang, L.T.M. Rapid accumulation of proline enhances salinity tolerance in Australian wild rice Oryza australiensis domin. *Plants* **2021**, *10*, 2044. [CrossRef] [PubMed]
72. Yadav, T.; Kumar, A.; Yadav, R.K.; Yadav, T.; Kushwaha, M. Salicylic acid and thiourea mitigate the salinity and drought stress on physiological traits governing yield in pearl millet-wheat. *Saudi J. Biol. Sci.* **2020**, *27*, 2010–2017. [CrossRef] [PubMed]
73. Zhang, Y.F.; Chen, T.S.; Fei, G.Q.; Zhang, M.X.; An, L.Z. Effects of Salinity on the Contents of Osmotica of Three Desert Plants. *J. Desert Res.* **2007**, *27*, 787–790.
74. Watanabe, S.; Kojima, K.; Ide, Y.; Sasaki, S. Effects of saline and osmotic stress on proline and sugar accumulation in Populus euphratica in vitro. *Plant Cell Tissue Organ. Cult.* **2000**, *63*, 199–206. [CrossRef]
75. Liu, S.G.; Zhu, D.Z.; Chen, G.H.; Gao, X.Q.; Zhang, X.S. Disrupted actin dynamics trigger an increment in the reactive oxygen species levels in the Arabidopsis root under salt stress. *Plant Cell Rep.* **2012**, *31*, 1219–1226. [CrossRef]
76. Daccord, N.; Celton, J.M.; Linsmith, G.; Becker, C.; Choisne, N.; Schijlen, E.; Van de Geest, H.; Bianco, L.; Micheletti, D.; Velasco, R. High-quality de novo assembly of the apple genome and methylome dynamics of early fruit development. *Nat. Genet.* **2017**, *49*, 1099–1106. [CrossRef] [PubMed]
77. Shi, Y.; Wen, Z.N.; Cong, S.H. Comparisons of relationships between leaf and fine root traits in hilly are of the loess Plateau, Yanne Biver basin. *Acta Ecol. Sin.* **2011**, *31*, 6805–6814.
78. Wang, Q.; Wu, C.; Xie, B.; Liu, Y.; Cui, J.; Chen, G.; Zhang, Y. Model analysing the antioxidant responses of leaves and roots of switchgrass to NaCl-salinity stress. *Plant Physiol. Biochem.* **2012**, *58*, 288–296. [CrossRef]
79. Jithesh, M.N.; Prashanth, S.R.; Sivaprakash, K.R.; Parida, A.K. Antioxidative response mechanisms in halophytes: Their role in stress defence. *J. Genet.* **2006**, *85*, 237–254. [CrossRef] [PubMed]
80. Withington, J.M.; Reich, P.B.; Oleksyn, J.; Eissenstat, D.M. Compari-sons of structure and life span in roots and leaves among temperate trees. *Ecol. Monogr.* **2006**, *76*, 381–397. [CrossRef]
81. Flores-Moreno, H.; Fazayeli, F.; Banerjee, A.; Datta, A.; Kattge, J.; Butler, E.E.; Atkin, O.K.; Wythers, K.; Chen, M.; Anand, M.; et al. Robustness of trait connections across environmental gradients and growth forms. *Glob. Ecol. Biogeogr.* **2019**, *28*, 1806–1826. [CrossRef]
82. Liu, C.C.; Li, Y.; Xu, L.; Chen, Z.; He, N.P. Variation in leaf morphological, stomatal, and anatomical traits and their relationships in temperate and subtropical forests. *Sci. Rep.* **2019**, *9*, 5803. [CrossRef]
83. Laughlin D, C. The intrinsic dimensionality of plant traits and its relevance to community assembly. *J. Ecol.* **2014**, *102*, 186–193. [CrossRef]

Disclaimer/Publisher's Note: The statements, opinions and data contained in all publications are solely those of the individual author(s) and contributor(s) and not of MDPI and/or the editor(s). MDPI and/or the editor(s) disclaim responsibility for any injury to people or property resulting from any ideas, methods, instructions or products referred to in the content.

forests

Article

The Influence of Intraspecific Trait Variation on Plant Functional Diversity and Community Assembly Processes in an Arid Desert Region of Northwest China

Lamei Jiang [1], Abudoukeremujiang Zayit [1], Kunduz Sattar [2], Shiyun Wang [1], Xuemin He [1], Dong Hu [3], Hengfang Wang [1] and Jianjun Yang [1,*]

[1] College of Ecology and Environment, Xinjiang University, Urumqi 830017, China; jianglam0108@126.com (L.J.)

[2] Xinjiang Uygur Autonomous Region, Forestry Planning Institute, Forestry and Grassland Bureau, Urumqi 830049, China

[3] College of Life Science, Northwest University, Xi'an 710069, China

* Correspondence: yjj@xju.edu.cn

Abstract: Exploring how functional traits vary along environmental gradients has long been one of the central questions of trait-based community ecology. Variation in functional traits includes both intraspecific trait variation (ITV) and interspecific trait variation (V_{inter}); however, the effects of ITV on functional diversity and community assembly remain to be explored. In this study, we compared functional diversity among three communities (i.e., riverbank, transition zone, and desert margin communities) at three spatial scales (i.e., 10 m × 10 m, 20 m × 20 m, and 50 m × 50 m) in the desert ecosystem of the Ebinur Lake basin in Xinjiang. We also analyzed the effects of ITV and environmental and spatial factors on functional diversity. Our results showed that incorporating ITV increased measurements of functional richness at the 10 m × 10 m scale in all three communities ($p < 0.01$). Rao's quadratic entropy (RaoQ) represents the differences in functional traits between different species. ITV significantly increased RaoQ at the 50 m × 50 m scale in the riverbank and desert margin community, whereas it significantly decreased RaoQ in the transitional zone community. Similarly, ITV significantly increased functional β-diversity at the 10 m × 10 m and 20 m × 20 m scales in the transitional zone community. Spatial factors mainly influenced functional diversity at smaller scales, whereas environmental factors were influential mainly at larger scales. After considering ITV, spatial factors had less of an effect on functional β-diversity, except for the 50 m × 50 m scale in the transitional zone and desert margin community, indicating that ITV can reduce the measured effect of dispersal on functional β-diversity. Considering ITV did not change the interpretation of the main ecological processes affecting functional diversity. However, it did change the extent to which environmental filtering and dispersal effects explained functional diversity.

Keywords: functional trait; intraspecific trait variation; functional diversity; community assembly; arid desert region

Citation: Jiang, L.; Zayit, A.; Sattar, K.; Wang, S.; He, X.; Hu, D.; Wang, H.; Yang, J. The Influence of Intraspecific Trait Variation on Plant Functional Diversity and Community Assembly Processes in an Arid Desert Region of Northwest China. *Forests* **2023**, *14*, 1536. https://doi.org/10.3390/f14081536

Academic Editor: Giovanbattista De Dato

Received: 7 May 2023
Revised: 19 July 2023
Accepted: 24 July 2023
Published: 27 July 2023

Copyright: © 2023 by the authors. Licensee MDPI, Basel, Switzerland. This article is an open access article distributed under the terms and conditions of the Creative Commons Attribution (CC BY) license (https://creativecommons.org/licenses/by/4.0/).

1. Introduction

Plant functional traits are a set of core plant attributes closely related to plant colonization, growth, survival, and mortality. They reflect the plant's response and adaptation to the growing environment, effectively linking individual plant characteristics to environmental conditions and influencing ecosystem function [1]. Plant functional traits can vary due to differences in the external environment. By exploring how functional plant traits vary in heterogeneous habitats, we can identify the main ecological processes of different environments [2]. The leaves are the main organ of photosynthesis and plant material production. They are closely related to the plant's resource acquisition capacity and utilization efficiency [3,4]. Leaves are sensitive to environmental changes during plant evolution, and their traits result from plants' adaptation to habitat heterogeneity and environmental changes [5]. Functional traits in leaves are easier to measure than other plant traits and are one of the main targets of current plant functional trait research.

Variations in plant functional traits within communities include interspecific (V_{inter}) and intraspecific trait variation (ITV). V_{inter} refers to the difference in functional traits between different species. ITV refers to the difference in traits between different individuals within the same species. However, most studies examining community assembly processes and the mechanisms of plant diversity have used the mean values of plant traits to calculate functional diversity. This approach has successfully determined factors leading to variation in community structure and ecosystem function [6–9]. Previous studies have shown that the magnitudes of ITV and V_{inter} in communities are comparable [6]. Furthermore, ITV has been found to account for 25% of the total trait variation within communities and 32% of the total trait variation between communities [10]. However, ecologists have recently become increasingly concerned about considering ITV when measuring functional diversity [11]. Ignoring intraspecific variation may obscure ecological patterns or lead to inaccurate interpretations [12]. Some studies have reported that intraspecific variation accounts for about 10% or even more than 50% of the total variation in some important functional traits [10,13]. Hulshof et al. (2013) found that ITV was greater than V_{inter} in specific leaf areas at low latitudes [14]. Considering ITV during secondary succession in semi-arid forest ecosystems may lead to higher values of functional α-diversity, while not significantly affecting the main process of forest succession. However, its consideration in top communities led to a divergence in the structure of the community's functional traits [15]. By contrast, Niu et al. showed that the importance of ITV in plant communities increased with environmental harshness in alpine meadows in Tibet [16]. One study using a large dataset of 10 leaf and petiole traits in temperate forest seedlings found that about 40% of trait variation was explained by intraspecific variation [17]. In conclusion, these findings suggest that using mean values for functional traits ignores actual trait variation and that the effect of intraspecific variation on functional diversity varies among ecosystems.

Previous studies have shown that salinity stress in arid desert areas causes leaf area and nitrogen traits to converge [18]. Soil organic carbon significantly affects functional α-diversity, whereas soil water content and salinity mainly drive changes in functional β-diversity [19]. Photosynthetic and carbohydrate-related traits adapted to drought stress in the same way in both small trees and shrubs, whereas leaf hydraulic traits adapted to environmental stress differently. This finding suggests that small trees and shrubs have different trait variations to adapt to drought due to genetic constraints and long-term ecological niche differentiation [20]. However, there are few studies on the effects of ITV on functional diversity in arid desert regions and the patterns of functional trait variation at different scales. Including ITV in an analysis can more accurately determine the causes of functional trait variation and functional diversity of desert plants in response to environmental changes. Based on this, we selected three communities at three different scales in the arid desert region to study ITV's effects on functional diversity and its magnitude in response to environmental and spatial factors. We also analyzed the phylogenetic signals of plant traits in the three communities to investigate the effects of genetic, environmental, and spatial factors on functional traits and diversity. To verify this hypothesis, this study aims to: (1) What are the characteristics of variations in plant functional traits among different plant communities? The worse the environment, the lower the plasticity of traits. (2) What is the relative contribution of variation sources of plant functional traits at different scales? (3) What is the impact of intraspecific variation on plant functional diversity and community-building mechanisms?

2. Materials and Methods

2.1. Study Area

The Ebinur Lake Wetland National Nature Reserve is located at the southwest edge of the Junggar Basin in the arid region of Northwest China. It is a good site for studying biodiversity in the arid desert regions of Xinjiang [18]. The region has a dry typical continental climate. The annual evaporation exceeds 1600 mm, the annual precipitation is about 100 mm, the extreme maximum temperature is 44 °C, and the extreme minimum temperature is −33 °C. The soil types are sandy soil, grey-brown desert soil, and grey desert soil, which are severely affected by desertification [19]. The vegetation structure is composed of annual herbs, perennial herbs, and shrubs.

2.2. Sample Layout

We set up one large sample plot of 1 hectare (100 m × 100 m) each on the river-bank, transition zone, and desert margin north of the Aqikesu River near the East Bridge Management Station in the Ebinur Lake Wetland National Nature Reserve. Each sample plot was divided into 400 sample squares of 5 m × 5 m each (Figure 1). The riverbank community mainly consisted of *Populus euphratica*, *Nitraria roborowskii*, *Lycium ruthenicum*, *Apocynum venetum*, *Halimodendron halodendron*, *Alhagi sparsifolia*, *Phragmites australis*, *Suaeda microphylla*, and other plants. The transition zone community mainly contained *A. venetum*, *P. euphratica*, *L. ruthenicum*, *Tamarix chinensis*, *Karelinia caspia*, *Suaeda glauca*, *A. sparsifolia*, *P. australis*, *Reaumuria soongarica*, *Halocnemum strobilaceum*, and other plants. The desert margin community mainly contained *Haloxylon ammodendron*, *S. glauca*, *Calligonum mongolicum*, *R. soongarica*, *N. roborowskii*, *A. sparsifolia*, *Seriphidium terrae-albae*, *Kali collinum*, and other plants.

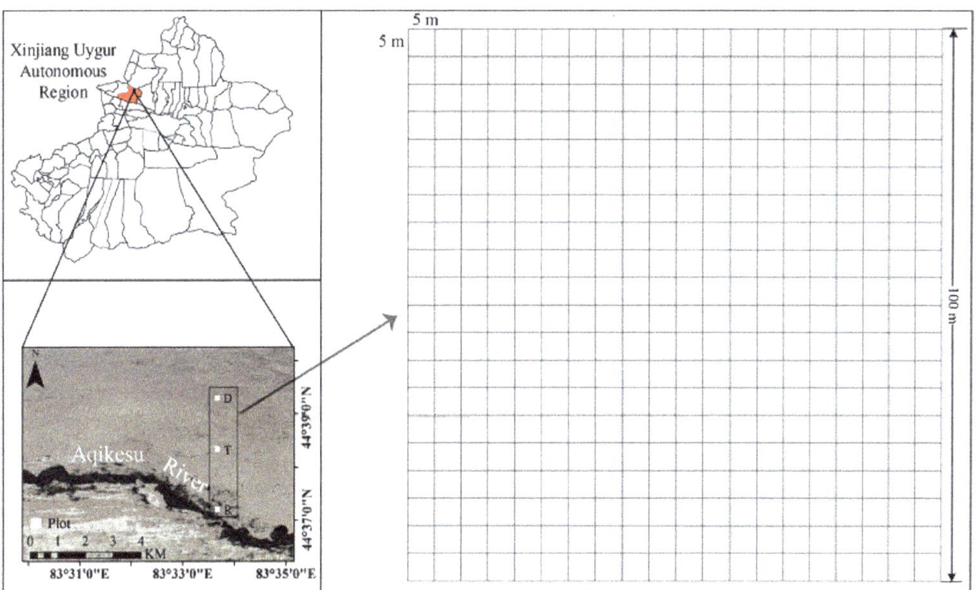

Figure 1. Schematic diagram of plot. Note: R, T, and D in the figure represent the riverbank community, transitional zone community, and desert margin community, respectively.

2.3. Collection and Measurement of Vegetation and Soil Samples

We investigated the plant community characteristics in each of the sample plots, including species composition, abundance, crown width, and maximum height of plants (H_{max}). The height of trees is measured using a laser altimeter, and the height of herbs and shrubs, as well as the crown width of all plants, were measured using a steel tape measure. We collected about 20 mature leaves from each plant in a 5 m sample square and measured the leaves' length (LL), leaves' width (LW), and thickness (LT) with vernier calipers. The leaf area was calculated using the photographed method (we laid the leaves flat on 1 mm^2 grid paper, photographed them, and uploaded the images to Photoshop 7.0 software). We weighed the fresh weight on a one-in-ten-thousand balance, placed it in an envelope, brought it back to the laboratory, dried it in an oven at 80 °C for 48 h, and immediately weighed it with an electronic balance with an accuracy of 0.001 g. Then, we calculated the dry weight. The leaves' dry weight was measured and used to calculate the specific leaf area (SLA) and dry matter content (LDMC). About 20 g of plant leaves were placed into paper envelopes, air-dried, and brought back indoors to determine the leaf carbon (LC), nitrogen (LN), and phosphorus (LP) contents. The LC, LN, and LP contents were determined using the potassium dichromate dilution heat method, the H_2SO_4-H_2O_2-Kjeldahl method, and the molybdenum antimony resistance colorimetric method [21].

In each 5 m × 5 m sample square, we used the diagonal method to select the center point. We took two samples from 0–20 cm of topsoil. One sample was collected in pre-weighed aluminum boxes, which were then numbered and weighed immediately after collection. It was then returned to the laboratory, dried in an oven, and weighed dry to calculate the soil water content. We returned the second sample to the laboratory and dried it naturally for later soil index determination.

We determined the soil salinity content (SA) and soil pH using the weighing method and a pH meter, respectively. The soil organic carbon (SOC), total nitrogen (TN), ammonium nitrogen (AN), and nitrate nitrogen (NN) were determined using the potassium dichromate dilution heat method, Kjeldahl digestion method, indophenol blue colorimetric method, and dual wavelength UV spectrophotometric method, respectively. The total phosphorus (TP) and available phosphorus (AP) were determined using the Mo–Sb colorimetric method. We determined the above physicochemical properties using soil agrochemical analysis [21]. The soil particle size was determined using a laser particle size meter.

2.4. Statistical Analysis

2.4.1. Functional Trait Variation and Relative Contribution

Interspecific (V_{inter})and intraspecific trait variation (ITV) constitute the main body of functional trait variation. In this study, the degree of interspecific (V_{inter}) and intraspecific trait variation (ITV) was quantified by the coefficient of variation: (CV) = standard deviation (σ)/mean (μ). The interspecific trait variation (V_{inter}) was calculated from the mean trait values of all species in the study area using the standard deviation and mean [22].

The total functional trait variation (V_{total}) is a measure of functional diversity calculated with the trait values of each individual of a species and includes both ITV and interspecific variation (V_{inter}) [23]. The calculation formulas are as follows. To measure the relative contribution of Vinter and ITV to community trait variation, we conducted variance decomposition analysis using the vegan package [24].

$$V_{total} = \sum_{i=1}^{n} A_{ij} T_{ij}$$

$$V_{inter} = \sum_{i=1}^{n} A_{ij} T_i$$

$$ITV = V_{total} - V_{inter}$$

Among them, n is the number of species, A_{ij} is the richness of the i-th species in plot j, T_{ij} is the trait value of species i in quadrat j, and Ti is the average of the trait values of species i in all plots.

2.4.2. Phylogenetic Signal of Functional Traits

Stochastic models of character evolution make a tree with any amount of hierarchical structure. Under such models as Brownian motion (stochastic models), evolutionary changes were simply added to values present in the previous generation or at the previous node on a phylogenetic tree. K is a descriptive statistic, to gauge the amount of phylogenetic signal. We used the K values of the Brownian motion evolutionary model to test whether functional traits were phylogenetically conserved [25]. K values and their significance were obtained using picante analysis in the statistical software R4.1.2(R Core Team; Vienna, Austria).

2.4.3. Calculation of Functional α- and β-Diversity

We calculated functional α-diversity indices using the *FD* package [26]. functional α-diversity indices included functional richness (FRic), functional evenness (FEve), functional dispersion (FDis), functional divergence (FDiv), and RaoQ. The calculation formula is as follows:

(1) FRic: Functional richness measures how many Ecological niche spaces are occupied by existing species in the community.

$$\text{Rts} = \max_{i \in s}[x_{its}] - \min_{i \in s}[x_{its}] = \int I_{st(x)}dx$$

$$\text{FRic} = \frac{U_{S \in S_c} R_{ts}}{U_{S \cup S_c} R_{ts}} = \frac{\int \max\limits_{s \in S_c}[I_{st(s)}]dx}{\int \max\limits_{s \cup S_c}[I_{st(s)}]dx}$$

In the above equation: S is the number of species, i is the individual of species S, t is the trait, x is the trait value, $I_{st}(x)$ is the indicator function of trait t of species s; S_c represents the community c to which species S belongs, and R_{ts} represents the range of traits t of species S.

(2) FEve: The functional evenness index measures the distribution pattern of species traits in the occupied trait space.

$$EW_I = \frac{\text{dist}(i,j)}{W_i + W_j}$$

$$PEW_I = \frac{EW_I}{\sum_{i=1}^{s-1} EW_I}$$

$$FEve = \frac{\sum_{i=1}^{s-1} \min\left(PEW_I \frac{1}{S-1}\right)}{1 - \frac{1}{S-1}}$$

In the above equation, S is the number of species, EW is the weight of evenness, dist (i, j) is the Euclidean distance between species i and j, Wi is the relative richness of species i, I is the branch length, and PEWI is the weight of branch length.

(3) FDiv: Functional divergence is the degree of aggregation of species along the trait axis.

$$g_k = \frac{1}{s} \sum_{i=1}^{s} x_{ik}$$

$$d_G = \sqrt{\sum_{k=1}^{T} (x_{ik} - g_k)^2}$$

$$\overline{dG} = \frac{1}{S} \sum_{I=1}^{S} dG_i$$

$$\Delta d = \sum_{i=1}^{s} w_i \times \left(dG_i - \overline{dG} \right)$$

$$\Delta |d| = \sum_{i=1}^{s} w_i \times \left| dG_i - \overline{dG} \right|$$

$$FD_{iv} = \frac{\Delta d + \overline{dG}}{\Delta |d| + \overline{dG}}$$

In the above equation, w_i is the abundance of species i, x_{ik} is the k trait value of species i, g_k is the center of gravity of trait k, S is the number of species, T is the number of traits, dG_i is the Euclidean distance between x_{ik} and the center of gravity, $\overline{dG}$ is the average distance between species i and the center of gravity, and d is the diversity weight dispersion.

(4)   FDis: Functional dispersion index measures the maximum statistical dispersion of the multi-degree distribution of community functional traits in the trait space.

$$c = [c_i] = \frac{\sum w_j x_{ij}}{\sum w_j}$$

$$FDis = \frac{\sum w_j z_j}{\sum w_j}$$

In the above equation, w_j is the relative abundance of species j, z_j is the weighted distance from species j to the center of gravity c, and x_{ij} is the value of the i-th trait of species j.

(5)   RaoQ: Rao uses the quadratic entropy equation to measure the differences in functional traits among different species.

$$RaoQ = \sum_{i=1}^{s} \sum_{j>1}^{s} d_{ij} w_i w_j$$

In the above equation, S represents the number of species, d_{ij} represents the Euclidean distance between species i and j in traits, and w_i and w_j represent the relative abundance of species i and j, respectively.

We calculated functional β-diversity as the dissimilarity among plots in community-weighted mean (CWM) trait values [25]. The calculation method of CWM is based on the formula in Section 2.4.1. We used Euclidean distances based on CWM trait values for all nine traits to represent functional β-diversity, both considering ITV and without considering ITV. We calculated β-diversity as the average distance-to-centroid, measured as the average distance. The β-diversity index calculation was carried out in the vegan package.

2.4.4. Impact of Trait Variation on Community Assembly Processes

(1)   Obtaining spatial factors

Calculate the spatial distance between different plots based on relative coordinates to quantify spatial factors. To obtain spatial factors (related to dispersal limitation), we used the R language adespatial package for MEM (Moran's Eigenvector Maps) analysis [27].

(2)   Distance-Based Redundancy Analysis (dbRDA) and Variance Decomposition

We used distance-based redundancy analysis and variance decomposition to explore the impact of different environmental and spatial factors on functional beta diversity. The variance decomposition analysis was conducted using the "varpart" function, The adjusted R^2 used in this study is more accurate [28].

3. Results and Analysis

3.1. Sources of Variation in Functional Traits

In this study, we found that leaf dry matter content variation in arid desert areas mainly stemmed from interspecific variation. The intraspecific variation in chemical element content in plant leaves was greater than the interspecific variation. There was also greater intraspecific variation in leaf area than in leaf dry matter content (Table S1).

Trait variation in the riverbank community was mainly influenced by ITV, with variation contribution rates of 70.04, 78.01, and 45.38% at the three scales, respectively. However, at the 10 m × 10 m and 20 m × 20 m scales in the transitional zone and desert margin communities, the impact of interspecific variation on community trait variation was greater than that of ITV (Figure 2).

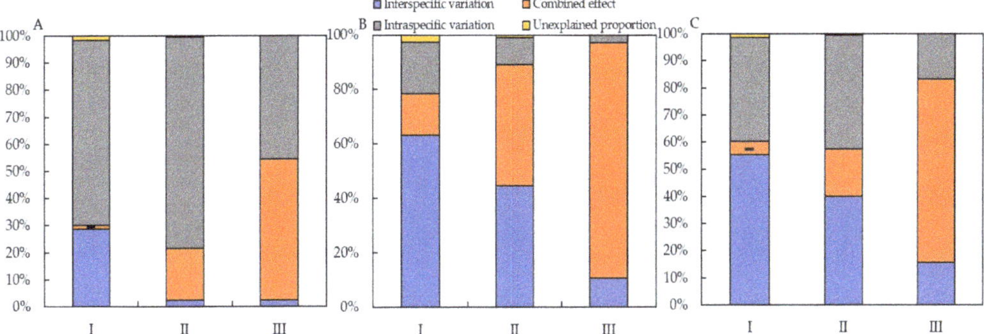

Figure 2. Impact of intraspecific and interspecific variation on functional trait variation in plant communities in the Ebinur Lake basin ((**A**) riverbank, (**B**) transitional zone, and (**C**) desert margin, —represents negative value. I, II and III represent the scale of 10 m × 10 m, 20 m × 20 m and 50 m × 50 m.).

The K values of leaf thickness and width in the three communities were all >1, indicating significant phylogenetic signals ($p < 0.05$). The leaf carbon content in the riverbank and transitional zone communities also showed strong phylogenetic signals ($p < 0.05$). However, traits such as the maximum plant height, specific leaf area and dry matter, nitrogen, and phosphorus contents did not show significant phylogenetic signals ($p > 0.05$). The plant functional traits in this area show a certain degree of phylogenetic conservation, and the influence of historical evolutionary factors varies for different functional traits. The overall influence is weak, except for the greater influence of leaf width, thickness, and carbon content (Table 1).

Table 1. The phylogenetic signals of plant functional traits.

Functional Trait	Riverbank		Transitional Zone		Desert Margin	
	K value	p	K value	p	K value	p
H_{max}	0.36	NS	0.56	NS	0.23	NS
LL	0.84	NS	0.82	*	0.53	NS
LW	1.23	**	1.08	**	1.44	*
LT	1.29	*	1.47	**	1.50	**
LDMC	0.71	*	0.85	NS	0.46	NS
SLA	0.34	NS	0.51	NS	0.31	NS
LC	1.37	**	1.15	0.001	0.79	NS
LN	0.67	NS	0.27	NS	0.45	NS
LP	0.60	NS	0.54	NS	0.36	NS

Note: * $p < 0.05$, ** $p < 0.01$, NS represents no significance.

3.2. Effect of Trait Variation on Functional Diversity

3.2.1. Effect of Intraspecific Variation on Functional α-Diversity

In the riverbank community, considering ITV greatly and significantly increased the values of functional richness (FRic) at the 10 m × 10 m and 20 m × 20 m scales ($p < 0.01$), and significantly decreased the functional evenness (FEve) at the 20 m × 20 m and 50 m × 50 m scales ($p < 0.01$). The RaoQ values at the 20 m × 20 m and 50 m × 50 m scales were significantly greater ($p < 0.01$) than those without ITV (Figure 3A–E).

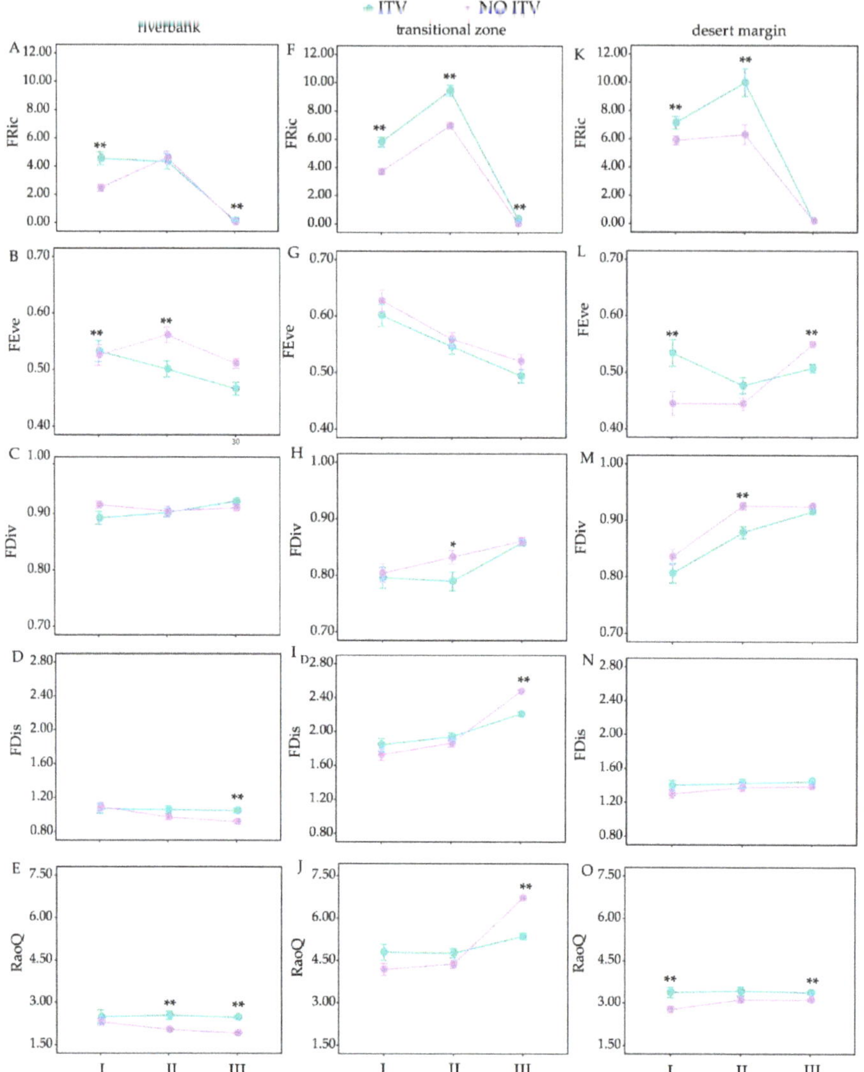

Figure 3. Effect of intraspecific variation on functional α-diversity in desert plant communities (* $p < 0.05$, ** $p < 0.01$, (**A–E**) represent the changes in FRic, FEve, FDis, FDiv and RaoQ index in river bank, (**F–J**) represent the changes in FRic, FEve, FDis, FDiv and RaoQ index in transitional zone, (**K–O**) represent the changes in FRic, FEve, FDis, FDiv and RaoQ index in desert margin. I, II, and III represent the scale of 10 m ×10 m, 20 m × 20 m, and 50 m × 50 m).

Considering ITV increased the values of functional richness ($p < 0.01$) and decreased the FEve at all scales in the transition community. It significantly decreased the functional dispersion (FDis) and RaoQ at the 50 m × 50 m scale. The functional divergence (FDiv) decreased at all scales when ITV was considered, but the difference was significant only at the 20 m × 20 m scale (Figure 3F–J).

Considering ITV in the desert margin community increased FRic significantly at the 10 m × 10 m and 20 m × 20 m scales ($p < 0.01$). FEve significantly decreased at 50 m × 50 m. Considering ITV decreased FDiv, the difference was highly significant at the 20 m × 20 m scale ($p < 0.01$). Considering intraspecific variation increased the functional dispersion at all scales, while RaoQ decreased. The RaoQ values obtained when considering ITV at the 10 m × 10 m and 50 m × 50 m scales were significantly greater than those obtained when ITV was not considered ($p < 0.01$) (Figure 3H–O).

3.2.2. Effect of Intraspecific Variation on Functional Beta Diversity

The incorporation of ITV significantly increased measures of functional β-diversity. The functional β-diversity increased significantly at all three scales in the riverbank and desert margin communities ($p < 0.01$) and the 10 m × 10 m and 20 m × 20 m scales in the transitional zone community. The functional β-diversity decreased with increasing scale, whereas it increased with increasing scale for riverbank communities when intraspecific variation was considered. It decreased with increasing scale without intraspecific variation. The functional β-diversity increased and then decreased with increasing scale both when considering and not considering intraspecific variability in desert margins, reaching a maximum at the 20 m scale. The functional β-diversity was significantly smaller when ITV was not considered (Figure 4).

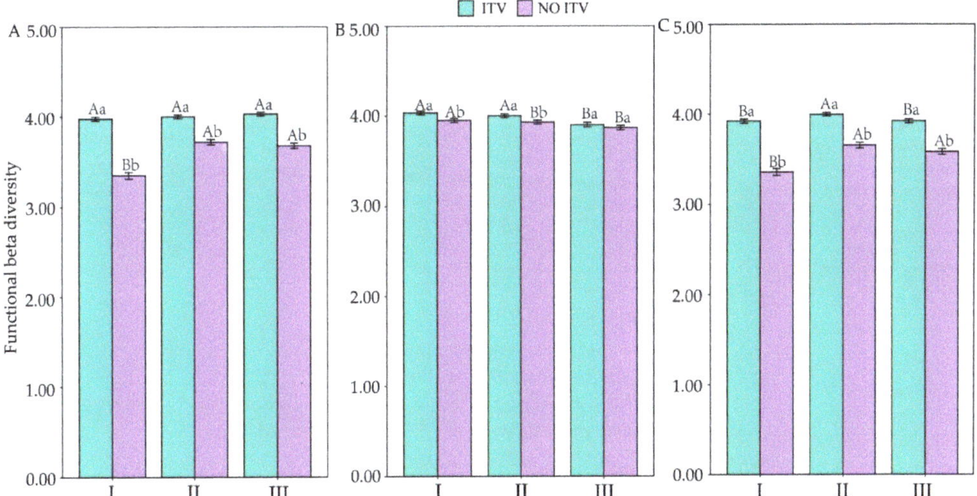

Figure 4. Effect of intraspecific variation on functional β-diversity in desert plant communities ((**A**) riverbank, (**B**) transitional zone and (**C**) desert margin, a and b means significant differences of plant function α diversity between ITV and NO-ITV at the same scale, A and B means significant differences of plant function α diversity between in different scales, the units of dissimilarity (functional beta diversity) is average difference in number of functional traits among plots. I, II, and III represent the scale of 10 m ×10 m, 20 m × 20 m, and 50 m × 50 m).

3.3. Effects of Intraspecific Variation on Community Assembly

3.3.1. Based on Functional α-Diversity

At the 10 m × 10 m scale, the spatial and environmental factors explained the changes in functional α-diversity of 0.28 and 0.05, respectively, and explanation ratios were 0.45 and 0.07 without considering ITV the riverbank community. The changes in functional α-diversity of 0.41 and 0.04 were explained, explanation ratios were 0.54 and 0.11 without considering ITV in the transitional community; The changes in functional diversity of 0.36 and 0.08 were explained, explanation ratios were 0.46 and 0.16 without considering ITV in the desert margin. At all scales, the proportion of unexplained variation was greater in all three communities when ITV was considered than when it was not.

As the scale increased, the environmental and spatialized environmental roles gradually increased, the role of spatial factors gradually decreased, and the proportion of unexplained variation also gradually decreased (Table 2).

Table 2. Relative contributions of environmental and spatial factors to the functional α-diversity of desert plant communities at different scales.

Community	Types	Scale	Soil Factors	Space Factors	Soil and Space	Unexplained Proportion
riverbank	ITV	10 m × 10 m	0.05	0.28	−0.01	0.69
		20 m × 20 m	0.07	0.44	0.12	0.28
		50 m × 50 m	0.49	0.23	0.12	0.16
	NO-ITV	10 m × 10 m	0.07	0.45	−0.06	0.54
		20 m × 20 m	0.03	0.50	0.18	0.20
		50 m × 50 m	0.07	0.13	0.79	0.01
transitional zone	ITV	10 m × 10 m	0.04	0.41	0.09	0.46
		20 m × 20 m	0.09	0.33	0.41	0.16
		50 m × 50 m	0.43	0.14	0.35	0.11
	NO-ITV	10 m × 10 m	0.11	0.54	−0.04	0.40
		20 m × 20 m	0.24	0.27	0.34	0.16
		50 m × 50 m	0.41	0.16	0.40	0.03
desert margin	ITV	10 m × 10 m	0.08	0.36	−0.06	0.62
		20 m × 20 m	0.18	0.37	0.18	0.27
		50 m × 50 m	0.03	0.21	0.68	0.08
	NO-ITV	10 m × 10 m	0.16	0.46	−0.10	0.48
		20 m × 20 m	0.11	0.61	0.12	0.16
		50 m × 50 m	0.07	0.10	0.80	0.03

3.3.2. Based on Functional β-Diversity

At the 10 × 10 m scale, the spatial and environmental factors explained the changes in functional β-diversity of 0.20 and 0.03, respectively, and explanation ratios were 0.37 and 0.05 without considering ITV the riverbank community. The changes in functional β-diversity of 0.09 and 0.02 were explained, explanation ratios were 0.48 and 0.07 without considering ITV in the transitional community; The changes in functional β-diversity of 0.29 and 0.10 were explained, explanation ratios were 0.45 and 0.08 without considering ITV in the desert margin. At the 10 m × 10 m scale, the functional β-diversity was more sensitive to spatial factors when ITV was not considered. The proportion of functional α-diversity explained by spatial factors with ITV was greater than that explained by functional diversity without ITV. Except for the transitional zone community at the 50 m × 50 m scale, the proportion of unexplained factors influencing functional diversity with ITV was greater than without ITV in all three communities (Table 3).

Table 3. Relative contributions of environmental and spatial factors to the functional β-diversity of desert plant communities at different scales.

Community	Types	Scale	Soil Factors	Space Factors	Soil and Space	Unexplained Proportion
riverbank	ITV	10 m × 10 m	0.03	0.20	−0.02	0.79
		20 m × 20 m	0.02	0.52	0.16	0.31
		50 m × 50 m	0.37	0.25	0.28	0.09
	NO-ITV	10 m × 10 m	0.05	0.37	−0.07	0.62
		20 m × 20 m	0.04	0.61	0.16	0.20
		50 m × 50 m	0.05	0.25	0.67	0.03
transitional zone	ITV	10 m × 10 m	0.02	0.09	0.10	0.80
		20 m × 20 m	0.12	0.36	0.30	0.22
		50 m × 50 m	0.24	0.18	0.47	0.11
	NO-ITV	10 m × 10 m	0.07	0.48	0.00	0.45
		20 m × 20 m	0.11	0.41	0.35	0.13
		50 m × 50 m	0.39	0.02	0.39	0.20
desert margin	ITV	10 m × 10 m	0.10	0.29	−0.03	0.64
		20 m × 20 m	0.05	0.49	0.08	0.39
		50 m × 50 m	0.11	0.18	0.62	0.08
	NO-ITV	10 m × 10 m	0.08	0.45	−0.16	0.53
		20 m × 20 m	0.10	0.66	0.10	0.14
		50 m × 50 m	0.06	0.10	0.81	0.04

4. Discussion

The importance of ITV when quantifying functional diversity has been emphasized [29], because ignoring it may strongly alter estimates of functional diversity and obscure ecological processes [11,12]. In particular, when assessing mechanisms of variation in functional diversity at local scales, ITV is expected to become increasingly important as the scale of study decreases [11]. On the other hand, increasing the scale tends to increase variability, so interspecific trait variation (V_{inter}) is expected to be relatively greater than ITV [11]. By contrast, there has been less focus on ITV during community assembly processes [30]. In this sense, ITV can potentially contribute to ecosystem functioning [31,32].

ITV has been reported for alpine grassland, scrubland, subtropical, and tropical rainforest communities [33,34]. However, relatively little research has been conducted on desert plant communities. The degree of trait variation represents the range of inherent characteristics and individual differences; the wider the range of resources available to plants, the greater the variability of plant functional traits [35]. Intraspecific variation in plant functional traits is influenced by a combination of genetic variation and environmental conditions [36]. Furthermore, its magnitude or degree can reflect plants' ability to adapt to environmental conditions [37]. Albert et al. conducted a study on 13 living species. They found that about 30% of trait variation came from within the species, with the highest intraspecific variability in leaf nitrogen content and carbon content [13]. ITV in specific stem densities of subtropical broadleaf evergreen forest plants in China explained up to 51.50% of total trait variation [38]. We found that the variation in leaf dry matter content in the arid desert region mainly originated from interspecific variation. This finding is consistent with Burton et al., who found greater interspecific variation in leaf dry matter content [39]. The intraspecific variation in the chemical element content of plant leaves was greater than the interspecific variation in this study, which is consistent with studies that found greater intra-species variation in plant leaf chemical element content [10,30,40]. The photosynthetic rate and nutrient cycling are correlated with the leaf element content. There may be a link between the large intraspecific variation in this study and intraspecific trait differentiation due to competition between species. In addition, ITV has been shown to have ecological effects similar to interspecific variation [41]. For example, controlled experiments in North Carolina scrub communities in the USA found that fire conditions significantly altered the community-weighted mean of specific leaf areas. Furthermore, changes in the community-weighted mean were primarily driven by intraspecific variation, i.e., plants adapted to fire disturbance through intraspecific variation in functional traits [41]. The specific leaf area and leaf dry matter content, which are also leaf economic-type spectral traits, showed greater intraspecific variation, possibly due to the greater genetic variation in specific leaf areas [42]. The larger intraspecific variation in leaf area allows for greater variation among individuals of the same species, which promotes species coexistence [43]. The lowest overall trait variation in the desert margin community may be due to its harsher habitat conditions than the other two communities. It has been suggested that species with low variation are more commonly found in harsh habitats. The low plasticity of traits may be due to specialization toward unfavorable habitats [43].

Inferring species coexistence mechanisms based on phylogenetic structures requires conserving functional traits on phylogenetic trees [44]. The phylogenetic and functional trait structures of a community are consistent when the functional traits of the community have a phylogenetic signal [44]. Examining leaf functional traits and phylogenetic signals in different desert communities revealed that more than half of the nine functional traits did not exhibit phylogenetic signals (Table 1). Therefore, the functional traits in those sites were not all evolutionarily conserved, and inconsistencies in the community's phylogenetic and functional trait patterns were observed. The K values of leaf thickness and width in all three communities were >1, showing significant phylogenetic signals. No significant phylogenetic signals were detected for most functional traits, suggesting that evolutionary history has not influenced plant communities in the Ebinur Lake basin with much intensity. This observation is consistent with Che's findings in subalpine meadow communities [45]. Due to long-term environmental adaptation, these functional traits do not always exhibit significant phylogenetic signals, which can lead to the convergent evolution of more distantly related species [46].

In this study, we found that considering ITV increased the values of functional richness and RaoQ. Greater ITV implied a wider distribution of traits and a higher complementarity of ecological niches within the community, which contributes to its stability [47]. These observations are consistent with those made of grassland plants in the Mediterranean, where ITV significantly influenced the size of FRic and RaoQ [48]. Furthermore, considering ITV significantly increased indices of functional β-diversity in our study. The effects of environmental and spatial factors on β-diversity have been controversial. Some studies have suggested that environmental factors such as climate, soil, and topography or spatial factors are more important for β-diversity [49,50]. By contrast, others have suggested that both roles are equally important [51]. Our results indicated that the relative importance of environmental and spatial factors on functional β-diversity varies with scale. On a small scale, spatial factors play a major role in functional diversity, while on a large scale, environmental factors and spatialized environmental effects play a role. The main reason is that as the sampling scale increases, the differences in species habitats become greater, and the role of environmental factors also becomes greater [52]. Dispersal limitation generally occurs at smaller scales, so the effect of spatial factors is more pronounced at smaller scales [53]. The results of this study are consistent with those of previous studies [53]. Therefore, dispersal limitation significantly impacts functional diversity at smaller scales, while environmental filters have a significant impact at larger scales. In recent years, more studies have shown that ITV plays an important role in species coexistence, niche differentiation, community assembly, and maintenance of ecosystem function [54–56]. In addition to explaining community trait variation, intraspecific trait variation has important implications for community assembly and ecosystem function [57]. A global meta-analysis showed that ITV accounted for 25% of the total variation within communities and 32% between communities, particularly in grassland community assembly [10,58]. Regarding community assembly, Jung et al. found that intraspecific trait variation helps more species pass through biotic and abiotic screens and promotes species coexistence [22]. Siefert also concluded that using a null model with ITV enhances the effects of environmental filtering and provides more comprehensive information on community assembly [59]. Spasojevic et al. found that variance decomposition analysis based on functional β-diversity without considering ITV may hinder the explanation of habitat filtering processes [60]. Spatial factors had less of an effect on functional β-diversity after considering ITV, except at the 50 m × 50 m scale in the transitional zone and desert margin communities. This finding indicated that ITV could reduce the effect of dispersal on functional diversity. Although considering ITV did not change the main ecological processes affecting functional diversity, it changed the extent to which its environmental filtering and dispersal effects explained functional diversity. Moreover, the proportion of unexplained factors in the variance decomposition of functional diversity, in which environmental and variance factors increased after considering ITV, suggests that there are other factors affecting ITV, such as genetic factors, ecological drift, and species extinction [61,62] or unmeasured spatial and environmental factors [63].

5. Conclusions

Including measurements of ITV significantly increased functional richness at the 10 m × 10 m scale in all three communities. It significantly increased RaoQ at the 50 m × 50 m scale in the riverbank and desert margin communities, whereas it significantly decreased RaoQ in the transition zone community. ITV significantly increased functional β-diversity except at the transition zone community's 50 m × 50 m scale. Considering ITV significantly increased functional β-diversity. The dispersal limitation significantly impacts functional diversity at small scales, while environmental filtering significantly impacts functional diversity at larger scales. Spatial factors had less of an effect on functional β-diversity after considering intraspecific variation at all scales, except at the 50 m × 50 m scale in the transitional zone and desert margin communities. This finding indicates that ITV can reduce the effect of dispersal on functional diversity. Although considering ITV did not change the interpretation of the main ecological processes affecting functional diversity, it changed the extent to which environmental filtering and dispersal effects explained functional diversity.

Supplementary Materials: The following supporting information can be downloaded at: https://www.mdpi.com/article/10.3390/f14081536/s1, Table S1: Coefficient of variation for interspecific and intraspecific functional traits in the Ebinur Lake basin.

Author Contributions: Investigation, L.J., S.W., K.S. and X.H.; Conceptualization, L.J. and J.Y.; Methodology, L.J.; Software, L.J., D.H. and H.W.; Writing—original draft preparation, L.J.; Writing—review and editing, L.J.; Supervision, J.Y. and A.Z.; Funding acquisition, J.Y. All authors have read and agreed to the published version of the manuscript.

Funding: This research was supported by the National Natural Science Foundation of China (42171026), Natural Science Foundation of Xinjiang (2022D01C42), and Xinjiang Uygur Autonomous Region Graduate Research, and Innovation Project (XJ2020G011).

Institutional Review Board Statement: Not applicable.

Informed Consent Statement: Not applicable.

Data Availability Statement: Not available.

Conflicts of Interest: The authors declare that they have no conflict of interest.

References

1. Webb, C.T.; Hoeting, J.A.; Ames, G.M.; Pyne, M.I.; Poff, N.L. A structured and dynamic framework to advance traits-based theory and prediction in ecology. *Ecol. Lett.* **2010**, *13*, 267–283. [CrossRef]
2. Shipley, B.; De Bello, F.; Cornelissen, J.H.C.; Laliberté, E.; Laughlin, D.C.; Reich, P.B. Reinforcing loose foundation stones in trait-based plant ecology. *Oecologia* **2016**, *180*, 923–931. [CrossRef] [PubMed]
3. Westoby, M. A leaf-height-seed (LHS) plant ecology strategy scheme. *Plant Soil* **1998**, *199*, 213–227. [CrossRef]
4. Vendramini, F.; Díaz, S.; Gurvich, D.E.; Wilson, P.J.; Thompson, K.; Hodgson, J.G. Leaf traits as indicators of resource-use strategy in floras with succulent species. *New Phytol.* **2002**, *154*, 147–157. [CrossRef]

5. Reich, P.B.; Walters, M.B.; Ellsworth, D.S. From tropics to tundra: Global convergence in plant functioning. *Proc. Natl. Acad. Sci. USA* **1997**, *94*, 13730–13734. [CrossRef]

6. Messier, J.; McGill, B.J.; Lechowicz, M.J. How do traits vary across ecological scales? A case for trait-based ecology. *Ecol. Lett.* **2010**, *13*, 838–848. [CrossRef] [PubMed]

7. Gross, N.; Bagousse-Pinguet, Y.L.; Liancourt, P.; Berdugo, M.; Gotelli, N.J.; Maestre, F.T. Functional trait diversity maximizes ecosystem multifunctionality. *Nat. Ecol. Evol.* **2017**, *1*, 132. [CrossRef]

8. Wang, J.; Chen, C.; Li, J.; Feng, Y.; Lu, Q. Different ecological processes determined the alpha and beta components of taxonomic, functional, and phylogenetic diversity for plant communities in dryland regions of Northwest China. *PeerJ* **2019**, *6*, e6220. [CrossRef]

9. Mensah, S.; Salako, K.V.; Assogbadjo, A.; Kakaï, R.G.; Sinsin, B.; Seifert, T. Functional trait diversity is a stronger predictor of multifunctionality than dominance: Evidence from an A fromontane forest in South Africa. *Ecol. Indic.* **2020**, *115*, 106415. [CrossRef]

10. Siefert, A.; Violle, C.; Chalmandrier, L.; Albert, C.H.; Taudiere, A.; Fajardo, A.; Aarssen, L.W.; Baraloto, C.; Carlucci, M.B.; Cianciaruso, M.V.; et al. A global meta-analysis of the relative extent of intraspecific trait variation in plant communities. *Ecol. Lett.* **2015**, *18*, 1406–1419. [CrossRef]

11. Albert, C.H.; de Bello, F.; Boulangeat, I.; Pellet, G.; Lavorel, S.; Thuiller, W. On the importance of intraspecific variability for the quantification of functional diversity. *Oikos* **2012**, *121*, 116–126. [CrossRef]

12. Ross, S.R.P.-J.; Hassall, C.; Hoppitt, W.J.E.; Edwards, F.; Edwards, D.P.; Hamer, K.C. Incorporating intraspecific trait variation into functional diversity: Impacts of selective logging on birds in Borneo. *Methods Ecol. Evol.* **2017**, *8*, 1499–1505. [CrossRef]

13. Albert, C.H.; Thuiller, W.; Yoccoz, N.G.; Douzet, R.; Aubert, S.; Lavorel, S. A multi-trait approach reveals the structure and the relative importance of intra- vs. *interspecific variability in plant traits. Funct. Ecol.* **2010**, *24*, 1192–1201. [CrossRef]

14. Hulshof, C.M.; Violle, C.; Spasojevic, M.J.; McGill, B.; Damschen, E.; Harrison, S.; Enquist, B.J. Intra-specific and inter-specific variation in specific leaf area reveal the importance of abiotic and biotic drivers of species diversity across elevation and latitude. *J. Veg. Sci.* **2013**, *24*, 921–931. [CrossRef]

15. Chai, Y.; Dang, H.; Yue, M.; Xu, J.; Zhang, L.; Quan, J.; Guo, Y.; Li, T.; Wang, L.; Wang, M.; et al. The role of intraspecific trait variability and soil properties in community assembly during forest secondary succession. *Ecosphere* **2019**, *10*, e02940. [CrossRef]

16. Niu, K.; Zhang, S.; Lechowicz, M.J. Harsh environmental regimes increase the functional significance of intraspecific variation in plant communities. *Funct. Ecol.* **2020**, *34*, 1666–1677. [CrossRef]

17. Jiang, F.; Cadotte, M.W.; Jin, G. Individual-level leaf trait variation and correlation across biological and spatial scales. *Ecol. Evol.* **2021**, *11*, 5344–5354. [CrossRef] [PubMed]

18. Gong, Y.; Ling, H.; Lv, G.; Chen, Y.; Guo, Z.; Cao, J. Disentangling the influence of aridity and salinity on community functional and phylogenetic diversity in local dryland vegetation. *Sci. Total. Environ.* **2019**, *653*, 409–422. [CrossRef]

19. Jiang, L.; Hu, D.; Wang, H.F.; Lv, G. Discriminating ecological processes affecting different dimensions of α- and β-diversity in desert plant communities. *Ecol. Evol.* **2022**, *12*, e8710. [CrossRef]

20. Yang, X.-D.; Anwar, E.; Zhou, J.; He, D.; Gao, Y.-C.; Lv, G.-H.; Cao, Y.-E. Higher association and integration among functional traits in small tree than shrub in resisting drought stress in an arid desert. *Environ. Exp. Bot.* **2022**, *201*, 104993. [CrossRef]

21. Bao, S.D. *Soil Agrochemical Analysis*, 3rd ed.; China Agricultural Press: Beijing, China, 2000.

22. Jung, V.; Violle, C.; Mondy, C.; Hoffmann, L.; Muller, S. Intraspecific variability and trait-based community assembly. *J. Ecol.* **2010**, *98*, 1134–1140. [CrossRef]

23. Spasojevic, M.J.; Grace, J.B.; Harrison, S.; Damschen, E.I. Functional diversity supports the physiological tolerance hypothesis for plant species richness along climatic gradients. *J. Ecol.* **2014**, *102*, 447–455. [CrossRef]

24. Oksanen, J.; Blanchet, F.G.; Friendly, M.; Kindt, R.; Legendre, P.; McGlinn, D.; Minchin, P.R.; O'Hara, R.B.; L. Simpson, G.; Solymos, P.; et al. vegan: Community Ecology Package. R package Version 2.5-7. 2020. Available online: https://CRAN.R-project.org/package=vegan (accessed on 24 April 2022).

25. Blomberg, S.P.; Garland, T.; Ives, A.R. Testing for phylogenetic signal in comparative data: Behavioral traits are more labile. *Evolution* **2003**, *57*, 717–745. [PubMed]

26. Villéger, S.; Mason, N.W.H.; Mouillot, D. New multidimensional functional diversity indices for a multifaceted framework in functional ecology. *Ecology* **2008**, *89*, 2290–2301. [CrossRef]

27. Borcard, D.; Gillet, F.; Legendre, P. *Numerical Ecology with R[M]. Numerical Ecology with R*; Springer: New York, NY, USA, 2011.

28. Peres-Neto, P.R.; Legendre, P.; Dray, S.; Borcard, D. Variation partitioning of species data matrices: Estimation and comparison of fractions. *Ecology* **2006**, *87*, 2614–2625. [CrossRef]

29. Violle, C.; Enquist, B.J.; McGill, B.J.; Jiang, L.; Albert, C.H.; Hulshof, C.; Jung, V.; Messier, J. The return of the variance: Intraspecific variability in community ecology. *Trends Ecol. Evol.* **2012**, *27*, 244–252. [CrossRef]

30. Herrera, C.M.; Medrano, M.; Bazaga, P. Continuous within-plant variation as a source of intraspecific functional diversity: Patterns, magnitude, and genetic correlates of leaf variability in *Helleborus foetidus* (Ranunculaceae). *Am. J. Bot.* **2015**, *102*, 225–232. [CrossRef]

31. Herrera, C.M. *Multiplicity in Unity: Plant Sub individual Variation and Interactions with Animals*, 1st ed.; University of Chicago Press: Chicago, IL, USA, 2009.

32. Arceo-Gómez, G.; Vargas, C.; Parra-Tabla, V. Selection on intra-individual variation in stigma-anther distance in the tropical tree *Ipomoea wolcottiana* (Convolvulaceae). *Plant Biol.* **2017**, *19*, 454–459. [CrossRef]
33. Kumordzi, B.B.; Wardle, D.A.; Freschet, G.T. Plant assemblages do not respond homogenously to local variation in environmental conditions: Functional responses differ with species identity and abundance. *J. Veg. Sci.* **2015**, *26*, 32–45. [CrossRef]
34. Carlucci, M.B.; Debastiani, V.J.; Pillar, V.D.; Duarte, L.D.S. Between- and within-species trait variability and the assembly of sapling communities in forest patches. *J. Veg. Sci.* **2015**, *26*, 21–31. [CrossRef]
35. Wang, L.R.; Zhu, G.R.; Fang, W.C. The evaluating criteria of some fruit quantitative characters of peach (*Prunus persica* L.) genetic resources. *Acta Hortic. Sin.* **2005**, *32*, 1–5.
36. Bolnick, D.I.; Svanbäck, R.; Fordyce, J.A.; Yang, L.H.; Davis, J.M.; Hulsey, C.D.; Forister, M.L. The ecology of individuals: Incidence and implications of individual specialiation. *Am. Nat.* **2003**, *161*, 1–28. [CrossRef]
37. Mitchell, R.M.; Bakker, J.D. Quantifying and comparing intraspecific functional trait variability: A case study with *Hypochaeris radicata*. *Funct. Ecol.* **2014**, *28*, 258–269. [CrossRef]
38. Tang, Q.; Huang, Y.; Ding, Y.; Zang, R. Interspecific and intraspecific variation in functional traits of subtropical evergreen and deciduous broad-leaved mixed forests. *Biodivers. Sci.* **2016**, *24*, 262–270. [CrossRef]
39. Burton, J.I.; Perakis, S.S.; McKenzie, S.C.; Lawrence, C.E.; Puettmann, K.J. Intraspecific variability and reaction norms of forest understorey plant species traits. *Funct. Ecol.* **2017**, *31*, 1881–1893. [CrossRef]
40. Kang, M.; Chang, S.X.; Yan, E.-R.; Wang, X.-H. Trait variability differs between leaf and wood tissues across ecological scales in subtropical forests. *J. Veg. Sci.* **2014**, *25*, 703–714. [CrossRef]
41. Mitchell, R.M.; Ames, G.M.; Wright, J.P. Intraspecific trait variability shapes leaf trait response to altered fire regimes. *Ann. Bot.* **2021**, *127*, 543–552. [CrossRef] [PubMed]
42. Wilson, P.J.; Thompson, K.; Hodgson, J.G. Specific leaf area and leaf dry matter content as alternative predictors of plant strategies. *New Phytol.* **1999**, *143*, 155–162. [CrossRef]
43. Grassein, F.; Till-Bottraud, I.; Lavorel, S. Plant resource-use strategies: The importance of phenotypic plasticity in response to a productivity gradient for two subalpine species. *Ann. Bot.* **2010**, *106*, 637–645. [CrossRef] [PubMed]
44. Yang, J.; Zhang, G.; Ci, X.; Swenson, N.G.; Cao, M.; Sha, L.; Li, J.; Baskin, C.C.; Slik, J.W.F.; Lin, L. Functional and phylogenetic assembly in a Chinese tropical tree community across size classes, spatial scales and habitats. *Funct. Ecol.* **2014**, *28*, 520–529. [CrossRef]
45. Che, Y.D.; Liu, M.X.; Li, L.R.; Jiao, J.I.; Xiao, W. Exploring the community assembly of subalpine meadow communities based on functional traits and community phylogeny. *Chin. J. Plant Ecol.* **2017**, *41*, 1157–1167.
46. Hao, S.-J.; Li, X.-Y.; Hou, M.-M.; Zhao, X.-H. Spatial variations of community functional traits at different successional stages in temperate forests of Changbai Mountains, Northeast China. *Chin. J. Plant Ecol.* **2019**, *43*, 208–216. [CrossRef]
47. Mason, N.W.; Peltzer, D.A.; Richardson, S.J.; Bellingham, P.J.; Allen, R.B. Stand development moderates effects of ungulate exclusion on foliar traits in the forests of New Zealand: Ungulate impacts on foliar traits. *J. Ecol.* **2010**, *98*, 1422–1433. [CrossRef]
48. Wong, M.K.L.; Carmona, C.P. Including intraspecific trait variability to avoid distortion of functional diversity and ecological inference: Lessons from natural assemblages. *Methods Ecol. Evol.* **2021**, *12*, 946–957. [CrossRef]
49. Jones, M.M.; Gibson, N.; Yates, C.; Ferrier, S.; Mokany, K.; Williams, K.J.; Manion, G.; Svenning, J.C. Underestimated effects of climate on plant taxonomic turnover in the Southwest Australian Floristic Region. *J. Biogeogr.* **2016**, *43*, 289–300. [CrossRef]
50. Zhao, M.; Wang, G.; Xing, K.; Wang, Y.; Xue, F.; Kang, M.; Luo, K. Patterns and determinants of species similarity decay of forest communities in the western Qinling Mountains. *Biodiversity* **2017**, *25*, 3–10. [CrossRef]
51. Qian, H.; Shimono, A. Effects of geographic distance and climatic dissimilarity on taxonomic turnover in alpine meadow communities across a broad spatial extent on the Tibetan Plateau. *Plant Ecol.* **2012**, *213*, 1357–1364. [CrossRef]
52. Weiher, E.; Freund, D.; Bunton, T.; Stefanski, A.; Lee, T.; Bentivenga, S. Advances, challenges and a developing synthesis of ecological community assembly theory. *Philos. Trans. R. Soc. B Biol. Sci.* **2011**, *366*, 2403–2413. [CrossRef]
53. Hu, D.; Jiang, L.; Hou, Z.; Zhang, J.; Wang, H.; Lv, G. Environmental filtration and dispersal limitation explain different aspects of beta diversity in desert plant communities. *Glob. Ecol. Conserv.* **2021**, *33*, e01956. [CrossRef]
54. Lasky, J.R.; Yang, J.; Zhang, G.; Cao, M.; Tang, Y.; Keitt, T.H. The role of functional traits and individual variation in the co-occurrence of Ficus species. *Ecology* **2014**, *95*, 978–990. [CrossRef]
55. Wright, J.P.; Ames, G.M.; Mitchell, R.M. The more things change, the more they stay the same? When is trait variability important for stability of ecosystem function in a changing environment. *Philos. Trans. R. Soc. B Biol. Sci.* **2016**, *371*, 20150272. [CrossRef]
56. Zhang, D.; Peng, Y.; Li, F.; Yang, G.; Wang, J.; Yu, J.; Zhou, G.; Yang, Y. Trait identity and functional diversity co-drive response of ecosystem productivity to nitrogen enrichment. *J. Ecol.* **2019**, *107*, 2402–2414. [CrossRef]
57. Benavides, R.; Carvalho, B.; Bastias, C.C.; López-Quiroga, D.; Mas, A.; Cavers, S.; Gray, A.; Albet, A.; Alía, R.; Ambrosio, O.; et al. The GenTree Leaf Collection: Inter- and intraspecific leaf variation in seven forest tree species in Europe. *Glob. Ecol. Biogeogr.* **2021**, *30*, 590–597. [CrossRef]
58. Fajardo, A.; Siefert, A. Intraspecific trait variation and the leaf economics spectrum across resource gradients and levels of organization. *Ecology* **2018**, *99*, 1024–1030. [CrossRef]
59. Siefert, A. Incorporating intraspecific variation in tests of trait-based community assembly. *Oecologia* **2012**, *170*, 767–775. [CrossRef] [PubMed]

60. Spasojevic, M.J.; Turner, B.L.; Myers, J.A. When does intraspecific trait variation contribute to functional beta-diversity? *J. Ecol.* **2016**, *104*, 487–496. [CrossRef]

61. Legendre, P.; Mi, X.; Ren, H.; Ma, K.; Yu, M.; Sun, I.-F.; He, F. Partitioning beta diversity in a subtropical broad-leaved forest of China. *Ecology* **2009**, *90*, 663–674. [CrossRef]

62. Segre, H.; Ron, R.; De Malach, N.; Henkin, Z.; Mandel, M.; Kadmon, R. Competitive exclusion, beta diversity, and deterministic vs. stochastic drivers of community assembly. *Ecol. Lett.* **2014**, *17*, 1400–1409. [CrossRef]

63. Myers, J.A.; Chase, J.M.; Jiménez, I.; Jørgensen, P.M.; Araújo-Murakami, A.; Paniagua-Zambrana, N.; Seidel, R. Beta-diversity in temperate and tropical forests reflects dissimilar mechanisms of community assembly. *Ecol. Lett.* **2013**, *16*, 151–157. [CrossRef]

Disclaimer/Publisher's Note: The statements, opinions and data contained in all publications are solely those of the individual author(s) and contributor(s) and not of MDPI and/or the editor(s). MDPI and/or the editor(s) disclaim responsibility for any injury to people or property resulting from any ideas, methods, instructions or products referred to in the content.

Article

Studies on the Correlation between δ^{13}C and Nutrient Elements in Two Desert Plants

Zhou Zheng [1,2,3], Xue Wu [1,2,3,4,*], Lu Gong [1,2,3], Ruixi Li [1,2,3], Xuan Zhang [1,2,3], Zehou Li [1,2,3] and Yan Luo [1,2,3,4]

[1] College of Ecology and Environment, Xinjiang University, Urumqi 830046, China; zhengzhouxd98@163.com (Z.Z.); gonglu721@163.com (L.G.); 18483639493@163.com (R.L.); 15939015185@163.com (X.Z.); luoyan505@xju.edu.cn (Y.L.)
[2] Key Laboratory of Oasis Ecology, Education Ministry, Urumqi 830046, China
[3] Xinjiang Jinghe Observation and Research Station of Temperate Desert Ecosystem, Ministry of Education, Jinghe 833300, China
[4] Ecological Postdoctoral Research Station, Xinjiang University, Urumqi 830046, China
* Correspondence: wuxue@xju.edu.cn

Citation: Zheng, Z.; Wu, X.; Gong, L.; Li, R.; Zhang, X.; Li, Z.; Luo, Y. Studies on the Correlation between δ^{13}C and Nutrient Elements in Two Desert Plants. Forests 2023, 14, 2394. https://doi.org/10.3390/f14122394

Academic Editor: Antonio Gallardo Correa

Received: 1 November 2023
Revised: 4 December 2023
Accepted: 6 December 2023
Published: 8 December 2023

Copyright: © 2023 by the authors. Licensee MDPI, Basel, Switzerland. This article is an open access article distributed under the terms and conditions of the Creative Commons Attribution (CC BY) license (https://creativecommons.org/licenses/by/4.0/).

Abstract: Stable carbon isotopes (δ^{13}C) and elemental stoichiometry characteristics are important ways to research the water and nutrient use strategies of plants. Investigating the variation patterns inof δ^{13}C and the major nutrient elements in different organs of plants and the correlation among them can reveal the ecological strategies of desert plants in extreme arid environments. In this study, two typical desert plants, *Alhagi sparsifolia* and *Karelinia caspia*, were studied in the Tarim Basin. By analyzing the changes in δ^{13}C, carbon (C), nitrogen (N), and phosphorus (P) and the ecological stoichiometry of their roots, stems, and leaves, the distribution patterns among different organs and their correlation with soil environmental factors were revealed. The results showed the following: (1) The δ^{13}C of the two plants differed significantly among different organs ($p < 0.01$). The root and stem of *Alhagi sparsifolia* had significantly greater δ^{13}C than the leave, while the δ^{13}C of *Karelinia caspia* showed a root > stem > leaf gradient; (2) the C content in the leaves of the two plants was significantly lower than that of the root ($p < 0.01$), whereas the N content showed the opposite trend ($p < 0.01$); (3) the average N:P of *Alhagi sparsifolia* was >16.00, indicating it was mainly limited by P elements, while the average N:P of *Karelinia caspia* was <14.00, suggesting it was mainly limited by N elements; (4) in the root, stem and leave of *Alhagi sparsifolia* and *Karelinia caspia*, the N content and C:N and the P content and C:P showed a significantly negative correlation ($p < 0.01$), and δ^{13}C was negatively correlated with C:P; (5) soil total phosphorus (TP) is an important soil environmental factor affecting δ^{13}C and the nutrient elements in *Alhagi sparsifolia* and *Karelinia caspia*. This study demonstrates that *Alhagi sparsifolia* and *Karelinia caspia* are able to effectively coordinate and regulate their water, N, and P use strategies in response to environmental stress. These results can provide scientific reference for the evaluation of plant physiological and ecological adaptations for ecological conservation in arid areas.

Keywords: desert plants; plant organs; stable carbon isotope; ecological stoichiometry

1. Introduction

The Tarim Basin in Xinjiang is a typical extreme arid region. The long-term effects of aridity, high temperatures, barrenness, and other stressors have resulted in plants in this region developing unique ecological adaptations to the extreme environment. Understanding the characteristics of plants in these arid desert areas is essential for the accurate prediction of fragile and sensitive ecosystems. Stable carbon isotopes (δ^{13}C) are a crucial indicator for plant water use efficiency (WUE) [1], which to some extent represents plant photosynthetic and transpiration rates [2,3]. Under aridity stress, plants enhance their adaptation by increasing their WUE [4]. C, N, and P are structural elements necessary for plant growth and development [5]. They participate in many plant physiological and biochemical processes [6] and are closely related to the synthesis of enzymes, amino acids, and other substances [7]. The elemental stoichiometric ratios of C, N, and P can reflect the dynamics of C accumulation, the efficiency of N and P utilization, and the pattern of nutrient limitation in the habitat of plants [8]. C:N and C:P are closely related to plant growth rate, while N:P can reflect N or P limitations in plant growth, which assist in the determination of plant nutrient utilization strategies. In recent years, more studies have focused on revealing plant WUE and nutrient use strategies by analyzing the changing patterns of plant δ^{13}C and nutrient elements [9–11]. The investigation of δ^{13}C and major nutrient elements in different organs of plants can help to understand their own resource allocation and ecological adaptation strategies in varying environments. However, studies exploring the trade-offs among different plant organs by combining water and nutrient use strategies are still limited.

Numerous scholars have worked on the δ^{13}C and nutrient characteristics of various research objects, including natural plants and cultivated plants [12–15]. Studies have verified that there are many factors affecting plants' δ^{13}C and nutrient elemental characteristics. The main factors are plant type, plant developmental stage, environmental changes, and human disturbances [16]. Diverse responses in the δ^{13}C and nutrient element stoichiometric characteristics of plants are triggered by changes in environmental factors such as elevation, atmospheric pressure, temperature, precipitation, fertilization, and aridity. Studies have shown that plant δ^{13}C is positively correlated with elevation [17] and negatively correlated with both precipitation and air temperature [18,19]; it also increases with increasing latitude [20]. At the species or community level, plant fine-root N and P contents are negatively correlated with mean annual temperature and annual precipitation [21]. Leaf N and P concentrations vary significantly among different life forms, usually showing a pattern of herb > shrub > tree, angiosperm > gymnosperm [22,23]. In arid desert regions, plants respond to water scarcity by altering their nutrient partitioning patterns.

Roots, stems, and leaves are key organs for the synthesis, uptake, transport, accumulation, and storage of nutrients in plants [24], and the ratio of nutrient distribution among them reflects the ability of plant resources to acquire, transport, and store nutrients [25]. Therefore, the distribution patterns of δ^{13}C and major nutrient elements among different organs of plants, including plant nutrient organs (roots, stems, leaves, etc.) and reproductive organs (flowers, fruits, seeds, etc.), have been given attention [26]. Numerous studies have shown that δ^{13}C in plant leaves is generally lower than in other organs. The N and P elements and their stoichiometric relationships in different plant organs depend critically on the growth and functional attributes of each organ. Due to the dilution of nutrients by water in the organs, N and P concentrations in plant stems and roots are lower than those in leaves and reproductive organs [27,28], which is confirmed by the global study of N and P's stoichiometric relationships in plant roots (i.e., N and P contents in fine roots are generally lower than in leaves) [22]. However, by analyzing plant and soil N and P elements, some scholars found that the elemental relationships among different organs may be decoupled or unchanged with environmental changes [29]. In extreme arid environments, the study of the changes in δ^{13}C and major nutrient content among different organs of desert plants can contribute to revealing their ecological adaptations.

Alhagi sparsifolia and *Karelinia caspia* are the dominant species in arid and semi-arid regions, with excellent tolerance to drought, salinity, and barrenness [30,31]. Previous studies mainly focused on their fixation of nutrient elements, the comparison of their ecological stoichiometric characteristics, and the changes in plant functional traits under adversity stress [32]. However, there are fewer studies on their δ^{13}C and nutrient elemental characteristics and the variations among organs. In this study, the Tarim Basin was taken as the study area, and *Alhagi sparsifolia* and *Karelinia caspia* were chosen as the research objects. We analyzed the stoichiometric characteristics of major nutrient elements (C, N, and P) and stable carbon isotopes (δ^{13}C) of plant roots, stems, and leaves and their correlations with soil physicochemical factors. The aim of this study is to enhance the understanding of the ecological adaptation strategies of plants in extreme environments, which could provide a theoretical basis for regional ecological environment protection and restoration.

2. Materials and Methods

2.1. Overview of the Study Area

The study area is located at the northern edge of the Taklamakan Desert in the Tarim Basin of Xinjiang, China, in the upper reaches of the Tarim River Basin. The region has a warm temperate continental arid climate characterized by perennial drought and low precipitation, with an average annual precipitation of only 75.00 mm. Moreover, it is prone to frequent wind and sandstorms. The temperature varies from a minimum of $-25\ °C$ to a maximum of $40\ °C$, and the frost-free period is 180−224 d. The main soil types are saline, wind−sand, brown desert, and scrub desert. There are a few vegetation types, dominated by arid desert plants such as *Alhagi sparsifolia*, *Karelinia caspia*, *Tamarix chinensis*, and *Populus euphratica* as the representative species. *Alhagi sparsifolia* and *Karelinia caspia* are perennial herbs of the *Leguminosae* and *Asteraceae* families, respectively. During the long evolutionary process, they have developed excellent resistance to drought, salinity, barrenness, wind, and sand. Therefore, *Alhagi sparsifolia* and *Karelinia caspia* are effective plants for desertification control, playing an important role in preventing sand erosion.

2.2. Sample Collection and Measurement

2.2.1. Sample Survey

In June and July 2021, during the plant-growing season, representative sample sites where there was less human influence and more vegetation cover were selected through investigation of the study area. Six 1.00×1.00 m herbaceous sample plots were set up with a distance of 20.00 m between different plots at each sample site. The number of *Alhagi sparsifolia* and *Karelinia caspia* plants in each sample plot was counted, and the plant height, crown width, and basal diameter were accurately measured and recorded. The specific information is shown in Table 1.

Table 1. *Alhagi sparsifolia* and *Karelinia caspia* parameters of plant height, crown width, basal diameter, and density (mean $\pm$ SE).

	Height (cm)	Crown Width (cm)	Basal Diameter (mm)	Density (No/m^2)
Alhagi sparsifolia	34.91 ± 11.85	22.67 ± 10.90	17.56 ± 9.00	5.18 ± 0.51
Karelinia caspia	28.63 ± 8.71	22.83 ± 11.13	17.86 ± 9.22	3.47 ± 0.41

2.2.2. Sample Collection

Within each sample plot, *Alhagi sparsifolia* and *Karelinia caspia* individuals of similar morphology, well established, and without obvious gnawing marks were selected.

Healthy and mature leaves of the two studied species from different orientations were collected and mixed, then put into kraft envelopes. One-year-old stems were cut using fruit pruning shears, stripped of leaves, and packed into kraft envelopes. A number of fine roots (diameter ranging from 0.05 to 0.30 cm and length of around 7.00 cm) were obtained by digging, then shaken off the soil on the root surface and put into kraft envelopes. In the center of each sample plot, soil samples ranging from 0 to 100.00 cm were collected according to the five-point sampling method (the soil depth included 0–20.00 cm, 20.00–40.00 cm, 40.00–60.00 cm, 60.00–80.00 cm, and 80.00–100.00 cm). After removing other plant roots, stones, and other debris, the collected soil samples were mixed evenly and divided into two portions. One portion was put into sample bags for determining the soil physicochemical properties, and the other portion was put into an aluminum box for determining the soil water content. Finally, all the collected samples were kept in a cryogenic ice box and brought back to the laboratory.

2.3. Sample Determination

2.3.1. Sample Processing

Collected plant roots, stems, and leaves were brought back to the laboratory and washed thoroughly, then placed into an oven, killed out at 105 °C for 15 min, and dried at 75 °C for 48 h until reaching a constant weight state. The dried plants were ground using a multifunctional pulverizer (NM200, Retsch, Haan, Germany) and passed through a 0.15 mm sieve, then put into sample bags for measurement. After the soil samples were brought back, one part of the samples was placed in an oven at 105 °C for 24 h to measure the soil moisture content, and the other part was air-dried in a cool place, ground, sieved using a 0.30 mm mesh sieve, and put into sample bags for measurement [33].

2.3.2. Stable Carbon Isotope Composition Determination (δ^{13}C)

The dried plants were ground and passed through a 0.15 mm sieve, then measured with a stable-isotope-ratio mass spectrometer (DELTA A V Advantage, Waltham, MA, USA). The δ^{13}C value of the sample was calculated according to the following formula:

$$\delta^{13}\text{C} \; (\text{‰}) = [(\text{R}_{\text{sample}} - \text{R}_{\text{standard}}) \div \text{R}_{\text{standard}}] \times 1000 \tag{1}$$

In the formula, $\text{R}_{\text{standard}}$ and R_{sample} are the ^{13}C/^{12}C standard sample and the ^{13}C/^{12}C sample, respectively; $\text{R}_{\text{standard}}$ is the adopted international standard substance, Vienna Pee Dee Belemnite (VPDB) [16].

2.3.3. Plant Nutrient Element Determination

Determination of plant C content was carried out using the potassium dichromate volumetric method (external heating method); determination of N content was performed using a fully automated Kjeldahl nitrogen analyzer (The Swedish FOSS, KjeltecTM8400, Copenhagen, Denmark); and P was determined using the aqua regia microwave-assisted digestion ICP−AES method.

2.3.4. Determination of Physical and Chemical Properties of Soil Samples

In the soil environment factors, the measurement characteristics included soil organic carbon (SOC), total nitrogen (TN) and total phosphorus (TP) elemental content (g·kg^{-1}), alkaline nitrogen content (AN, mg·kg^{-1}), quick-acting phosphorus content (AP, mg·kg^{-1}), soil water content (SWC, %), electrical conductivity (EC, $\mu\text{S·cm}^{-1}$), and pH. The SOC, TN, and TP contents of the soil were determined using the potassium dichromate volumetric method, Kjeldahl nitrogen determination, and molybdenum antimony colorimetric method, respectively. The soil AN and AP contents were determined using the alkaline diffusion method and sodium bicarbonate method, respectively. Soil SWC was determined using the drying method. Soil EC was determined using the electrical conductivity method (DDS−307, Shanghai Yidian Scientific Instrument Co., Ltd., Shanghai, China). Soil pH was determined using a pH meter (PHS−3C, Shanghai Yidian Scientific Instrument Co., Ltd., Shanghai, China). Specific experimental methods are described in the literature [34].

2.4. Data Processing

Differences in δ^{13}C and nutrient contents among different organs of *Alhagi sparsifolia* and *Karelinia caspia* were examined using One-Way ANOVA and the Multiple Comparison Tukey method. Pearson analysis was used to analyze the correlation between δ^{13}C and nutrient element contents across different organs of the two desert plants. Redundancy Analysis (RDA) was used to study the relationships among the whole plant δ^{13}C, nutrient element contents, and soil environmental factors. Data processing, statistical analysis, and drawing were completed using SPSS 26 (IBM, Chicago, IL, USA), Canoco 5 (CanocoLab, Microsoft, Redmond, Washington, DC, USA), and Origin 2021 (OriginLab, Nothampton, MA, USA).

3. Results and Analysis

3.1. Characteristics of $\delta^{13}C$ and Major Nutrient Elements in Alhagi sparsifolia and Karelinia caspia

As shown in Table 2, except for the P content, the average δ^{13}C and major nutrient element contents of *Karelinia caspia* were greater than those of *Alhagi sparsifolia*. The stoichiometric ratios showed that the C:P and N:P of *Karelinia caspia* were greater than those of *Alhagi sparsifolia*, while C:N was lower than that of *Alhagi sparsifolia*. The coefficients of variation in δ^{13}C and the C content were within 6.00% for both plants, which was low; the coefficients of variation in N, P, C:N, C:P, and N:P in *Karelinia caspia* and P and C:P in *Alhagi sparsifolia* ranged from 20.00% to 40.00%, which was relatively stable. However, the coefficients of variation in N, C:N, and N:P in *Karelinia caspia* ranged from 50.00% to 60.00%, showing relatively high variability.

Table 2. Characteristics of $\delta^{13}C$ and major nutrient element contents of *Alhagi sparsifolia* and *Karelinia caspia*.

	Index	Maximum	Minimum	Average	Standard Error	Coefficient of Variation (%)
Alhagi sparsifolia	δ^{13}C (‰)	−24.77	−28.76	−26.49	1.24	1.09
	C (g·kg^{-1})	467.06	416.83	445.78	1.55	2.71
	N (g·kg^{-1})	20.66	5.67	12.23	1.13	39.00
	P (g·kg^{-1})	1.05	0.43	0.72	0.05	28.54
	C:N	73.49	20.92	42.19	3.91	39.34
	C:P	1039.09	413.17	676.11	48.36	30.34
	N:P	24.73	8.08	17.36	1.27	29.95
Karelinia caspia	δ^{13}C (‰)	−25.51	−30.27	−27.98	1.39	1.18
	C (g·kg^{-1})	455.69	371.80	412.57	12.78	5.65
	N (g·kg^{-1})	14.46	3.01	7.51	0.93	52.34
	P (g·kg^{-1})	1.64	0.74	1.24	0.07	22.35
	C:N	143.46	27.90	73.18	9.39	54.45
	C:P	596.76	259.17	352.38	22.73	27.37
	N:P	10.83	2.18	6.40	0.78	51.71

3.2. Characteristics of $\delta^{13}C$ and Major Nutrient Elements in Different Organs of Alhagi sparsifolia and Karelinia caspia

The δ^{13}C values ranged from −30.00 to −25.00‰ for *Alhagi sparsifolia* and *Karelinia caspia*, and the mean δ^{13}C values for the two plants were −26.49‰ and −27.98‰, respectively. As shown in Figure 1, the results of the One-Way ANOVA indicated that the difference in δ^{13}C among different organs for *Alhagi sparsifolia* and *Karelinia caspia* was significant ($p < 0.05$). The δ^{13}C of *Alhagi sparsifolia* showed a pattern of root > stem > leaf, with both root and stem δ^{13}C values significantly higher than that of the leaf. The δ^{13}C of *Karelinia caspia* showed a root > stem > leaf pattern; the δ^{13}C of the root was significantly higher than that of the stem, and the δ^{13}C of the stem was significantly higher than that of the leaf.

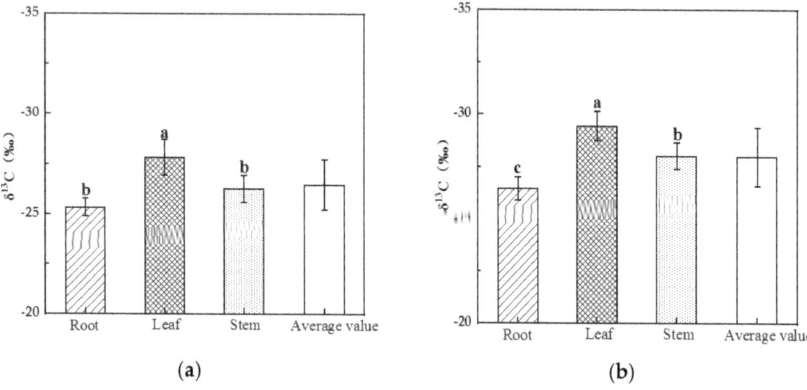

Figure 1. Characteristics of $\delta^{13}C$ in different organs of *Alhagi sparsifolia* (**a**) and *Karelinia caspia* (**b**) (mean ± SE). Note: Different lowercase letters indicate a significant difference ($p < 0.01$).

The major nutrient contents and stoichiometric ratio characteristics across different organs of *Alhagi sparsifolia* and *Karelinia caspia* are shown in Figure 2. Except for the C content in *Alhagi sparsifolia*, the other nutrient contents and stoichiometry of the two studied species were significantly different among the root, stem, and leaf. For *Karelinia caspia*, the C contents of the root and stem were significantly higher than that of the leaf. The N content in all organs of *Alhagi sparsifolia* and *Karelinia caspia* showed a pattern of leaf > root > stem. The leaf P content in *Alhagi sparsifolia* was significantly higher than in the root and stem, whereas the P element content in *Karelinia caspia* was much higher in the stem and leaf compared to the root ($p < 0.01$). In addition, the variation patterns in C:N among organs for both *Alhagi sparsifolia* and *Karelinia caspia* were consistent (stem > root > leaf ($p < 0.01$)). The C:P values in the root and stem were significantly higher than in the leaf in *Alhagi sparsifolia*, and the C:P value in the root was significantly higher than in the leaf and stem in *Karelinia caspia* ($p < 0.01$). For *Alhagi sparsifolia*, leaf N:P was higher than root N:P, and root N:P was significantly higher than stem N:P ($p < 0.05$).

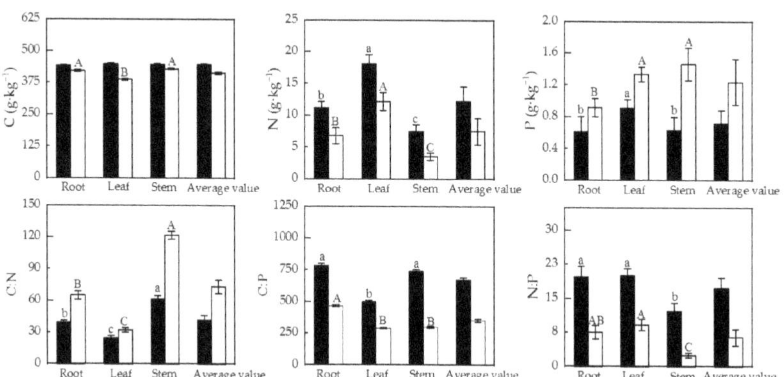

Figure 2. Characteristics of main nutrient element contents and stoichiometric ratios in different organs of *Alhagi sparsifolia* and *Karelinia caspia* (mean ± SE). Note: Black columns represent *Alhagi sparsifolia*; white columns represent *Karelinia caspia*. Note that lowercase letters indicate a significant difference in *Alhagi sparsifolia*, while capital letters indicate a significant difference in *Karelinia caspia*.

3.3. Correlation between $\delta^{13}C$ and Major Nutrient Elements in Different Organs of Alhagi sparsifolia and Karelinia caspia

As shown in Table 3, only *Alhagi sparsifolia*'s leaf $\delta^{13}C$ showed a significant positive correlation with the C content ($p < 0.05$). There was no significant correlation between $\delta^{13}C$ and major nutrient elements in other organs. However, the major nutrient elements and their stoichiometric ratios exhibited close relationships, as the N content and C:N and the P content and C:P of the three organs of *Alhagi sparsifolia* all showed extremely significant negative correlations. There was also an extremely significant negative correlation between root P content and N:P, and an extremely significant positive correlation between root C:P and N:P ($p < 0.01$).

Table 3. Correlation between $\delta^{13}C$ and major nutrient elements in different organs of *Alhagi sparsifolia*.

Organ	Index	$\delta^{13}C$ (‰)	C (g·kg^{-1})	N (g·kg^{-1})	P (g·kg^{-1})	C:N	C:P	N:P
	$\delta^{13}C$ (‰)	1.00						
	C (g·kg^{-1})	−0.09	1.00					
	N (g·kg^{-1})	0.69	0.02	1.00				
Root	P (g·kg^{-1})	0.46	−0.39	0.16	1.00			
	C:N	−0.71	0.14	−0.99 **	−0.24	1.00		
	C:P	−0.60	0.37	−0.31	−0.99 **	0.39	1.00	
	N:P	−0.42	0.35	−0.02	−0.99 **	0.10	0.96 **	1.00
	$\delta^{13}C$ (‰)	1.00						
	C (g·kg^{-1})	0.05	1.00					
	N (g·kg^{-1})	−0.32	0.46	1.00				
Stem	P (g·kg^{-1})	0.05	−0.29	0.05	1.00			
	C:N	0.45	−0.29	−0.98 **	−0.16	1.00		
	C:P	−0.01	0.55	−0.01	−0.95 **	0.16	1.00	
	N:P	−0.26	0.60	0.63	0.74	−0.52	0.76	1.00
	$\delta^{13}C$ (‰)	1.00						
	C (g·kg^{-1})	0.88 *	1.00					
	N (g·kg^{-1})	−0.32	−0.55	1.00				
Leaf	P (g·kg^{-1})	0.04	−0.30	0.58	1.00			
	C:N	0.47	0.69	−0.98 **	−0.53	1.00		
	C:P	0.10	0.47	−0.60	−0.98 **	0.58	1.00	
	N:P	−0.39	−0.24	0.43	−0.48	−0.46	0.45	1.00

Note: ** indicates extremely significant correlation ($p < 0.01$); * indicates significant correlation ($p < 0.05$). The same notation applies below.

Based on Table 4, it can be found that the stem $\delta^{13}C$ of *Karelinia caspia* showed an extremely significant positive correlation with the N content ($p < 0.01$), but the relationship with C:N showed an extremely significant negative correlation. The N content and C:N and the P content and C:P of the three organs all showed significantly negative correlations. Furthermore, the C:N and N:P of the root and leaf and the N and C:P of the stem also displayed significantly negative correlations ($p < 0.05$). There was a significantly positive correlation observed between N:P and P in the root and leaf.

Table 4. Correlation between δ^{13}C and major nutrient elements in different organs of *Karelinia caspia*.

Organ	Index	δ^{13}C (‰)	C (g·kg^{-1})	N (g·kg^{-1})	P (g·kg^{-1})	C:N	C:P	N:P
	δ^{13}C (‰)	1.00						
	C (g·kg^{-1})	0.01	1.00					
	N (g·kg^{-1})	0.28	0.36	1.00				
Root	P (g·kg^{-1})	0.67	−0.56	−0.28	1.00			
	C:N	−0.24	−0.21	−0.99 **	0.18	1.00		
	C:P	0.18	0.71	0.34	−0.98 **	−0.29	1.00	
	N:P	−0.13	0.57	0.88 *	−0.70	−0.85 *	0.75	1.00
	δ^{13}C (‰)	1.00						
	C (g·kg^{-1})	−0.36	1.00					
	N (g·kg^{-1})	0.92 **	−0.15	1.00				
Stem	P (g·kg^{-1})	0.64	0.42	−0.75	1.00			
	C:N	−0.93 **	0.44	−0.95 **	−0.57	1.00		
	C:P	−0.78	−0.15	−0.87 *	−0.96 **	0.76	1.00	
	N:P	−0.06	−0.75	−0.14	−0.75	−0.08	0.58	1.00
	δ^{13}C (‰)	1.00						
	C (g·kg^{-1})	−0.275	1.00					
	N (g·kg^{-1})	−0.192	0.726	1.00				
Leaf	P (g·kg^{-1})	−0.405	0.354	0.041	1.00			
	C:N	0.097	−0.511	−0.959 **	0.081	1.00		
	C:P	−0.546	0.160	0.342	−0.866*	−0.357	1.00	
	N:P	−0.343	0.551	0.924 **	−0.344	−0.929 **	0.654	1.00

Note: ** indicates extremely significant correlation ($p < 0.01$); * indicates significant correlation ($p < 0.05$).

3.4. Relationships among δ^{13}C, Major Nutrient Elements, and Soil Physicochemical Factors in Alhagi sparsifolia and Karelinia caspia

According to the results of the RDA, the interpretation amounts of soil environmental factors to the variations in δ^{13}C, major nutrient elements, and stoichiometric ratios of *Alhagi sparsifolia* in axes I and II were 44.47% and 0.45%, respectively. Those of *Karelinia caspia* were explained at 59.15% and 14.31%, respectively. The sum of the explanations of the two axes was 44.93% and 73.46% for *Alhagi sparsifolia* and *Karelinia caspia*, respectively. Therefore, the first two axes could effectively reflect the relationship between δ^{13}C, the major nutrient elements, and the soil physicochemical factors, with axis I playing the more important role.

Figure 3a,b show the two-dimensional ordination maps of RDA for *Alhagi sparsifolia* and *Karelinia caspia*. The solid lines represent δ^{13}C, the major nutrient elements, and the ecological stoichiometry ratios; the dotted lines represent the soil physicochemical factors of their habitats. The length of the arrow line indicates the strength of the correlation, with a longer arrow representing a greater correlation and vice versa. The angle of the arrow to the sorting axis indicates a positive or negative correlation, with an acute or obtuse angle representing a positive or negative correlation, respectively. In the figure, the length of the arrows of the TP content in the soil physicochemical factors was the longest, indicating that soil TP played a crucial role in explaining the δ^{13}C, major nutrient elements, and stoichiometric variability in the two studied species. Soil TP was negatively correlated with the δ^{13}C values of *Alhagi sparsifolia* and *Karelinia caspia* and most highly correlated with N, C:N, and N:P.

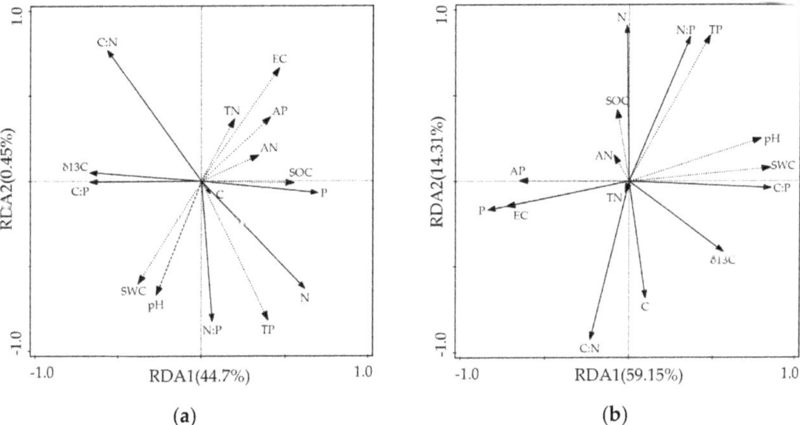

Figure 3. RDA two-dimensional ordination plots of soil physicochemical factors for δ^{13}C and major nutrient elements in *Alhagi sparsifolia* and *Karelinia caspia*. (**a**) Correlation plots for δ^{13}C, major nutrient elements, and soil physicochemical factors in *Alhagi sparsifolia*. (**b**) Correlation plots for δ^{13}C, major nutrient elements, and soil physicochemical factors in *Karelinia caspia*. Note: Solid lines indicate plants. Dashed lines indicate soil physicochemical factors.

4. Discussion

4.1. Variation in δ^{13}C Characteristics among Organs of Alhagi sparsifolia and Karelinia caspia

Water use is a crucial process for photosynthetic carbon sequestration in plants. Its efficiency reflects the plant's ability to adapt to stressful environments. Traditional instantaneous WUE refers to the ratio of photosynthetic rate (A) to transpiration rate (E), while δ^{13}C is a key indicator of long-term WUE in plants, which is important for revealing the relationship between plants and their long-term role in the environment. Due to the fractionation effect of δ^{13}C, carbon isotope discrimination (Δ) in plant tissue was developed as a method for determining transpiration efficiency. It has been verified that it is negatively correlated with A/E at the leaf level [35]. By reviewing numerous studies, it was found that the most important factor affecting δ^{13}C characteristics is the type of plant photosynthesis (C$_3$, C$_4$, and CAM). In terrestrial ecosystems, the δ^{13}C of C$_3$, C$_4$, and CAM plants ranges from -34.00% to $-21.00‰$, -19.00% to $-9.00‰$, and -38.00% to $-1.30‰$, respectively [36]. In the current study, both *Alhagi sparsifolia* and *Karelinia caspia* were identified as C$_3$ plants. Their average δ^{13}C values were $-26.49‰$ and $-27.98‰$, respectively, which are relatively higher among C$_3$ plants. This may be due to the fact that they are distributed in the extreme arid zone, where maximum temperatures can reach 39 °C and annual precipitation is less than 100.00 mm, with evaporation exceeding 2500.00 mm. These higher temperatures and scarcer precipitation resulted in relatively high δ^{13}C values in *Alhagi sparsifolia* and *Karelinia caspia* (Table 2).

Previous studies have verified that δ^{13}C content tends to differ among plant organs. The δ^{13}C values across different organs of plants can reveal the physiological and ecological characteristics of plants and their responses to environmental changes (e.g., climate, history, geography, etc.) [37,38]. A study of 10 *Glycine max* by Feng et al. found that δ^{13}C presented a pattern of roots > stems > seeds > leaves [39]. The δ^{13}C pattern of different organs in *Populus tomentosa* was root > twig > leaf [40]. The current study showed that the δ^{13}C of the root and stem of *Alhagi sparsifolia* was significantly greater than that of the leaf, while the δ^{13}C of *Karelinia caspia* displayed a trend of root > stem > leaf, which is consistent with most scholars' findings. There may be several explanations for the discrepancy among different organs. Firstly, their chemical and physiological properties vary greatly [41]. Secondly, since water metabolism in plants begins with water uptake by the root system, soil water, atmospheric precipitation, surface runoff, or groundwater can enter a plant's roots under the combined effect of root pressure and transpiration tension. Then, it is transported upwards along the ducts or tubular cells to the leaves and other tissues. On the one hand, it finally dissipates through transpiration or exhalation. On the other hand, it is involved in life activities such as respiration and photosynthesis in the cells of various tissues. These complex processes could lead to potential fractionation during water transport in plants. Thirdly, under arid conditions, dry matter accumulation is reduced to varying degrees in all plant organs, with leaves being more susceptible to stress than stems or roots [42]. Finally, as ^{12}C is a lighter isotope than ^{13}C, leaves and stems are more likely to uptake ^{12}C during growth. Therefore, the heavier ^{13}C is enriched in its position closer to the soil, thus resulting in a maximum $\delta^{13}C$ value in the roots [43–45].

4.2. Variation in Major Nutrient Element Content and Stoichiometric Ratio among Organs of Alhagi sparsifolia and Karelinia caspia

Plant growth and development are closely related to the environment in which they are located. Changes in elemental content and its stoichiometry could reflect the ability of plants to absorb and accumulate soil nutrients. Hence, they can characterize their response to changes in environmental conditions [46–48]. Therefore, investigating the ecological stoichiometric characteristics of different organs can provide novel insight into the distribution of the C, N, and P elements in desert plants and the ability of plants to regulate their own nutrients in order to adapt to the environment [49]. *Alhagi sparsifolia* and *Karelinia caspia* are dominant species in arid desert areas with infertile soils, scarce water resources, and severe saline intrusion. Their capacity to sustain the relative stability of chemical composition plays a determining role, and they have developed unique nutrient utilization strategies that are compatible with arid desert environments [50,51].

Through the analysis of the major nutrient element contents in different organs of the two studied species, we found that the leaf C content in *Karelinia caspia* was significantly lower than in the root and stem. The leaf N content in the two species was significantly higher than in the root and stem, and the leaf P content was significantly higher than in the root ($p < 0.01$). The main reason for this phenomenon may be that plants absorb or transport C, N, and P in different ways. The C element is mainly fixed into the plant by the leaves through photosynthesis, while N and P are mainly absorbed from the soil through the fine roots [52]. Secondly, desert plants are susceptible to water stress, which contributes to the increase in free amino acids in the leaves to maintain intracellular osmotic balance [53]; therefore, the leaf N content increases. This is a sign of the active adaptation employed by plants to promote photosynthesis and a response to prevent the loss of organic C due to salt stress [12]. Meanwhile, the higher P content in the leaf relative to the root may be due to the strategy adopted by plants to cope with high soil salinity and low water availability [54]. According to the nutrient limitation theory proposed by Koerselman et al. [55], the average N:P of *Alhagi sparsifolia* was greater than 16, suggesting it is mainly limited by the P element. The probable reason for this is that the sampling occurred during the peak plant growth season, when the uptake of nutrients by the plant was not as fast as the rate of cell expansion, resulting in an insufficient supply of the P element [56]. In contrast, the average N:P in *Karelinia caspia* was below 14, indicating that it was mainly N-limited. This may be because water stress significantly reduces nitrate reductase activity, slowing down nitrate reduction and ammonia assimilation. The uptake of nitrate N by *Karelinia caspia*'s roots is reduced [54,57]. Meanwhile, as *Alhagi sparsifolia* belongs to the leguminous family, Koerselman et al.'s study reported that dryland legumes may enhance their adaptability to saline environments by both building symbiotic associations with rhizobium and utilizing readily available N when salt stress is moderate. It was proven that non-leguminous plants could also obtain nitrogen through their interaction with leguminous plants, but nitrogen mobility was generally low [31,58]. Therefore, *Karelinia caspia*, the non-leguminous plant, is limited by the N element.

4.3. Relationships among $\delta^{13}C$, Major Nutrient Elements, and Soil Physicochemical Factors in Alhagi sparsifolia and Karelinia caspia

The study of plant–soil correlations is conducive to reducing ecological niche overlap and rationalizing species allocation. The results of the RDA indicated that soil TP content had the greatest effect on $\delta^{13}C$, major nutrient elements, and stoichiometry in the two studied plants; thus, it was the main driver affecting $\delta^{13}C$ and the major nutrient elements of *Alhagi sparsifolia* and *Karelinia caspia*. In summarizing a large number of studies, we found that the relationships between plant $\delta^{13}C$ and soil TP were inconsistent. Harris et al. found a negative correlation between $\delta^{13}C$ and soil TP in the vegetation of Xilingol grassland; they thought the uptake of soil P by plants was related to the mass-flow effect [59]. But some scholars, through experiments, found a positive correlation between plant $\delta^{13}C$ and soil P content under conditions of severe water deficit and the addition of different N sources [60]. The uptake of P mainly comes from the transpiration force, which drives soil-soluble P to the root surface. The greater the transpiration accompanied by lower WUE, the lower the corresponding $\delta^{13}C$ value; thus, there is an inverse relationship between $\delta^{13}C$ and soil P content. Our results were in line with this conclusion. The average annual precipitation in the study area is within 75.00 mm, but the annual evaporation exceeds 2500.00 mm, which indicates that moisture is a critical limiting factor for plant growth and survival in this area. Affected by strong transpiration, $\delta^{13}C$ was negatively correlated with soil TP content in *Alhagi sparsifolia* and *Karelinia caspia*. However, plant $\delta^{13}C$ and soil TP content are not necessarily negatively correlated in high-salt areas. This is because salinity changes plant $\delta^{13}C$ mainly by affecting stomatal conductance, photosynthesis, and other physiological activities in C_3 plant leaves [61]. The process of photosynthetic product synthesis in the saline environment is affected by P, which in turn affects leaf $\delta^{13}C$. High-salt environments reduce soil-soluble P, resulting in the inability of plants to enrich more P by increasing transpiration.

5. Conclusions

This study has revealed that the $\delta^{13}C$ values in the organs of *Alhagi sparsifolia* and *Karelinia caspia* show a varying pattern of root > stem > leaf. The N and P contents in the leaf were higher than those in the root. The results showed that the $\delta^{13}C$ of the two species was negatively correlated with C:P and N:P, indicating that they may promote C and N fixation under conditions of low water-use efficiency by increasing the efficiency of P utilization. Aridity and infertility are the main environmental factors that limit plant growth and vegetation recovery in the Tarim Basin. *Alhagi sparsifolia* and *Karelinia caspia* have developed adaptive strategies that involve coordinating water use efficiency and nutrient use efficiency. The findings of this study could provide a theoretical basis for the protection and restoration of vegetation in desert ecosystems.

Author Contributions: Conceptualization, Z.Z. and X.W.; methodology, Z.Z. and X.W.; software, Z.Z., X.Z. and Z.L.; validation, Z.Z.; formal analysis, Z.Z.; investigation, X.W., R.L., Y.L. and X.Z.; resources, Z.Z., X.W., R.L. and X.Z.; data curation, Z.Z.; writing—original draft preparation, Z.Z.; writing—review and editing, Z.Z., X.W. and L.G.; visualization, Z.Z.; supervision, X.W.; project administration, Z.Z. and X.W.; funding acquisition, X.W. All authors have read and agreed to the published version of the manuscript.

Funding: National Natural Sciences Foundation of China (No. 32001145); Xinjiang Uygur Autonomous Region Education Department Basic Scientific Project (No. XJEDU2023P005); Xinjiang Uygur Autonomous Region Education Department Tianchi Doctoral Research Project (No. TCBS202054); Xinjiang University Doctoral Science Foundation (No. 2020670018); and Sciences Foundation of Xinjiang Uygur Autonomous Region (No. 2020D01C053).

Data Availability Statement: Data are contained within the article.

Conflicts of Interest: The authors declare no conflict of interest.

References

1. Shi, R.L.; Zhang, Q.F.; Li, M.; Li, Q.Q.; Zhang, M.M. Application of Plant Carbon Isotope Fractionation in the Study of Water Use Efficiency. *Chin. Agric. Sci. Bull.* **2022**, *38*, 15–20. [CrossRef]
2. Gavito, M.E.; Jakobsen, I.; Mikkelsen, T.N.; Mora, F. Direct evidence for modulation of photosynthesis by an arbuscular mycorrhiza-induced carbon sink strength. *New Phytol.* **2019**, *223*, 896–907. [CrossRef] [PubMed]
3. Chen, P.; Zhang, J.S.; Meng, P.; He, C.X.; Jia, C.X.; Li, J.Z. Feasibility analysis on the determination of WUE by stable carbon isotope: *Cassia obtusifolia* L. as an example. *Acta Ecol. Sin.* **2014**, *34*, 5453–5459. [CrossRef]
4. Deng, X.X.; Shi, Z.; Zeng, L.X.; Lei, L.; Pei, X.X.; Wu, S.; Xiao, W.F. Effects of drought and shading on instantaneous water use efficiency and $\delta^{13}C$ of *Pinus massoniana* seedlings. *Chin. J. Ecol.* **2014**, *33*, 2736–2742.
5. Warren, C.R.; Adams, M.A.; Chen, Z.L. Is photosynthesis related to concentrations of nitrogen and Rubisco in leaves of Australian native plants? *Funct. Plant Biol.* **2000**, *27*, 407–416. [CrossRef]
6. Ma, F.; Liang, W.Y.; Zhou, Z.N.; Xiao, G.J.; Liu, J.L.; He, J.; Jiao, B.Z.; Xu, T.T. Spatial Variation in Leaf Stable Carbon Isotope Composition of Three Caragana Species in Northern China. *Forests* **2018**, *9*, 297. [CrossRef]
7. Walker, A.P.; Beckerman, A.P.; Gu, L.H.; Kattge, J.; Cernusak, L.A.; Domingues, T.F.; Scales, J.C.; Wohlfahrt, G.; Wullschleger, S.D.; Woodward, F.I. The relationship of leaf photosynthetic traits—V_{cmax} and J_{max}—To leaf nitrogen, leaf phosphorus, and specific leaf area: A meta-analysis and modeling study. *Ecol. Evol.* **2014**, *4*, 3218–3235. [CrossRef] [PubMed]
8. Dijkstra, F.A.; Carrillo, Y.; Aspinwall, M.J.; Maier, C.; Canarini, A.; Tahaei, H.; Choat, B.; Tissue, D.T. Water, nitrogen and phosphorus use efficiencies of four tree species in response to variable water and nutrient supply. *Plant Soil* **2016**, *406*, 187–199. [CrossRef]
9. Xia, D.J.; Liu, Q.R.; Zou, L.L.; Ge, Z.W.; Xue, X.H.; Peng, S.L. Foliar δ13C correlates with elemental stoichiometey in halophytes of coastal wetlands. *Acta Ecol. Sin.* **2020**, *40*, 2215–2224.
10. Knelman, J.E.; Schmidt, S.K.; Lynch, R.C.; Darcy, J.L.; Castle, S.C.; Cleveland, C.C.; Nemergut, D.R.; Anthony, G.J. Nutrient Addition Dramatically Accelerates Microbial Community Succession. *PLoS ONE* **2014**, *9*, e102609. [CrossRef]
11. Castle, S.C.; Sullivan, B.W.; Knelman, J.; Hood, E.; Nemergut, D.R.; Schmidt, S.K.; Cleveland, C.C. Nutrient limitation of soil microbial activity during the earliest stages of ecosystem development. *Oecologia* **2017**, *185*, 429–511. [CrossRef] [PubMed]
12. Su, Y.H.; Song, X.Q.; Zheng, J.W.; Zhang, Z.H.; Tang, Z.H. Ecological Stoichiometric Characteristics of C, N and P and Their Relationship with Soil Factors from Different Organs of the Halophytic *Chenopodiaceae* Plants in Hulunbuir. *Bull. Bot. Res.* **2022**, *42*, 910–920. [CrossRef]
13. Li, W.Z.; Gao, Y.; Yang, L.; Jiang, Z.Y.; Wang, X.W. Carbon, nitrogen, and phosphorus stoichiometry of recently senesced larch leaves in response to environmental factors across an entire growing season. *Chin. J. Ecol.* **2020**, *39*, 2832–2841. [CrossRef]
14. Wang, R.Z.; Luo, L.Y.; Sun, J.W.; Gu, H.B.; Wang, G.J. Seasonal dynamics of leaf, branch and root C:N:P ecological stoichiometry of mature *Cunninghamia lanceolata* in Huitong. *J. Cent. South Univ. For. Technol.* **2020**, *40*, 64–71. [CrossRef]
15. Yang, Y.; Liu, B.R.; An, S.S. Ecological stoichiometry in leaves, roots, litters and soil among different plant communities in a desertified region of Northern China. *Catena* **2018**, *166*, 328–338. [CrossRef]
16. Liu, Y.; Li, P.; Shen, B.; Feng, C.H.; Liu, Q.; Zhang, Y. Effects of drought stress on *Bothriochloa ischaemum* water-use efficiency based on stable carbon isotope. *Acta Ecol. Sin.* **2017**, *37*, 3055–3064. [CrossRef]
17. Zhou, C.L.; Li, Y.K.; Cao, G.M.; Peng, C.J.; Song, M.H.; Xu, X.L.; Zhou, H.K.; Lin, L. Carbon and nitrogen stable isotopes technology in the researches on alpine meadow ecosystem in Qinghai-Tibet Plateau: Progress and prospect. *Chin. J. Appl. Ecol.* **2020**, *31*, 3568–3578. [CrossRef]
18. Li, J.Z. Stable Carbon Isotope Compositions in C3, C4 Herbaceous Plants and Their Responce to Changing Temperature. Master's Thesis, Ludong University, Yantai, China, 2009.
19. Yang, S.Y.; Zhao, X.N.; Gao, X.D.; Yu, L.Y. Difference of Water Use Efficiency Between Ecological and Economic Forest and Its Response to Environment Using Carbon Isotopeithe Loess Plateau of China. *J. Soil Water Conserv.* **2022**, *36*, 7. [CrossRef]
20. Haliguli, A.; Yiliminuer; Guan, W.K.; Abudourexii, R. Resposeof Leaf $\delta^{13}C$ and $\delta^{15}N$ Environmental Factorsin Different Habitats of Populus euphraticai. *Acta Bot. Boreal. Occident. Sin.* **2020**, *40*, 1031–1042. [CrossRef]
21. Ma, Y.Z.; Zhong, Q.L.; Jin, B.J.; Lu, H.D.; Guo, B.Q.; Zheng, Y.; Li, M.; Cheng, D.L. Spatial changes and influencing factors of fine root carbon, nitrogen and phosphorus stoichiometry of plants in China. *Chin. J. Plant Ecol.* **2015**, *39*, 159–166. [CrossRef]
22. Tian, D.; Yan, Z.B.; Niklas, K.J.; Han, W.X.; Kattge, J.; Reich, P.B.; Luo, Y.K.; Chen, Y.H.; Tang, Z.Y.; Hu, H.F. Global leaf nitrogen and phosphorus stoichiometry and their scaling exponent. *Natl. Sci. Rev.* **2017**, *5*, 728–739. [CrossRef]
23. Tian, D.; Yan, Z.B.; Fang, J.Y. Review on characteristics and main hypotheses of plant ecological stoichiometry. *Chin. J. Plant Ecol.* **2021**, *45*, 682–713. [CrossRef]
24. Iii, F.; Mooney, S. The Ecology and Economics of Storage in Plants—Annual Review of Ecology and Systematics. *Annu. Rev. Ecol. Syst.* **1990**, *21*, 423–447.
25. Jiang, G.F.; Li, S.Y.; Li, Y.C.; Roddy, A.B. Coordination of hydraulic thresholds across roots, stems, and leaves of two co-occurring mangrove species. *Plant Physiol.* **2023**, *4*, 4. [CrossRef]
26. Badeck, F.W.; Tcherkez, G.; Salvador, N.; Clément, P.; Ghashghaie, J. Post-photosynthetic fractionation of stable carbon isotopes between plant organs—A widespread phenomenon. *Rapid Commun. Mass Spectrom.* **2005**, *19*, 1381–1391. [CrossRef]

27. Tang, Z.Y.; Xu, W.T.; Zhou, G.Y.; Bai, Y.F.; Li, J.X.; Tang, X.L.; Chen, D.M.; Liu, Q.; Ma, W.H.; Xiong, G.M.; et al. Patterns of plant carbon, nitrogen, and phosphorus concentration in relation to productivity in China's terrestrial ecosystems. *Proc. Natl. Acad. Sci. USA* **2018**, *115*, 4033. [CrossRef]

28. Zhao, N.; Yu, G.R.; He, N.P.; Xia, F.C.; Wang, Q.F.; Wang, R.L.; Xu, Z.W.; Jia, Y.L. Invariant allometric scaling of nitrogen and phosphorus in leaves, stems, and fine roots of woody plants along an altitudinal gradient. *J. Plant Res.* **2016**, *129*, 647–657. [CrossRef]

29. He, M.Z.; Dijkstra, F.A.; Zhang, K.; Li, X.R.; Tan, H.J.; Gao, Y.H.; Li, G. Leaf nitrogen and phosphorus of temperate desert plants in response to climate and soil nutrient availability. *Sci. Rep.* **2014**, *4*, 6932. [CrossRef]

30. Xing, R.X.; Zhang, B.; Guo, P.L.; Zhang, Z.H.; Huang, C.B.; Zeng, F.J. The ecological stoichiometric characteristics of *Alhagi sparsifolia* and *Karelinia caspia* in different habitats. *Chin. J. Ecol.* **2020**, *39*, 733–740. [CrossRef]

31. Guo, P.L.; Liu, B.; Zhang, Z.H.; Xing, R.X.; Zhang, B.; Zeng, F.J. Effects of interaction between *Alhagi sparsifolia* and *Karelinia caspia* on nitrogen fixation and rhizosphere microorganisms. *Acta Ecol. Sin.* **2020**, *40*, 6632–6643.

32. Zhang, L. Mechanism of the effects of typical temperate desert plant diversity on ecosystem function. Master's Thesis, Xingjiang University, Urumqi, China, 2019.

33. Gang, C.; Sheng, L. *Plant Physiology Lab*; Higher Education Press (HEP): Beijing, China, 2016.

34. Dong, M. *Observation and Analysis of Terrestrial Biocommunities*; Standards Press of China: Beijing, China, 1997.

35. Lin, Z.; Xing, X. Review on influential factors of plant water use efficiency. *Agric. Res. Arid Areas* **2005**, *23*, 208–213. [CrossRef]

36. Zhou, R.C.; Zhang, W.B.; Cheng, X.L.; Xu, X.Y. A Review on the Responses of Plant and Soil Carbon Stable Isotope. Composition to Environmental Change. *Environ. Sci.* **2019**, *32*, 565–572. [CrossRef]

37. Ge, T.D.; Wang, D.D.; Zhu, Z.K.; Wei, L.; Wei, X.M.; Wu, J.S. Tracing technology of carbon isotope and its applications to studies of carbon cycling in terrestrial ecosystem. *Chin. J. Plant Ecol.* **2020**, *44*, 360–372. [CrossRef]

38. Feng, Q.H.; Shi, Z.M.; Dong, L.L. Response of Plant Functional Traits to Environment and Its Application. *Acta Ecol. Sin.* **2008**, *44*, 125–131. [CrossRef]

39. Feng, H.Y.; Chen, T.; Xu, S.J.; An, L.Z.; Wang, X.L. Effect of enhanced UV-B radiation on growth, yield and stable carbon isotope composition in Glycine max cultivars (SCI). *Acta Bot. Sin.* **2001**, *43*, 709–713. [CrossRef]

40. Fang, X.J.; Li, J.Y.; Nie, L.B.; Shen, Y.B.; Zhang, Z.Y. The characteristics of stable carbon isotope and water use efficiency for Populus tomentosa hybrid clones. *Ecol. Environ. Ences* **2009**, *18*, 2267–2271. [CrossRef]

41. Greitner, C.S.; Winner, W.E. Increases in δ^{13} values of radish and soybean plants caused by ozone. *New Phytol.* **2010**, *108*, 489–494. [CrossRef]

42. Li, Y.Y. Application of carbon isotope technique on the study of water water use efficiency of crops. *Acta Agric. Nucl. Sin.* **2000**, *14*, 115–121. [CrossRef]

43. Quan, X.L.; Duan, Z.H.; Qiao, Y.M.; Pei, H.K.; Chen, M.C.; He, G.F. Variations in soil carbon and nitrogen stable isotopes and densitu among different alpine meadows. *Acta Pratac. Sin.* **2016**, *25*, 27–34. [CrossRef]

44. Li, M.C.; Liu, H.H.; Yi, X.F.; Li, L.X. Characterization of photosynthetic pathway of plant species growing in the eastern Tibetan plateau using stable carbon isotope composition. *Photosynthetica* **2006**, *44*, 102–108. [CrossRef]

45. Zhang, X.Y.; Sun, X.Y.; Zhang, L.; Li, S.Y. A Study on the Characteristics of Soil Stable Carbon Isotope Composition in Different Types of Grassland in Northwest China. *Chin. J. Soil Sci.* **2013**, *44*, 348–354. [CrossRef]

46. Sun, L.; Gong, L.; Zhu, M.L.; Xie, L.N.; Li, H.L.; Luo, Y. Leaf stoichiometric characteristics of typical desert plants and their relationships to soil environmental factors in the northern margin of the Tarin Basin. *Chin. J. Ecol.* **2017**, *36*, 1208–1214. [CrossRef]

47. Wang, Z.N.; Yang, H.M. Response of ecological stoichiometry of carbon, nitrigen and phosphorus in plants to abiotic envrionmental factors. *Pratac. Sci.* **2013**, *30*, 927–934.

48. Abliz, A.; Lü, G.H.; Zhang, X.N.; Gong, Y.M. Carbon, nitrogen and phosphorus stoichiometry of photosynthetic organs across Ebinur Lake Wetland Natural Reserve of Xinjiang. *Chin. J. Ecol.* **2015**, *34*, 2123–2130. [CrossRef]

49. Wang, S.Q.; Yu, G.R. Ecological stoichiometry characteristics of ecosystem carbon, nitrogen and phosphorus elements. *Acta Ecol. Sin.* **2008**, *28*, 3937–3947. [CrossRef]

50. Drenovsky, R.E.; James, J.J.; Richards, J.H. Variation in nutrient resorption by desert shrubs. *J. Arid. Environ.* **2010**, *74*, 1564–1568. [CrossRef]

51. Xiang, X.M.; De, K.J.; Feng, T.X.; Lin, W.S.; Qian, S.W.; Wei, X.J.; Wang, W.; Xu, C.T.; Zhang, L.; Geng, X.P. Effect of Exogenous Nitrogen Addition on inter-monthly Variation of Plant-soil Nutrients in Alpine Meadow. *Acta Agrestia Sin.* **2022**, *30*, 1836–1845. [CrossRef]

52. Zhang, J.F.; Zhang, X.D.; Zhou, J.X.; Franz, M. Effects of salinity stress on poplars seedling growth and soil enzyme activity. *Chin. J. Appl. Ecol.* **2005**, *16*, 426–430.

53. Zhang, J.X.; Ge, S.F.; Wu, Y.H.; Yang, Y.F.; Xu, G.D.; Liu, P. Effects of Drought Stress on Carbon and Nitrogen Metabolism of Ardisia japonical Leaves. *J. Soil Water Conserv.* **2015**, *29*, 278–282. [CrossRef]

54. Zhang, J.; Qi, X.X.; Liu, D.; Zhao, H.Y.; Xie, H.J.; Cao, J.J. Stoichiometric responses of Phragmites australis organs to soil factors in a wetland from arid area. *Chin. J. Ecol.* **2021**, *40*, 701–711. [CrossRef]

55. Koerselman, W.; Meuleman, A.F. The Vegetation N:P Ratio: A New Tool to Detect the Nature of Nutrient Limitation. *J. Appl. Ecol.* **1996**, *33*, 1441. [CrossRef]

56. Li, H.L.; Gong, L.; Hong, Y. Seasonal variations in C, N, and P stoichiometry of roots, stems, and leaves of Phragmites australis in the Keriya Oasis. *Acta Ecol. Sin.* **2016**, *36*, 6547–6555. [CrossRef]
57. Su, Y.H.; Song, X.Q.; Zheng, J.W.; Zhang, Z.H.; Tang, Z.H. Comparison of stoichiometric characteristics of main nutrient elements in different organs from four *Chenopodiaceae* species. *Chin. J. Ecol.* **2022**, *42*, 1–12. [CrossRef]
58. Li, M.M.; Petrie, M.D.; Tariq, A.; Zeng, F.J. Response of nodulation, nitrogen fixation to salt stress in a desert legume *Alhagi sparsifolia*. *Environ. Exp. Botany* **2021**, *183*, 104348. [CrossRef]
59. Zhou, Y.C.; Cheng, X.L.; Fan, J.W.; Harris, W. Relationships between foliar carbon isotope composition and elements of C-3 species in grasslands of Inner Mongolia, China. *Plant Ecol.* **2016**, *217*, 883–897. [CrossRef]
60. Chen, K.L. Soil phosphorus levels in relation to crop stable carbon isotope fractionation and biological yield. *For. Pract. Technol.* **2003**, *6*, 15–16. [CrossRef]
61. Wang, W.W. Summary on Relationship Between Stable Carbon Isotope Composition of Plants and Soil Salinity. *J. Anhui Agric. Sci.* **2012**, *40*, 431–436. [CrossRef]

Disclaimer/Publisher's Note: The statements, opinions and data contained in all publications are solely those of the individual author(s) and contributor(s) and not of MDPI and/or the editor(s). MDPI and/or the editor(s) disclaim responsibility for any injury to people or property resulting from any ideas, methods, instructions or products referred to in the content.

forests

MDPI

Article

Divergent Nitrogen, Phosphorus, and Carbon Concentrations among Growth Forms, Plant Organs, and Soils across Three Different Desert Ecosystems

Alamgir Khan, Xu-Dong Liu, Muhammad Waseem, Shi-Hua Qi, Shantwana Ghimire, Md. Mahadi Hasan and Xiang-Wen Fang *

State Key Laboratory of Grassland Agro-Ecosystems, College of Ecology, Lanzhou University, Lanzhou 730000, China; alamgir14389@gmail.com (A.K.); liuxd19@lzu.edu.cn (X.-D.L.); muhad@lzu.edu.cn (M.W.); qishh21@lzu.edu.cn (S.-H.Q.); shantwana@lzu.edu.cn (S.G.); hasanmahadikau@gmail.com (M.M.H.)
* Correspondence: fangxw@lzu.edu.cn

Abstract: Quantifying the dryland patterns of plant carbon (C), nitrogen (N), and phosphorus (P) concentrations and their stoichiometric values along environmental gradients is crucial for understanding ecological strategies. To understand the plant adaptive strategies and ecosystem nutrient concentrations across three desert ecosystems (e.g., desert, steppe desert, and temperate desert), we compiled a dataset consisting of 1295 plant species across three desert ecosystems. We assessed the element concentrations and ratios across plant growth forms, plant organs, and soils and further analysed the leaf vs. root N, P, and N:P scaling relationships. We found that the leaf N, P, and C concentrations were significantly different only from those of certain other growth forms and in certain desert ecosystems, challenging the generality of such differences. In leaves, the C concentrations were always greater than the N and P concentrations and were greater than those in soils depending on the soil chemistry and plant physiology. Thus, the element concentrations and ratios were greater in the organs than in the soils. The values in the leaf versus the root N, P, and N:P scaling relationships differed across the three desert ecosystems; for example, αN (1.16) was greater in the desert, αP (1.10) was greater in the temperate desert ecosystem, and αN:P (2.11) was greater in the desert ecosystem. The mean annual precipitation (MAP) and mean annual temperature (MAT) did not have significant effects on the leaf elemental concentrations or ratios across the desert ecosystems. This study advances our understanding of plant growth forms and organs, which support resource-related adaptive strategies that maintain the stability of desert ecosystems via divergent element concentrations and environmental conditions.

Keywords: desert ecosystems; plant organs; growth forms; elemental concentrations; climatic variables

Citation: Khan, A.; Liu, X.-D.; Waseem, M.; Qi, S.-H.; Ghimire, S.; Hasan, M.M.; Fang, X.-W. Divergent Nitrogen, Phosphorus, and Carbon Concentrations among Growth Forms, Plant Organs, and Soils across Three Different Desert Ecosystems. *Forests* 2024, 15, 607. https://doi.org/10.3390/f15040607

Academic Editor: Romà Ogaya

Received: 11 February 2024
Revised: 8 March 2024
Accepted: 11 March 2024
Published: 27 March 2024

Copyright: © 2024 by the authors. Licensee MDPI, Basel, Switzerland. This article is an open access article distributed under the terms and conditions of the Creative Commons Attribution (CC BY) license (https://creativecommons.org/licenses/by/4.0/).

1. Introduction

Nitrogen (N), phosphorus (P), and carbon (C) are essential limiting elements in terrestrial ecosystems that often constrain plant growth and ecological functions [1–3]. They are important for the function and biogeochemical C cycles of terrestrial ecosystems [4,5] and help sustain the stability of ecosystems worldwide [2,6]. However, plant responses to increased carbon dioxide (CO_2) can be constrained and modulated by other growth conditions, including insufficient nutrient levels [7]. Additionally, plant leaf C:N and C:P ratios help to assimilate C in plants under N or P accumulation [8] and thereby help increase plant growth rates. Exploring the dynamics of N and P in plant life history strategies can advance our understanding of the requirements of plant growth, which represents an important nutrient-use adaptation strategy to help fulfil plant defence strategies [9], influencing soil–plant nutrient cycling and ecosystem processes [10,11]. Hence, the N:P ratio is considered to determine nutrient limitations in plant populations and communities [5]. Therefore, plant C:N:P stoichiometry is considered to play an important role in the study of environmental changes, with an impact on ecosystem functions and nutrient limitations for biogeochemical cycling [10,12], which highlights the importance of environmental changes in plant tissue nutrients. It is important to understand how and why plant growth patterns and organs respond differently across three desert ecosystems, namely, deserts, steppe deserts, and temperate deserts, especially in terms of nutrient concentrations and climatic factors, at scales ranging from the regional scale to the global scale [13–15].

Previous studies have focused mainly on N and P absorption within a single organ type, including leaves [16,17], stems [16,18], and fine roots [19,20]. According to the leaf economics spectrum (LES), leaves are a basic part of plants that are sensitive to the external environment; therefore, plant functional traits have the potential to explain the adaptive strategies of species and their responses to environmental changes [21]. Leaves are sensitive to C fixation [22], while roots help water and nutrient transport [23] and soil C chemistry [24,25]. For example, plants allocate limited amounts of P and N to stabilize their metabolic rate by changing their allocations under water stress and soil nutrient limitations. Such coordination of different plant characteristics, also called the whole-plant economic spectrum (PES), occurs during slow and fast growth [26], suggesting that the fundamental constraints on fast and slow growth depend upon the coordination of different organs [26]. For example, if plant N and P allocation strategies also depend on the PES, it would be reasonable to assume that a scaling relationship between the N and P contents exists for all organs because nutrients and water are integrated across different organs to achieve an appropriate growth rate [27].

Thus, an increase in the distribution of N to leaves is an adaptive response to low photosynthetic rates and reduced stomatal conductance, which enhances water use efficiency in desert conditions [28,29]. Moreover, plants absorb atmospheric CO_2 through photosynthesis, by which CO_2 is converted into biomass; hence, C accumulates in plant leaves and roots [30]. Thus, plant root systems can easily increase water and nutrient uptake from the soil environment, which has a positive effect on stomatal conductance and photosynthesis [7]; therefore, an increase in CO_2 decreases stomatal conductance more in water-limited plants [7]. However, in the soil, more P is allocated to leaves with increasing soil P, and vice versa; when soil N increases, a larger proportion of N is allocated to roots [31]. The LES explains that plant species adopt ecological strategies, such as those used by conservative species (slow-growing, i.e., N-poor leaves) and acquisitive species (fast-growing, i.e., N-rich leaves), which could cause basic changes in ecosystem stability when resource availability is altered [13,32].

Studies have revealed clear geographical patterns in the compositions of leaf N and P contents in terrestrial plants at regional/global scales [13,33]. Therefore, biogeographic N and P gradients could occur because of the effect of temperature on plant physiology [13,34], while other climatic variables, such as precipitation, could also influence the changes in N and P [13] in plant organs. Dryland covers approximately 45% of the global terrestrial area and is expected to increase because of climate change [2]. Reports on drylands have shown that aridity has no negative or positive influence on the C, N, or P content in plant organs. Aridity does not affect foliar N or P concentrations in temperate desert plants [35]. In addition, it was reported that aridity does not affect the C and N concentrations in terrestrial ecosystems that are composed of grasses, N-fixing shrubs, or trees [36]. Based on the results from various reports, it remains controversial whether plant C contents decrease with increasing aridity in grasslands in China [37]; in contrast, several studies have reported that the C, N, and P concentrations in plants are positively correlated with aridity. This inconsistency in the positive and negative effects of aridity on C, N and P concentrations is possibly due to differences in geographic areas, plant types, soil types, and climatic conditions [2]. Therefore, herbaceous species require more nutrients for faster growth, and woody species need less nutrients for slow growth under dryland conditions [38]. There is a need for a better understanding of whether the shifts in these nutrient concentrations are because of the growth forms and their organs across desert ecosystems.

The log–log scaling relationships between N and P and the stoichiometric relationships between these elements in plant organs are important for understanding plant ecology and desert ecosystem patterns at the local and global levels. Earlier studies have provided evidence that the foliar N versus P scaling relationships are similar at local, regional, and global scales, although changes occur because of environmental differences; that is, the scaling exponent for the N versus P relationship has been reported to range from 0.62 to 0.78 [39,40]. Based on a wide-ranging study (i.e., approximately 2500 species), a general 2/3 scaling exponent was proposed [39] for the global foliar N versus P scaling relationship, with the relationships for plant functional types and vegetation biomes all having a similar scaling exponent. Additionally, based on 763 terrestrial plant species, a 0.82-power "law" was proposed for the global root N vs. P contents across different plant groups and ecosystems [19]. According to this sparse information, we hypothesized that the N vs. P and N:P ratio scaling relationships for roots and leaves will differ for the three different desert ecosystems, namely, desert, steppe desert, and temperate desert.

The leaf C, N, and P contents and their ratios in dryland plants have been extensively studied; however, the variations in C, N, and P across three desert ecosystems, namely, desert, steppe desert, and temperate desert ecosystems, especially among the growth forms and their organs, are unclear. Thus, we hypothesized that shifts in nutrient concentrations possibly cause plant growth forms and their organs to adapt, mediating a shift in plant adaptive strategies across the three desert ecosystems. Here, we compiled 1295 different plant growth forms (e.g., herbs, shrubs, trees, and grasses) along with soil and environmental conditions that were identified in previous studies across three desert ecosystems, namely, desert, steppe desert and temperate desert, focusing only on "'drylands". Our main objectives were (1) to quantify the changes in N, P, and C and their ratios in plant organs (leaves versus roots) and the soil and across the growth forms in three desert ecosystems; (2) to test whether differences in stoichiometric scaling exponents exist between root and leaf N, P, and N:P ratios across desert ecosystems; (3) to analyse the effects of temperature and precipitation on the leaf N, P, and C contents and their stoichiometry across desert ecosystems; and (4) to further quantify the element concentrations between deciduous and evergreen species and between N-fixing and non-N-fixing species from all the pooled data for the three different deserts.

2. Materials and Methods

2.1. Data Sources

Using the Web of Science (http://apps.webofknowledge.com, accessed on 10 February 2024) and Google Scholar (http://scholar.google.com, accessed on 10 February 2024), we surveyed studies that were published from 2005 to 2022, and additional references that were published between 1982 and 1991 were also checked for potential data (because some of the published literature used datasets covering 1982–1991). The following keywords were used: "nitrogen (N)", "phosphorus (P)", "carbon (C)", "stoichiometry", "N:P:C"; "leaves" and "roots"; and "soil". We searched for these key terms using Google Scholar and the Web of Science, each in combination with the following terms: "deserts", "mean annual temperature (MAT)" and "mean annual precipitation (MAP)", which represented climatic variables. We compiled a dataset of a total of 1295 desert plant species (Supplementary Materials) from the literature (sources and literature information) can be found in the Supplementary Materials). After compiling the dataset, we divided the desert vegetation into (1) desert (D) (496 species), (2) steppe desert (SD) (363 species), and (3) temperate desert (TD) (436 species) based on the descriptions of the whole dataset that were provided in the original articles (Supplementary Table S1). The geographic locations, climatic variables, element concentrations and stoichiometric ratios, plant growth forms, and plant species collected from different desert ecosystems were compiled from the dataset. A complete list of information is provided in Supplementary Materials. The selected sites have different climatic conditions, with MAPs varying from 20 to 502 mm and MATs ranging from -2.17 to 13.9 °C; the geographical locations cover longitudes varying from 1.6° E to 129.18° E and latitudes varying from 11.18° N to 57.32° N in China. A complete description of the geographical locations and climatic variables for each desert vegetation type is provided in Supplementary Table S1 if this information was not specified in the original research papers. Basic climatic parameter values (e.g., MAP and MAT) were obtained from the global biodiversity information facility (gbif.org, USA) based on the site locations if site climate information was not provided in the source papers.

The datasets were extracted from Supplementary Materials, and the data in the figures were extracted using GetData Graph Digitizer 2.26 (http://getdata-graph-digitizer.com, accessed on 10 February 2024) software. If the element contents (in milligrams per gram) were provided, rather than the element ratios, in table text, or figure form, we then calculated the element ratios (i.e., N:P, C:N, and C:P) for the leaves, roots, and soil. As a result, the differences in the literature were also checked for data (sources and the literature information in the Supplementary Materials [14,25,41–56]). Plant species were explored for N fixation (N-fixing and non-N-fixing) and growth (herbs, shrubs, trees, and grass), and plant species were further categorized according to their growth habit, i.e., deciduous or evergreen. The data pertain to desert vegetation and therefore cover an enormous range of environmental conditions and ecosystem types. Some publications did not provide root and soil data; in those cases, we simply used Student's t-tests (between two groups) and one-way ANOVA (more than two groups). Therefore, all the elemental concentrations and their stoichiometries are expressed as means $\pm$ standard errors (means $\pm$ SEs), with lowercase letters indicating significant differences (Supplementary Table S1). We also compared the N and P contents and their stoichiometric ratios in the leaves and roots between N-fixing and non-N-fixing plants and the leaf N, P, and C concentrations between deciduous and evergreen plants (Figures S1 and S2).

2.2. Statistical Analysis

We used SPSS version 19.0 (Armonk, NY, USA) for statistical analysis, and the figures were created with SigmaPlot 12.5. The differences in elemental concentrations and their ratios in plant organs and soils and among the growth forms across three different desert vegetation categories were tested using the least significant difference (LSD) test following one-way ANOVA (Figures 1 and 2). Furthermore, the data were log-transformed prior to analyses of the effects of climatic factors on element contents and their ratios among the different desert types. The correlations between the leaf element concentrations and environmental variables (MAP and MAT) were tested using simple linear regressions to adjust the r^2 and p-values. We also performed Student's t-tests (between two groups) to assess the differences between N-fixing and non-N-fixing plant organs and the element concentrations between deciduous and evergreen species.

To compare the numerical values of the scaling exponents (α) for the leaf and root element contents and their ratios among the three desert vegetation categories, we performed reduced major axis (RMA) regressions using the lmodel2 function of the R package LMODEL2 (Figure 3, Table 1) [1,57]. The data were log-transformed prior to performing the RMA regressions (Figure 3 and Table 1) to determine the scaling relationships of leaf vs. root N, P, and N:P for the three different desert ecosystems. The allocation of N and P concentrations among plant organs was assessed using a scaling relationship that may be described as a log–log scaling relationship, taking the form:

$$Y = \beta X^{\alpha} \tag{1}$$

Taking the log on both sides results in the following:

$$\log Y = \log \beta + \alpha \log X \tag{2}$$

where α and β represent the slopes of the Y versus X regression curves (the scaling exponent), respectively [18,29]. The scaling exponent α explains the allocation of N with respect to the allocation of P in plant organs. This simple equation provides a useful empirical model in which α can be used to predict plant and ecosystem functioning [10]. Therefore, the numerical values of the scaling exponents for the plant growth forms and ecosystems were determined using RMA regressions as explained above.

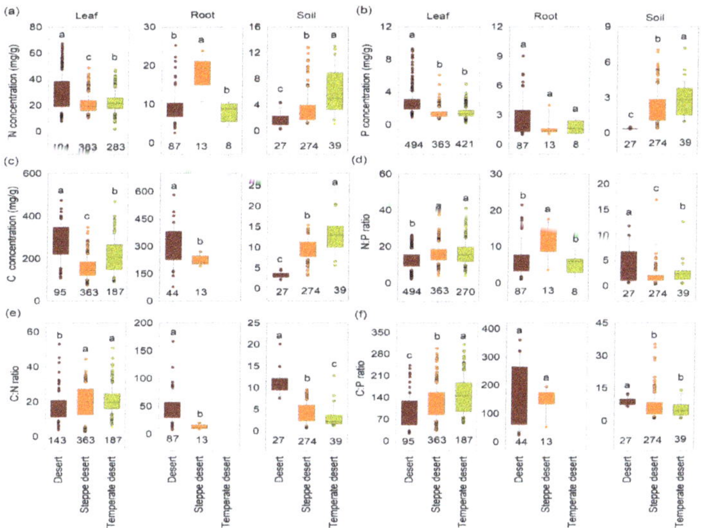

Figure 1. Comparisons of the nitrogen (N), phosphorus (P), and carbon (C) concentrations and their stoichiometric ratios in the leaves, roots, and soil of three ecosystem types (e.g., desert, steppe desert, and temperate desert). Here, (**a**–**c**) show the N concentrations, P concentrations, and C concentrations and (**d**–**f**) show the N:P, C:N, and C:P ratios, respectively, in two organs (e.g., leaves and roots) and in soil when all growth forms (e.g., herbs, shrubs, trees, and grasses) were pooled together. The different letters in each graph indicate significant differences among the three ecosystem types and differences in elements among leaves, roots, and soil (one-way ANOVA, $p < 0.05$). The numbers (n) under the box plot represent the number of species in each group.

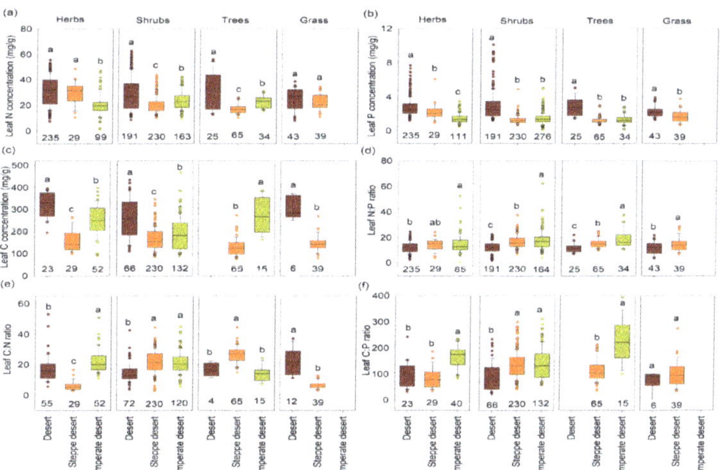

Figure 2. Comparisons of the nitrogen (N), phosphorus (P), and carbon (C) concentrations and their stoichiometric ratios in the leaves of plants with different growth forms (e.g., herbs, shrubs, trees, and grasses) among the three ecosystem types (e.g., desert, steppe desert, and temperate desert). Here, (**a**–**c**) show the N concentrations, P concentrations, and C concentrations and (**d**–**f**) show the N:P, C:N, and C:P ratios, respectively, in leaves. The different letters in each graph indicate significant differences in the elements across the three ecosystem types and among growth forms (one-way ANOVA, $p < 0.05$). The numbers (n) under the box plot indicate the number of species in each group.

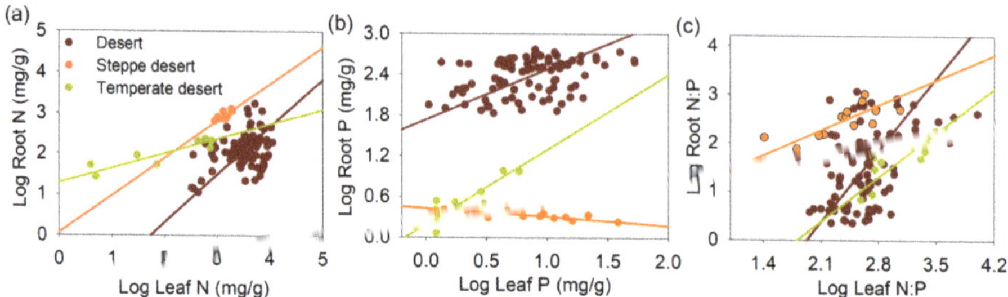

Figure 3. Scaling relationships of the nitrogen (N) and phosphorus (P) concentrations and N:P ratios between roots and leaves among the three vegetative types (e.g., desert, steppe desert, and temperate desert). The results of the reduced major axis (RMA) regressions are shown as comparisons between root and leaf N (**a**), between root and leaf P (**b**), and between root and leaf N:P ratios (**c**) across desert, steppe desert, and temperate desert ecosystems. All the results are described in Table 1.

Table 1. Summary of the reduced major axis (RMA) regressions of the log-transformed leaf nitrogen (N), phosphorus (P), and N:P ratios pooled across all species. α_{RMA} and β_{RMA} represent the scaling exponent and intercept, respectively. n represents the number of individuals. Lowercase letters indicate the differences among the three ecosystem types (shown by the exponents) based on likelihood-ratio tests at $p < 0.05$.

Ecosystem Type	n	α_{RMA} (95% CLs)	β_{RMA} (95% CLs)	r^2	p
N					
All	108	0.817 (0.677; 0.985)	−0.559 (−1.090; −0.029)	0.039	=0.041
Desert	87	1.169 (0.956; 1.424) [a]	−2.022 (−2.847; −1.197)	0.147	<0.01
Steppe desert	13	0.912 (0.572; 1.455) [a]	0.061 (−1.319; 1.441)	0.469	<0.01
Temperate desert	8	0.360 (0.237; 0.548) [b]	1.279 (0.943; 1.615)	0.812	<0.01
P					
All	108	2.102 (1.749; 2.528)	0.290 (−0.076; 0.655)	0.074	<0.01
Desert	87	0.742 (0.609; 0.905) [a]	1.735 (1.595; 1.875)	0.145	<0.01
Steppe desert	13	−0.128 (−0.195; −0.084) [b]	0.434 (0.381; 0.486)	0.578	<0.01
Temperate desert	8	1.106 (0.665; 1.838) [a]	0.190 (−0.064; 0.443)	0.718	<0.01
N:P					
All	108	1.987 (1.653; 2.389)	−3.606 (−4.589; −2.622)	0.077	<0.01
Desert	87	2.111 (1.734; 2.568) [a]	−4.060 (−5.174; −2.946)	0.161	<0.01
Steppe desert	13	0.740 (0.484; 1.130) [b]	0.713 (−0.067; 1.493)	0.568	<0.01
Temperate desert	8	1.307 (0.747; 2.289) [ab]	−2.362 (−4.674; −0.050)	0.651	=0.016

3. Results

3.1. Patterns of Element Concentrations and Stoichiometry among Plant Organs across Growth Forms and in the Soil across Three Desert Ecosystems

3.1.1. Changes in Concentrations and Ratios among Leaves, Roots, and Soils

According to our results, leaves had higher N and P concentrations and N:P ratios than roots and soils. Across the three desert ecosystems, plants in desert ecosystems had higher N, P, and C contents than those in the steppe and temperate desert ecosystems. Surprisingly, the lowest N, P, and C contents were found in the soils of the desert ecosystem, while the desert ecosystem had higher leaf and root C contents than did the other two desert ecosystems. However, the steppe and temperate desert ecosystems had higher P and C contents in their soils than did the plant organs. Among all the element contents and their ratios, the highest C storage was found in organs and soils. Furthermore, the N:P, C:N, and C:P ratios in the leaves, roots, and soils varied significantly across the three different desert ecosystems (Figure 1d). The soil N:P ratios were significantly greater in the desert ecosystem than in the other ecosystems. The soils had considerably greater

C:N ratios than the leaves and roots in the desert ecosystem. In the desert ecosystem, compared with those in the soils and roots, the C:P ratios in the leaves were significantly lower, but the differences were not significant. Finally, substantial variations were found in the ratios across organs and roots and across the desert ecosystems. However, within the plant organs and soils, almost all the leaf and root element contents and ratios significantly differed between the plant species that were associated with the three desert ecosystems (Figure 1), indicating that plant traits are also dependent on soil status and maintain different ecosystem stabilities under such conditions.

The shrubs in the different desert ecosystems had varying element concentrations and ratios as follows: desert ecosystem (means $\pm$ SEs) N (5.75 $\pm$ 0.71), P (3.20 $\pm$ 0.31), and C (13.93 $\pm$ 0.95) contents; steppe desert ecosystem N (3.31 $\pm$ 0.16), P (2.18 $\pm$ 0.09), and C (9.20 $\pm$ 0.16); and desert ecosystem N (1.21 $\pm$ 0.3), P (0.34 $\pm$ 0.01), and C (3.04 $\pm$ 0.19). Regarding the ratios, shrubs in the desert ecosystem had means $\pm$ SEs of N:P (3.49 $\pm$ 0.87), C:N (10.65 $\pm$ 0.3), and C:P (8.51 $\pm$ 0.39); those in the steppe desert had means of N:P (1.86 $\pm$ 0.1), C:N (3.99 $\pm$ 0.15), and C:P (6.37 $\pm$ 0.34); and those in the temperate desert had means of N:P (2.22 $\pm$ 0.28), C:N (3.30 $\pm$ 0.5), and C:P (4.73 $\pm$ 0.67) (Supplementary Table S2). The results also demonstrated that the means $\pm$ SEs of the element concentrations and their ratios were greater in the roots of different growth forms than in the soils associated with the growth forms across the different ecosystems.

3.1.2. Changes in Leaf Concentrations and Ratios among Growth Forms

We found that trees (woody), herbs, shrubs (woody), and grasses had higher N contents, while woody species had higher P contents than herbs and grasses in the desert ecosystem. Higher leaf N and P contents were observed in the herbaceous species than in the grasses and woody species (shrubs and trees) of the steppe desert. Additionally, lower leaf N and P contents were found in grass species than in herbaceous and woody species in the desert ecosystem. We observed that woody species exhibited higher N:P ratios than herbaceous species and grasses. We determined that the plants in the desert ecosystem had the lowest P concentrations and highest C concentrations, which demonstrates the importance of different growth forms and their organs in desert ecosystems for nutrient storage and photosynthetic activity under harsh conditions.

We found significant relationships between N-fixing and non-N-fixing plants in terms of the leaf and root N and P concentrations, except for the root and leaf N:P ratios, which were not significant (Supplementary Figure S1). For example, the leaf N and P concentrations were significantly greater for N-fixing species than for non-N-fixing species, while the leaf N:P ratios were not significantly greater for N-fixing species than for non-N-fixing species across the three desert ecosystems. The root N and P concentrations were significantly greater for N-fixing species than for non-N-fixing species, and the root N:P ratios were not significantly greater for non-N-fixing species than for N-fixing species (Supplementary Figure S1). Across all the pooled data for the three desert ecosystems, the leaf N contents were greater for deciduous species than for evergreen species, and the leaf P contents were greater for evergreen species than for deciduous species, although the difference was not significant; however, the leaf C contents were significantly greater for evergreen species than for deciduous species (Supplementary Figure S2).

3.2. N vs. P Scaling Exponents across Desert Ecosystems

According to the reduced major regression (RMA) analysis of log-transformed Equation (1) (Figure 3 and Table 1), we pooled all the data for the three desert ecosystems, where the leaf N vs. root N and leaf P vs. root P scaling exponents (αN and αP, respectively) were 0.81 (95% CIs = 0.677–0.985; Table 1) and 2.10 (95% CIs = 1.749–2.528; Table 1), respectively, and the leaf N:P versus root N:P scaling exponents (αN:P) were 1.98 (95% CIs = 1.653–2.389; Table 1).

We also observed that the αN, αP, and αN:P values varied across the three different desert ecosystems; for example, αN was greater (1.16) in the desert ecosystem than in the steppe desert (0.91) and temperate desert (0.36), whereas αP was greater (1.10) in the temperate desert than in the desert (0.74) and steppe desert (0.12), and αN:P was greater (2.11) in the desert than in the temperate desert (1.30) and steppe desert (0.74). There were significant differences among the three desert ecosystems in terms of the scaling exponents (αN, αP, and αN:P) based on the RMA regressions ($p < 0.05$). Furthermore, we observed that αN, αP, and αN:P were weakly correlated across all the pooled data; no correlations were found in the desert ecosystem ($r^2 = 0.039$ to $r^2 = 0.0747$), but a weak positive correlation was found among the three desert ecosystems ($r^2 = 0.14$ to $r^2 = 0.81$). The ratio of N to P among the three desert ecosystems was generally statistically greater than 2/3 (or 0.66), which is close to/greater than 1, whereas N was statistically less than 2/3 for temperate desert (αN = 0.36) and steppe desert (αP = 0.12) ecosystems.

3.3. Effects of Climate on Leaf Elements and Stoichiometry

The overall effects of climatic factors (mean annual precipitation (MAP) and mean annual temperature (MAT)) explained the significant variations in leaf N, P, and C contents and their ratios (Figure 4) that were found by using data for all species and plant types across the three desert ecosystems. Overall, the leaf N concentrations increased with increasing MAP from high to low in the following order: desert > steppe desert > temperate desert; leaf N concentrations decreased with increasing MAT from high to low as follows: desert > steppe desert ≈/or > temperate desert; leaf P concentrations increased with increasing MAP from high to low as follows: desert ≈ steppe desert > temperate desert; and leaf P concentrations decreased with decreasing MAT from high to low as follows: desert > steppe desert > temperate desert. The ranking of leaf C contents showed negative relationships with increased MAP in the desert, and the order in which leaf C contents increased with increasing MAP was similar to the relationship of steppe desert ≈ temperate desert. While the leaf C contents increased with increasing MAT in the desert, the ranking of leaf C content decreased with increasing MAT from high to low in the following order: temperate desert > steppe desert. Leaf N, leaf P, and temperate desert had lower slope relationships ($r^2 = 0.11$, $p < 0.43$ and $r^2 = 0.002$, $p < 0.32$) with MAT than did the other two desert ecosystems. The steppe desert leaf C content had the lowest values, with no significant difference in the relationship with MAT ($r^2 = 0.004$, $p < 0.21$) when compared to the other two ecosystems.

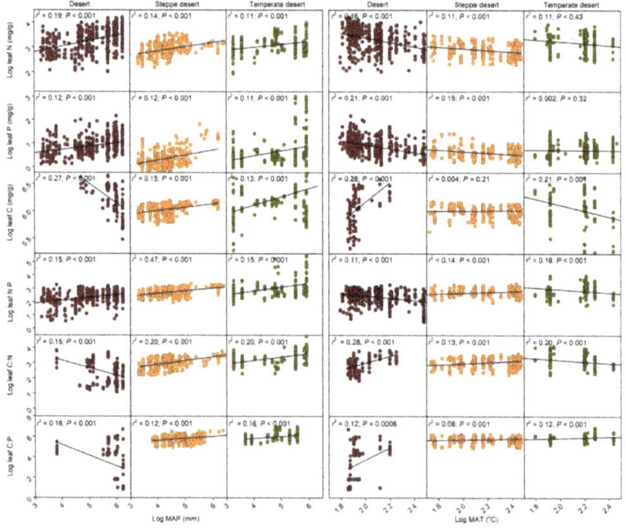

Figure 4. Comparison of the leaf nitrogen (N), phosphorus (P), and carbon (C) concentrations and stoichiometric ratios (C:N and C:P) in relation to the mean annual precipitation (mm) (MAP) and mean annual temperature (°C) (MAT) among desert, steppe desert, and temperate desert ecosystems. The fitting results (r^2 and p-values) are shown in Figure 4.

Furthermore, compared with those of MAP, the leaf N:P ratios were weakly related to MAT across the three desert ecosystems. The leaf C:N and C:P ratios showed negative relationships with MAP and positive relationships with MAT in the desert, while positive relationships of leaf C:N and C:P ratios with MAP and the leaf C:N and C:P ratios remained constant with increasing temperature in the steppe desert and temperate desert. The steppe desert showed weak relationships ($r^2 = 0.06$, $p < 0.001$) compared with the other desert ecosystems.

4. Discussion

By using a dataset of species from three desert ecosystems and a scaling approach, we found variations in stoichiometric scaling exponents across desert ecosystems for the leaf N, root N, and root P contents, reflecting the influences of plant traits and the stability strategies of species found in various geographic locations on nutrient allocation. From an ecological and evolutionary perspective, the nutrient distribution patterns across growth forms and desert ecosystems improve the understanding of species adaptation and ecosystem functioning to overcome climatic challenges. Therefore, we divided the typical desert into three different desert ecosystems, despite earlier studies having focused only on typical deserts.

4.1. Variations in Elements and Stoichiometry among Growth Forms of Organs According to Soil Factors across Three Desert Ecosystems

Our results support previous studies [27,58] showing that leaves display greater N and P concentrations and N:P ratios than roots and soils, possibly due to the metabolic requirements for photosynthetic activities under limited water and nutrient concentrations. These results support previous studies of the steppe region in Xinjiang, China [59], including Northwest China [2,11]. The results from this study showed that leaf traits and N and P patterns possibly cause basic changes across the three desert ecosystems when resource availability and climate conditions are altered, which follow the predictions of the LES. These results also support those of previous studies [29,60]. Furthermore, these results support the hypothesis that desert plants allocate more N and P to leaves than roots, supporting previous findings [59], probably because plants specifically allocate more N for photosynthetic activity [27,58]. Our results showed that the N:P ratios in leaves and roots were affected more by the N content than by the P content, so a total decrease in P in soils was found across terrestrial ecosystems in China [11,43]; this is why N:P ratios are considered good indicators of soil N and P. Furthermore, in desert ecosystems, the C contents were greater in leaves than in roots, consistent with the findings of a previous study showing that high proportions of protein and other C-rich compounds were enriched in leaves [61]. These results are inconsistent with a previous report on a steppe desert [61] in which a lower C content was observed in the leaves than in the roots. This suggested that temperate ecosystems follow a conservative strategy with investment in structure and defence [12,62]. Our analyses revealed that roots, soils, and both growth forms (Figures 1b and 2b) had the lowest P concentrations in the steppe and temperate desert ecosystems of Northeast China, which is consistent with the findings of low leaf P concentrations in previous reports, possibly reflecting low soil P concentrations in Chinese desert ecosystems [2,43]. Overall, studies have shown that the concentrations of N and especially P are lower in Chinese dryland regions. However, these harsh conditions increase the stability of certain general plant organ growth characteristics that could be applied to a wide range of habitats of various species in different desert ecosystems.

Higher N concentrations were detected in the leaves of trees (woody), herbs, shrubs, (woody) and grasses, while higher P concentrations were detected in woody species than in herbs and grasses in the desert ecosystem. The results for the N concentrations are inconsistent with the findings for trees, consistent with the findings for woody shrubs [58], and inconsistent with the findings for P concentrations, which were greater in woody species than in herbaceous plants [2,58] and grass species in desert ecosystems. Compared with herbaceous species, a positive relationship was reported between the soil and organ N contents for woody species; indeed, woody species use deep roots to acquire soil N, which is why organ N levels may reflect soil N contents [19]. Higher N and P concentrations were detected in the leaves of herbaceous species than in those of grasses or woody species in the steppe desert ecosystem, which is consistent with previous reports [2]. The N:P ratios in the leaves of woody species were greater than those in the leaves of herbaceous plants and grasses, which indicates that different species are variably susceptible to N and P limitations. Compared with those of herbaceous and woody species, lower N and P concentrations were observed for grass species in desert ecosystems, which supports the findings of a previous report [19,62] suggesting that the roots of grasses exhibit somewhat low metabolic activity and slow absorption rates [63,64]. This finding demonstrates the importance of trade-offs in determining the P and C patterns in plants in global desert ecosystems (drylands). Across the different desert ecosystems, lower P concentrations were observed with higher C concentrations; such plants showed conservative nutrient strategies and displayed lower P contents but higher C contents [62,63].

We found variations in the leaf and root N and P concentrations, which were greater in N-fixing species than in non-N-fixing species (Supplementary Materials, Figure S1). These results support the finding [65,66] that the maximum metabolic cost of N fixation is correlated with higher N concentrations [67] and, indirectly, that N-rich cellulose phosphatase leads to increased P concentrations [68]. However, the N:P ratios in leaves and roots were not significantly greater in N-fixing species than in non-N-fixing species (Supplementary Materials, Figure S1), possibly due to the close correlation of N with P contents. These results are in accordance with previous findings in terrestrial ecosystems [69] because N and P are required for plant metabolic activities [13,17]. Additionally, higher N and lower P and C contents were observed in deciduous species than in evergreen species (Supplementary Materials, Figure S2), which supports the findings of previous studies showing higher N contents but inconsistent P contents [1,5]; these results also depend on the lifespans and growth rates of species. It is suggested that the P contents were limited across the different desert ecosystems, which is consistent with previous reports in arid regions [37,70], following the "plant physiology hypothesis" [13,71]. In contrast to the findings of previous studies, especially in desert conditions, evergreen plants employ expensive strategies with high P and C gains [29], and deciduous species may follow conservative strategies and invest more in leaf formation [38]. Thus, this study expands our understanding of the effects of the dynamics of three different desert ecosystems (drylands) with limited element concentrations and ratios on nutrient allocation patterns, and the results are helpful for determining future nutrient limitations and climatic trajectories.

4.2. Variations in Leaf vs. Root N, P, and N:P Scaling Exponents in Different Desert Ecosystems

Our results revealed divergent root and leaf N and P concentrations and N:P ratio allocation strategies across three different desert ecosystems and overall, based on RMA, where αN is >1 (i.e., more N is allocated to roots than to leaves). Our compiled data also revealed that αP is <1 (i.e., less P is allocated to roots than to leaves). The changes in the numerical values of the scaling exponent that were reported among studies are expected given the differences in sample sizes [39]. The higher αN (1.16) and lower αP (0.74) in the desert ecosystem indicate that more N is allocated to roots than to leaves, while less P is allocated to roots than to leaves. Our results showed a greater metabolic rate in leaves than in roots because of the greater requirement of P for photosynthesis and C partitioning in light–dark cycling [72]; therefore, the leaf and root differences in αN and αP likely explain their divergent requirements for N and P in functional and physiological processes [73,74]. Our observations that αP < 1 in the desert ecosystem agree with the findings of [75], who demonstrated a disproportionate increase in leaf P content in relation to the increases in root P content. These results agreed with the "growth rate hypothesis", which states that plants with greater growth rates need disproportionately more P than N to support rapid protein synthesis [15], thus leading to an N versus P scaling exponent that is less than 1.0 (α < 1) [76]. Overall, the values for the three desert ecosystems (αN = 0.81, αP = 2.10, and αN:P = 1.98) differed from those reported previously, with slightly higher values than those reported at the global scale (0.76) [74] and the regional scale (0.79) [77] but lower values than those reported at the local scale (0.87) [40]. Furthermore, the higher leaf and root N:P ratios in desert (2.11) and temperate desert (1.30) ecosystems may support the soil substrate-age hypothesis, where greater precipitation in areas that are geologically older leads to lower fertility rates and areas that are more highly depleted of P than other areas [34,78]. The influence of soil substrate age on soil P and N availability depends on soil fertility in young and old soils [79], thereby enhancing N:P ratios in the leaves and roots of desert and temperate desert ecosystems.

The αN, αP, and αN:P ratios differed among the three desert ecosystems. For example, we found different values for αN (>1), αP (<1), and the αN:P ratio (>1) in desert ecosystems, indicating changes in the allocations of N and P to leaves and roots because of different combinations of stressors, e.g., high temperature and low-P soils [13]; as a result, physiological stressors became the cause of the N and P investments in leaves and roots, respectively. Therefore, N:P ratios are considered indicators of nutrient availability and limited N and P in terrestrial ecosystems [5] and are often considered to be positively correlated with stress tolerance. The relatively high N:P and C:P ratios among plants under dryland conditions reflect slow growth rates and high stress tolerance, possibly because these plants are limited by P compared with other nutrients and drought conditions. In agreement with our results, plants under desert conditions are possibly more limited by P than by C and N with greater aridity [2,19]. Although the N and P concentrations are low in plant metabolic organs (e.g., leaves and roots) in desert ecosystems, they are greater than those in steppe desert ecosystems, which is consistent and inconsistent, respectively, with the finding that both plant N and P contents are low in temperate deserts [35].

In some cases, the N and P concentrations and N:P ratios differed among plant organs, and the metabolic organs (e.g., leaves and roots) had increased nutrient concentrations and N:P ratios compared with structural organs (e.g., stems and coarse roots), which has been reported by previous researchers [27,80]. However, the possible reason that the leaves and roots are involved in the plant photosynthesis, respiration, and nutrient uptake processes is the need for greater nutrient concentrations to sustain plant physiological activity [70]. Consistent with the prediction of PES theory, our results revealed a scaling relationship between N and P in plant organs (e.g., leaves, stems, and roots). This suggested that the nutrient-use strategies of all the organs (surprisingly) varied across the growth forms in the three different desert ecosystems. Therefore, the results of both the present study and of previous studies demonstrate that the PES is applicable for water-related as well as C- and nutrient (i.e., N and P)-related traits across scales [26,27]. Our results indicate that similar and dissimilar isometric relationships between N and P occur among plant leaves and root organs, although previously reported observations showed that N increases rapidly with P in particular organs [29,39]. Thus, heterogeneous environmental conditions and various trade-off strategies caused different leaf and root N and P contents and different leaf and root N:P ratios across the three desert ecosystems.

4.3. Effects of Environmental Conditions on Element Concentrations and Stoichiometry

Shifts in species ranges and abundances are expected to occur with climate change. We hypothesized that along with nutrient concentrations, MAP and MAT are also key drivers of the shifts in and stabilization of elemental concentrations and their ratios in desert ecosystems; for example, climate variables and nutrient limitations in soils influence the biogeographic patterns of root and leaf N and P allocations [13,74]. However, due to the lack of data on the C, N, and P concentrations and their ratios in the roots and soils in our dataset, our understanding of the N and P relationships in the roots and soils with climatic factors is still severely limited. We found that the leaf N and P contents increased with MAP and that the leaf N and P contents decreased with increasing MAT (except in the temperate desert ecosystem, Figure 4, r^2 = 0.002, p = 0.32). However, because the temperatures in temperate desert ecosystems (40–265 °C) are lower than those in the other two desert ecosystems, there is no support for the hypothesis that N and P contents increase with MAT; for example, low temperatures suppress biogeochemical processes [13], and interactions with many other processes may limit the degree to which this increase occurs, especially at all but the lowest temperatures [81]. This result showed that biogeographic variations, along with climate variables, had different effects on the total nutrient contents across the desert ecosystems.

According to our results, the three desert ecosystems were limited by P, and these findings were consistent with the low precipitation and high evaporation present in the studied areas [82]. This suggested that these desert plants, through morphological and physiological adjustments, decreased their metabolism rates and resource demands [59] to cope with harsh desert conditions. This result is consistent with the hypothesis that leaf N contents decrease/increase with decreasing/increasing MAP [71] because of the higher leaf N contents of plants in desert ecosystems. These results are widely reported to indicate that plants increase their leaf N contents to increase light availability while decreasing stomatal conductance and transpiration rates [29,59]. Similarly, roots play a key role in nutrient cycling in desert habitats, but leaves are considered to be more important than roots and stems for supporting higher nutrient levels [35,60]. Our analyses showed that in all the studied desert plants, the leaves maintained greater N and P concentrations than did the roots and soil.

With increasing MAP in desert species, the leaf C contents decreased with increasing C:N and C:P ratios, which indicates that the plant leaf C contents in desert ecosystems are highly correlated with MAP in steppe desert and temperate desert ecosystems. At the leaf level, species displayed different responses to climatic variables in the different desert ecosystems. The finding that the C:N and C:P ratios increased (and the N and P contents decreased) with increasing MAT is consistent with previous plant physiology hypotheses that C:N and C:P ratios increase with increasing temperature (T) [13,83]. Surprisingly, the lowest P concentration was detected in roots, followed by leaves, but the P concentration was greater in the soils of the temperate desert ecosystem (Figure 1). Our results are consistent with those of previous studies [13,43]; that is, we did not observe a decrease in leaf N concentrations with decreasing MAT (Figure 4). However, MAT has a significant effect on leaf P concentrations in desert ecosystems, and the leaf P patterns are sensitive to temperature changes [13]; thus, temperature causes a shift in decreasing leaf P contents [27]. Our analyses showed that the leaf N:P ratios increased with MAP across the three desert ecosystems and that this increase was driven by leaf P concentrations, consistent with previous results [84].

The desert species had relatively high leaf N contents, which is explained by the adaptation of these plants to desert conditions and the accumulation of N compounds in their leaves to maintain the water balance [71]. The results demonstrated that roots are responsible for nutrient absorption and transport to leaves, resulting in the maintenance of metabolic activity, which is limited among desert plants by the nutrient contents [59]. This finding agrees with the hypothesis that N limits photosynthesis more than P [85]. Geographic cycles, climatic variables, and divergent N, P, and C contents and their ratios in plant organs and soil maintain plant distributions and ecosystem stability across different desert ecosystems. These results suggest that MAT and MAP play key roles in allocation strategies in leaves across desert ecosystems. Furthermore, the results indicate that leaf and root N, P, and N:P ratios play key roles in explaining variations in the numerical values of scaling exponents, which explains the shift in nutrients in organs of plant growth forms across the three deserts. The three desert ecosystems investigated in this study result from the collective influence of several drivers that led to a shift in nutrient allocation strategies across space. Through different adaptations to limited nutrients in the soil and environmental conditions, different desert plants have developed special survival strategies that help to determine species distributions and desert ecosystem stability. These results have important implications for understanding how plant species and their organs maintain nutrient allocation strategies across different geographic scales in deserts and highlight the resource and photosynthetic capacities of plant roots and leaves in response to environmental variables.

5. Conclusions

To our knowledge, this is the first study to address the nutrient allocation strategies in plant organs and soil across three different desert ecosystems rather than focusing only on "drylands". For instance, the plant species extracted from previous studies showed that even by using only three elements, namely, the N, P, and C concentrations in plants and soil, deserts can be differentiated from steppe desert and temperate desert ecosystems and characterized by the dependency of these elements on the organs, the soil, and the MAP and MAT. This study further revealed that the values of root and leaf N vs. P and N:P ratio scaling relationships vary for species growing across three desert ecosystems, which helps us to understand plant growth and ecosystem dynamics. This phenomenon needs to be tested in more diverse desert ecosystems where plant nutrient cycling is sensitive to limited soil nutrients, increased CO_2 concentrations, and climatic factors.

Supplementary Materials: The following supporting information can be downloaded at: https://www.mdpi.com/article/10.3390/f15040607/s1, Figure S1: Changes in nitrogen (N), phosphorus (P) concentrations and their stoichiometric ratios in the leaves and roots between N-fixing and Non-N-fixing. Here, (a)–(c) show the N concentrations, P concentrations and N:P ratios in leaves between N-fixing and non-N-fixing, and (d)–(f) show the N and P concentrations and N:P ratios in roots between N-fixing and non-N-fixing. The different letters in each graph indicates significant differences between N-fixing and Non-N-fixing in leaves and roots. Significant difference ($p < 0.05$) was calculated by Student's t-test. The numbers (n) under the box-plot are the number of species for each group; Figure S2: Changes in nitrogen (N), phosphorus (P) and carbon (C) concentrations in the leaves between deciduous and evergreen species when all pooled data were together. Here, (a) show the comparison of N concentrations, (b) P concentrations, and (c) the C concentrations. Different letters in each graph indicates the significant differences between deciduous and evergreen species (Student's *t*-test, $p < 0.05$). The numbers (n) under the box-plot are the number of species for each group; Table S1: Complete the basic information for the three different desert ecosystem and related climatic variables; Table S2: Comparison of the three different ecosystem types for different elements in root and soil for growth forms. References [14,25,41–56] are cited in the Supplementary Materials.

Author Contributions: A.K. and X.-W.F. designed the research ideas; A.K., M.W. and S.G. collected the data; A.K., X.-D.L., M.W., S.-H.Q., M.M.H. and S.G. analysed the data; A.K. led the writing of first draft of this manuscript. All authors substantially contributed to drafts and additional text. All authors have read and agreed to the published version of the manuscript.

Funding: This research was partially supported by the National Natural Science Foundation of China (Nos 32325036, 32171491), Gansu Science and Technology Major Project (22ZD6FA052, 22ZD6NA007, 23JRRA1037, 22JR5RA531) and Excellent Doctoral Project (23JRRA1117), and top leading talents in Gansu Province.

Data Availability Statement: The data presented in this study are available upon request from the corresponding authors. The data are not publicly available due to confidentiality.

Acknowledgments: We thank the Core facility of the School of Life Sciences, Lanzhou University.

Conflicts of Interest: The authors declare that they have no known competing financial interests or personal relationships that could have appeared to influence the work reported in this paper.

References

1. Wang, Z.; Gong, H.; Sardans, J.; Zhou, Q.; Deng, J.; Niklas, K.J.; Hu, H.; Li, Y.; Ma, Z.; Mipam, T.D.; et al. Divergent nitrogen and phosphorus allocation strategies in terrestrial plant leaves and fine roots: A global meta-analysis. *J. Ecol.* **2022**, *110*, 2745–2758. [CrossRef]
2. Xiong, J.; Dong, L.; Lu, J.; Hu, W.; Gong, H.; Xie, S.; Zhao, D.; Zhang, Y.; Wang, X.; Deng, Y.; et al. Variation in plant carbon, nitrogen and phosphorus contents across the drylands of China. *Funct. Ecol.* **2022**, *36*, 174–186. [CrossRef]

3. Penuelas, J.; Janssens, I.A.; Ciais, P.; Obersteiner, M.; Sardans, J. Anthropogenic global shifts in biospheric N and P concentrations and ratios and their impacts on biodiversity, ecosystem productivity, food security, and human health. *Glob. Chang. Biol.* **2020**, *26*, 1962–1985. [CrossRef] [PubMed]
4. Elser, J.J.; Bracken, M.E.; Cleland, E.E.; Gruner, D.S.; Harpole, W.S.; Hillebrand, H.; Ngai, J.T.; Seabloom, E.W.; Shurin, J.B.; Smith, J.E. Global analysis of nitrogen and phosphorus limitation of primary producers in freshwater, marine and terrestrial ecosystems. *Ecol. Lett.* **2007**, *10*, 1135–1142. [CrossRef] [PubMed]
5. Güsewell, S. N : P ratios in terrestrial plants: Variation and functional significance. *New Phytol.* **2004**, *164*, 243–266. [CrossRef]
6. Chapin, F.S., III. The Mineral Nutrition of Wild Plants. *Annu. Rev. Ecol. Syst.* **1980**, *11*, 233–260. [CrossRef]
7. Ofori-Amanfo, K.K.; Klem, K.; Veselá, B.; Holub, P.; Agyei, T.; Juráň, S.; Grace, J.; Marek, M.V.; Urban, O. The effect of elevated CO_2 on photosynthesis is modulated by nitrogen supply and reduced water availability in Picea abies. *Tree Physiol.* **2023**, *43*, 925–937. [CrossRef] [PubMed]
8. Li, L.; Liu, B.; Gao, X.; Li, X.; Li, C. Nitrogen and phosphorus addition differentially affect plant ecological stoichiometry in desert grassland. *Sci. Rep.* **2019**, *9*, 18673. [CrossRef] [PubMed]
9. Vrede, T.; Dobberfuhl, D.R.; Kooijman, S.A.L.M.; Elser, J.J. Fundamental connections among organism C:N:P stoichiometry, macromolecular composition, and growth. *Ecology* **2004**, *85*, 1217–1229. [CrossRef]
10. Elser, J.J.; Fagan, W.F.; Kerkhoff, A.J.; Swenson, N.G.; Enquist, B.J. Biological stoichiometry of plant production: Metabolism, scaling and ecological response to global change. *New Phytol.* **2010**, *186*, 593–608. [CrossRef]
11. Tang, Z.; Xu, W.; Zhou, G.; Bai, Y.; Li, J.; Tang, X.; Chen, D.; Liu, Q.; Ma, W.; Xiong, G.; et al. Patterns of plant carbon, nitrogen, and phosphorus concentration in relation to productivity in China's terrestrial ecosystems. *Proc. Natl. Acad. Sci. USA* **2018**, *115*, 4033–4038. [CrossRef]
12. Hu, Y.K.; Liu, X.Y.; He, N.P.; Pan, X.; Long, S.Y.; Li, W.; Zhang, M.Y.; Cui, L.J. Global patterns in leaf stoichiometry across coastal wetlands. *Glob. Ecol. Biogeogr.* **2021**, *30*, 852–869. [CrossRef]
13. Reich, P.B.; Oleksyn, J. Global patterns of plant leaf N and P in relation to temperature and latitude. *Proc. Natl. Acad. Sci. USA* **2004**, *101*, 11001–11006. [CrossRef] [PubMed]
14. He, J.S.; Wang, L.; Flynn, D.F.; Wang, X.; Ma, W.; Fang, J. Leaf nitrogen: Phosphorus stoichiometry across Chinese grassland biomes. *Oecologia* **2008**, *155*, 301–310. [CrossRef]
15. Elser, J.; Sterner, R.; Gorokhova, E.; Fagan, W.; Markow, T.; Cotner, J.; Harrison, J.; Hobbie, S.; Odell, G.; Weider, L. Biological stoichiometry from genes to ecosystems. *Ecol. Lett.* **2000**, *3*, 540–550. [CrossRef]
16. Niklas, K.J.; Cobb, E.D. N, P, and C stoichiometry of *Eranthis hyemalis* (Ranunculaceae) and the allometry of plant growth. *Am. J. Bot.* **2005**, *92*, 1256–1263. [CrossRef] [PubMed]
17. Tian, D.; Yan, Z.; Niklas, K.J.; Han, W.; Kattge, J.; Reich, P.B.; Luo, Y.; Chen, Y.; Tang, Z.; Hu, H. Global leaf nitrogen and phosphorus stoichiometry and their scaling exponent. *Natl. Sci. Rev.* **2018**, *5*, 728–739. [CrossRef]
18. Wang, Z.; Niklas, K.J.; Ma, Z.; Jiang, D.; Deng, J. The 2/3 scaling of twig nitrogen to phosphorus in woody plants. *For. Ecosyst.* **2022**, *9*, 100049. [CrossRef]
19. Wang, Z.; Yu, K.; Lv, S.; Niklas, K.J.; Mipam, T.D.; Crowther, T.W.; Umaña, M.N.; Zhao, Q.; Huang, H.; Reich, P.B. The scaling of fine root nitrogen versus phosphorus in terrestrial plants: A global synthesis. *Funct. Ecol.* **2019**, *33*, 2081–2094. [CrossRef]
20. Zhao, M.; Luo, Y.; Chen, Y.; Shen, H.; Zhao, X.; Fang, J.; Hu, H. Varied nitrogen versus phosphorus scaling exponents among shrub organs across eastern China. *Ecol. Indic.* **2021**, *121*, 107024. [CrossRef]
21. Westoby, M.; Wright, I.J. Land-plant ecology on the basis of functional traits. *Trends Ecol. Evol.* **2006**, *21*, 261–268. [CrossRef] [PubMed]
22. Pallardy, S.G. *Physiology of Woody Plants*; Academic Press: Cambridge, MA, USA, 2010.
23. Laliberté, E. Below-ground frontiers in trait-based plant ecology. *New Phytol.* **2017**, *213*, 1597–1603. [CrossRef] [PubMed]
24. McCormack, M.L.; Dickie, I.A.; Eissenstat, D.M.; Fahey, T.J.; Fernandez, C.W.; Guo, D.; Helmisaari, H.S.; Hobbie, E.A.; Iversen, C.M.; Jackson, R.B. Redefining fine roots improves understanding of below-ground contributions to terrestrial biosphere processes. *New Phytol.* **2015**, *207*, 505–518. [CrossRef]
25. Akram, M.A.; Wang, X.; Hu, W.; Xiong, J.; Zhang, Y.; Deng, Y.; Ran, J.; Deng, J. Convergent variations in the leaf traits of desert plants. *Plants* **2020**, *9*, 990. [CrossRef] [PubMed]
26. Reich, P.B. The world-wide 'fast–slow' plant economics spectrum: A traits manifesto. *J. Ecol.* **2014**, *102*, 275–301. [CrossRef]
27. Chen, X.; Wang, M.; Li, M.; Sun, J.; Lyu, M.; Zhong, Q.; Cheng, D. Convergent nitrogen–phosphorus scaling relationships in different plant organs along an elevational gradient. *AoB Plants* **2020**, *12*, plaa021. [CrossRef]
28. Palmroth, S.; Katul, G.G.; Maier, C.A.; Ward, E.; Manzoni, S.; Vico, G. On the complementary relationship between marginal nitrogen and water-use efficiencies among Pinus taeda leaves grown under ambient and CO_2-enriched environments. *Ann. Bot.* **2013**, *111*, 467–477. [CrossRef]
29. Wright, I.J.; Reich, P.B.; Westoby, M.; Ackerly, D.D.; Baruch, Z.; Bongers, F.; Cavender-Bares, J.; Chapin, T.; Cornelissen, J.H.C.; Diemer, M.; et al. The worldwide leaf economics spectrum. *Nature* **2004**, *428*, 821–827. [CrossRef]
30. Deng, W.; Liu, X.; Hu, H.; Liu, Z.; Ge, Z.; Xia, C.; Wang, P.; Liang, L.; Zhu, Z.; Sun, Y. Carbon storages and densities of different ecosystems in Changzhou city, China: Subtropical forests, urban green spaces, and wetlands. *Forests* **2024**, *15*, 303. [CrossRef]
31. Ågren, G.I. Stoichiometry and nutrition of plant growth in natural communities. *Annu. Rev. Ecol. Evol. Syst.* **2008**, *39*, 153–170. [CrossRef]

32. Wei, L.; Hong, H.; Bee, M.Y.; Wu, Y.; Ndayambaje, P.; Yan, C.; Kao, S.-J.; Chee, P.S.; Wang, Y. Different adaptive strategies of three mangrove species to nutrient enrichment. *Plant Ecol.* **2022**, *223*, 1093–1102. [CrossRef]
33. Ordoñez, J.C.; Bodegom, P.M.v.; Witte, J.-P.M.; Wright, I.J.; Reich, P.B.; Aerts, R. A global study of relationships between leaf traits, climate and soil measures of nutrient fertility. *Glob. Ecol. Biogeogr.* **2009**, *18*, 137–149. [CrossRef]
34. Chadwick, O.A.; Derry, L.A.; Vitousek, P.M.; Huebert, B.J.; Hedin, L.O. Changing sources of nutrients during four million years of ecosystem development. *Nature* **1999**, *397*, 491–497. [CrossRef]
35. He, M.; Dijkstra, F.A.; Zhang, K.; Li, X.; Tan, H.; Gao, Y.; Li, G. Leaf nitrogen and phosphorus of temperate desert plants in response to climate and soil nutrient availability. *Sci. Rep.* **2014**, *4*, 6932. [CrossRef] [PubMed]
36. Delgado-Baquerizo, M.; Eldridge, D.J.; Maestre, F.T.; Ochoa, V.; Gozalo, B.; Reich, P.B.; Singh, B.K. Aridity decouples C: N: P stoichiometry across multiple trophic levels in terrestrial ecosystems. *Ecosystems* **2018**, *21*, 459–468. [CrossRef]
37. Luo, W.; Li, M.H.; Sardans, J.; Lü, X.T.; Wang, C.; Peñuelas, J.; Wang, Z.; Han, X.G.; Jiang, Y. Carbon and nitrogen allocation shifts in plants and soils along aridity and fertility gradients in grasslands of China. *Ecol. Evol.* **2017**, *7*, 6927–6934. [CrossRef]
38. Poorter, L.; McDonald, I.; Alarcón, A.; Fichtler, E.; Licona, J.C.; Peña-Claros, M.; Sterck, F.; Villegas, Z.; Sass-Klaassen, U. The importance of wood traits and hydraulic conductance for the performance and life history strategies of 42 rainforest tree species. *New Phytol.* **2010**, *185*, 481–492. [CrossRef] [PubMed]
39. Reich, P.B.; Oleksyn, J.; Wright, I.J.; Niklas, K.J.; Hedin, L.; Elser, J.J. Evidence of a general 2/3-power law of scaling leaf nitrogen to phosphorus among major plant groups and biomes. *Proc. R. Soc. B Biol. Sci.* **2010**, *277*, 877–883. [CrossRef] [PubMed]
40. Zhao, N.; Yu, G.; He, N.; Xia, F.; Wang, Q.; Wang, R.; Xu, Z.; Jia, Y. Invariant allometric scaling of nitrogen and phosphorus in leaves, stems, and fine roots of woody plants along an altitudinal gradient. *J. Plant Res.* **2016**, *129*, 647–657. [CrossRef]
41. Chen, Z.; Huang, D.; Zhang, H. The characteristics of element chemistry of 122 plants on the Xilin river valley, Inner Mongolia. *Res. Grassl. Ecosyst.* **1985**, *1*, 112–131.
42. Mo, D.-L.; Wu, J.-X. A Study on the Characteristics of the Chemical Composition and the Interrelationship Between the Elements in the Plants of 86 Species in Hainan Island. *Chin. J. Plant Ecol.* **1988**, *12*, 51–62.
43. Han, W.; Fang, J.; Guo, D.; Zhang, Y. Leaf nitrogen and phosphorus stoichiometry across 753 terrestrial plant species in China. *New Phytol.* **2005**, *168*, 377–385. [CrossRef] [PubMed]
44. Hou, X. *Chinese Vegetable Geography and Chemical Elements: Analyses of the Dominant Plant Species*; Science Press: Beijing, China, 1982.
45. Huang, J.; Wang, P.; Niu, Y.; Yu, H.; Ma, F.; Xiao, G.; Xu, X. Changes in C: N: P stoichiometry modify N and P conservation strategies of a desert steppe species Glycyrrhiza uralensis. *Sci. Rep.* **2018**, *8*, 12668. [CrossRef] [PubMed]
46. Huang, J.-H.; Chen, L.-Z. A study of chemical contents in a mixed shrubland near Baihuashan mountain in Beijing. *Chin. J. Plant Ecol.* **1991**, *15*, 224.
47. Jiang, X.; Jia, X.; Gao, S.; Jiang, Y.; Wei, N.; Han, C.; Zha, T.; Liu, P.; Tian, Y.; Qin, S. Plant nutrient contents rather than physical traits are coordinated between leaves and roots in a desert shrubland. *Front. Plant Sci.* **2021**, *12*, 734775. [CrossRef] [PubMed]
48. Luo, Y.; Lian, C.; Gong, L.; Mo, C. Leaf Stoichiometry of Halophyte Shrubs and Its Relationship with Soil Factors in the Xinjiang Desert. *Forests* **2022**, *13*, 2121. [CrossRef]
49. Nurbolat, S.; Guanghui, L.; Lamei, J.; Lei, Z. Convergent Variation in the Leaf Traits of Desert Plants in the Ebinur Lake Basin. *Front. Environ. Sci.* **2022**, *10*, 927572. [CrossRef]
50. Tao, Y.; Qiu, D.; Gong, Y.-M.; Liu, H.-L.; Zhang, J.; Yin, B.-F.; Lu, H.Y.; Zhou, X.B.B.; Zhang, Y.M. Leaf-root-soil N: P stoichiometry of ephemeral plants in a temperate desert in Central Asia. *J. Plant Res.* **2022**, *135*, 55–67. [CrossRef] [PubMed]
51. Tao, Y.; Zhou, X.-B.; Zhang, Y.M.; Yin, B.F.; Li, Y.G.; Zang, Y.-X. Foliar C: N: P stoichiometric traits of herbaceous synusia and the spatial patterns and drivers in a temperate desert in Central Asia. *Glob. Ecol. Conserv.* **2021**, *28*, e01620. [CrossRef]
52. Li, J.-Q.; Gong, W.-G. Characteristic Analysis of Nutrient Contents of Major Tree Species in Northeast China. *Chin. J. Plant Ecol.* **1991**, *15*, 380–385.
53. Yin, H.; Zheng, H.; Zhang, B.; Tariq, A.; Lv, G.; Zeng, F.; Graciano, C. Stoichiometry of C: N: P in the roots of alhagi sparsifolia is more sensitive to soil nutrients than aboveground organs. *Front. Plant Sci.* **2021**, *12*, 698961. [CrossRef]
54. Zhang, K.; Li, M.; Su, Y.; Yang, R. Stoichiometry of leaf carbon, nitrogen, and phosphorus along a geographic, climatic, and soil gradients in temperate desert of Hexi Corridor, northwest China. *J. Plant Ecol.* **2020**, *13*, 114–121. [CrossRef]
55. Zhang, X.; Guan, T.; Zhou, J.; Cai, W.; Gao, N.; Du, H.; Jiang, L.; Lai, L.; Zheng, Y. Community characteristics and leaf stoichiometric traits of desert ecosystems regulated by precipitation and soil in an arid area of China. *Int. J. Environ. Res. Public Health* **2018**, *15*, 109. [CrossRef] [PubMed]
56. Zhu, Z.X.; Harris, A.; Nizamani, M.M.; Thornhill, A.H.; Scherson, R.A.; Wang, H.F. Spatial phylogenetics of the native woody plant species in Hainan, China. *Ecol. Evol.* **2021**, *11*, 2100–2109. [CrossRef] [PubMed]
57. Legendre, P. *lmodel2: Model II Regression*, R Package v.2018; Elsevier Science BV: Amsterdam, The Netherlands, 2018.
58. Kerkhoff, A.J.; Fagan, W.F.; Elser, J.J.; Enquist, B.J. Phylogenetic and growth form variation in the scaling of nitrogen and phosphorus in the seed plants. *Am. Nat.* **2006**, *168*, E103–E122. [CrossRef] [PubMed]
59. Luo, Y.; Peng, Q.; Li, K.; Gong, Y.; Liu, Y.; Han, W. Patterns of nitrogen and phosphorus stoichiometry among leaf, stem and root of desert plants and responses to climate and soil factors in Xinjiang, China. *Catena* **2021**, *199*, 105100. [CrossRef]
60. Khan, A.; Yan, L.; Hasan, M.M.; Wang, W.; Xu, K.; Zou, G.; Liu, X.-D.; Fang, X.-W. Leaf traits and leaf nitrogen shift photosynthesis adaptive strategies among functional groups and diverse biomes. *Ecol. Indic.* **2022**, *141*, 109098. [CrossRef]

61. Ma, S.; He, F.; Tian, D.; Zou, D.; Yan, Z.; Yang, Y.; Zhou, T.; Huang, K.; Shen, H.; Fang, J. Variations and determinants of carbon content in plants: A global synthesis. *Biogeosciences* **2018**, *15*, 693–702. [CrossRef]
62. Freschet, G.T.; Valverde-Barrantes, O.J.; Tucker, C.M.; Craine, J.M.; McCormack, M.L.; Violle, C.; Fort, F.; Blackwood, C.B.; Urban-Mead, K.R.; Iversen, C.M. Climate, soil and plant functional types as drivers of global fine-root trait variation. *J. Ecol.* **2017**, *105*, 1182–1196. [CrossRef]
63. Freschet, G.T.; Kichenin, E.; Wardle, D.A. Explaining within-community variation in plant biomass allocation: A balance between organ biomass and morphology above vs below ground? *J. Veg. Sci.* **2015**, *26*, 431–440. [CrossRef]
64. Roumet, C.; Birouste, M.; Picon-Cochard, C.; Ghestem, M.; Osman, N.; Vrignon-Brenas, S.; Cao, K.F.; Stokes, A. Root structure-function relationships in 74 species: Evidence of a root economics spectrum related to carbon economy. *New Phytol.* **2016**, *210*, 815–826. [CrossRef]
65. Wang, Z.; Huang, H.; Yao, B.; Deng, J.; Ma, Z.; Niklas, K.J. Divergent scaling of fine-root nitrogen and phosphorus in different root diameters, orders and functional categories: A meta-analysis. *For. Ecol. Manag.* **2021**, *495*, 119384. [CrossRef]
66. Ning, Z.-Y.; Li, Y.-L.; Yang, H.-L.; Sun, D.-C.; Bi, J.-D. Carbon, nitrogen and phosphorus stoichiometry in leaves and fine roots of dominant plants in Horqin Sandy Land. *Chin. J. Plant Ecol.* **2017**, *41*, 1069.
67. Wardle, D.; Greenfield, L. Release of mineral nitrogen from plant root nodules. *Soil Biol. Biochem.* **1991**, *23*, 827–832. [CrossRef]
68. Venterink, H.O. Legumes have a higher root phosphatase activity than other forbs, particularly under low inorganic P and N supply. *Plant Soil* **2011**, *347*, 137–146. [CrossRef]
69. McGroddy, M.E.; Daufresne, T.; Hedin, L.O. Scaling of C: N: P stoichiometry in forests worldwide: Implications of terrestrial redfield-type ratios. *Ecology* **2004**, *85*, 2390–2401. [CrossRef]
70. Liu, G.; Ye, X.; Huang, Z.; Dong, M.; Cornelissen, J.H. Leaf and root nutrient concentrations and stoichiometry along aridity and soil fertility gradients. *J. Veg. Sci.* **2019**, *30*, 291–300. [CrossRef]
71. Yang, X.; Chi, X.; Ji, C.; Liu, H.; Ma, W.; Mohhammat, A.; Shi, Z.; Wang, X.; Yu, S.; Yue, M. Variations of leaf N and P concentrations in shrubland biomes across northern China: Phylogeny, climate, and soil. *Biogeosciences* **2016**, *13*, 4429–4438. [CrossRef]
72. Lambers, H.; Chapin, F.S.; Pons, T.L. *Plant Physiological Ecology*; Springer: Berlin/Heidelberg, Germany, 2008; Volume 2.
73. Sardans, J.; Grau, O.; Chen, H.Y.; Janssens, I.A.; Ciais, P.; Piao, S.; Peñuelas, J. Changes in nutrient concentrations of leaves and roots in response to global change factors. *Glob. Chang. Biol.* **2017**, *23*, 3849–3856. [CrossRef]
74. Yuan, Z.; Chen, H.Y.; Reich, P.B. Global-scale latitudinal patterns of plant fine-root nitrogen and phosphorus. *Nat. Commun.* **2011**, *2*, 344. [CrossRef]
75. Zhao, H.; He, N.; Xu, L.; Zhang, X.; Wang, Q.; Wang, B.; Yu, G. Variation in the nitrogen concentration of the leaf, branch, trunk, and root in vegetation in China. *Ecol. Indic.* **2019**, *96*, 496–504. [CrossRef]
76. Sterner, R.W.; Elser, J.J. *Ecological Stoichiometry: The Biology of Elements from Molecules to the Biosphere*; Princeton University Press: Princeton, NJ, USA, 2003.
77. Geng, Y.; Wang, L.; Jin, D.; Liu, H.; He, J.-S. Alpine climate alters the relationships between leaf and root morphological traits but not chemical traits. *Oecologia* **2014**, *175*, 445–455. [CrossRef] [PubMed]
78. Vitousek, P.M.; Porder, S.; Houlton, B.Z.; Chadwick, O.A. Terrestrial phosphorus limitation: Mechanisms, implications, and nitrogen-phosphorus interactions. *Ecol. Appl.* **2010**, *20*, 5–15. [CrossRef] [PubMed]
79. Wardle, D.A.; Walker, L.R.; Bardgett, R.D. Ecosystem properties and forest decline in contrasting long-term chronosequences. *Science* **2004**, *305*, 509–513. [CrossRef] [PubMed]
80. Zhang, J.; He, N.; Liu, C.; Xu, L.; Yu, Q.; Yu, G. Allocation strategies for nitrogen and phosphorus in forest plants. *Oikos* **2018**, *127*, 1506–1514. [CrossRef]
81. Hobbie, S.E.; Nadelhoffer, K.J.; Högberg, P. A synthesis: The role of nutrients as constraints on carbon balances in boreal and arctic regions. *Plant Soil* **2002**, *242*, 163–170. [CrossRef]
82. Yang, X.; Tang, Z.; Ji, C.; Liu, H.; Ma, W.; Mohhamot, A.; Shi, Z.; Sun, W.; Wang, T.; Wang, X. Scaling of nitrogen and phosphorus across plant organs in shrubland biomes across Northern China. *Sci. Rep.* **2014**, *4*, 5448. [CrossRef] [PubMed]
83. Oleksyn, J.; Reich, P.; Zytkowiak, R.; Karolewski, P.; Tjoelker, M. Nutrient conservation increases with latitude of origin in European Pinus sylvestris populations. *Oecologia* **2003**, *136*, 220–235. [CrossRef]
84. Zheng, S.; Shangguan, Z. Spatial patterns of leaf nutrient traits of the plants in the Loess Plateau of China. *Trees* **2007**, *21*, 357–370. [CrossRef]
85. Reich, P.; Walters, M.; Ellsworth, D.; Uhl, C. Photosynthesis-nitrogen relations in Amazonian tree species: I. Patterns Among Species Communities. *Oecologia* **1994**, *97*, 62–72. [CrossRef]

Disclaimer/Publisher's Note: The statements, opinions and data contained in all publications are solely those of the individual author(s) and contributor(s) and not of MDPI and/or the editor(s). MDPI and/or the editor(s) disclaim responsibility for any injury to people or property resulting from any ideas, methods, instructions or products referred to in the content.

Article

Patterns in Tree Cavities (Hollows) in Euphrates Poplar (*Populus euphratica*, Salicaceae) along the Tarim River in NW China

Tayierjiang Aishan [1,2], Reyila Mumin [3], Ümüt Halik [1,2,*], Wen Jiang [1], Yaxin Sun [1], Asadilla Yusup [4] and Tongyu Chen [1]

[1] College of Ecology and Environment, Xinjiang University, Urumqi 830046, China; tayirjan@xju.edu.cn (T.A.); jiangwen202111@163.com (W.J.); sun_yx1003@163.com (Y.S.); chentongyu05@163.com (T.C.)
[2] Ministry of Education Key Laboratory of Oasis Ecology, Xinjiang University, Urumqi 830046, China
[3] Institute of Microbiology, School of Ecology and Nature Conservation, Beijing Forestry University, Beijing 100083, China; ramilla@163.com
[4] Institute of Ecology, College of Urban and Environmental Sciences, Peking University, Beijing 100871, China; asadilla@xju.edu.cn
* Correspondence: halik@xju.edu.cn

Abstract: *Populus euphratica* Oliv., an indicator species for eco-environmental change in arid areas, plays a key role in maintaining the stability of fragile oasis–desert ecosystems. Owing to human interference as well as to the harshness of the natural environment, *P. euphratica* forests have suffered severe damage and degradation, with trunk cavities (i.e., hollows) becoming increasingly pronounced, and thus posing a great threat to the growth, health, and survival of the species. Currently, there is a gap in our understanding of cavity formation and its distribution in *P. euphratica*. Here, cavities in the trunks and branches of a *P. euphratica* in a typical transect (Arghan) along the lower Tarim River were studied based on field positioning observations combined with laboratory analysis. The results revealed a large number of hollow-bearing *P. euphratica* stands in the study area; indeed, trees with hollows accounted for 56% of the sampled trees, with approximately 159 trees/ha. Sixty-six percent of hollow trees exhibited large (15 cm $\leq$ cavity width (CW) < 30 cm) or very large (CW > 30 cm) hollows. The main types of cavities in the trees were trunk main (31.3%), trunk top (20.7%), branch end (19.5%), and branch middle (19.5%). Tree parameters, such as diameter at breast height (DBH), tree height (TH), east–west crown width (EWCW), height under branches (UBH), and crown loss (CL) were significantly different between hollow and non-hollow trees. Both cavity height and width were significantly and positively correlated with DBH and CL, as well as with average crown width (ACW) ($p < 0.001$) and the distance from the tree to the river. The proportion of *P. euphratica* trees with cavities showed an overall increasing trend with increasing groundwater depth. Our findings show that cavities in *P. euphratica* varied with different tree architectural characteristics. Water availability is a major environmental factor influencing the occurrence of hollowing in desert riparian forests. The results provide scientific support for the conservation and sustainable management of existing desert riparian forest ecosystems.

Keywords: functional degradation; tree hollow; adaptation strategy; environmental stress; *Populus euphratica*

Citation: Aishan, T.; Mumin, R.; Halik, Ü.; Jiang, W.; Sun, Y.; Yusup, A.; Chen, T. Patterns in Tree Cavities (Hollows) in Euphrates Poplar (*Populus euphratica*, Salicaceae) along the Tarim River in NW China. *Forests* **2024**, *15*, 421. https://doi.org/10.3390/f15030421

Academic Editor: Akira Itoh

Received: 21 December 2023
Revised: 8 February 2024
Accepted: 19 February 2024
Published: 22 February 2024

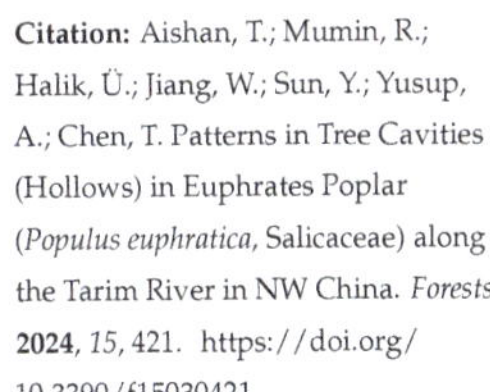

Copyright: © 2024 by the authors. Licensee MDPI, Basel, Switzerland. This article is an open access article distributed under the terms and conditions of the Creative Commons Attribution (CC BY) license (https://creativecommons.org/licenses/by/4.0/).

1. Introduction

Tree stems connect the roots and crown and have major transporting and supporting functions [1,2]. Cavities or hollows in the trunk of a tree reduce its mechanical stability [3]. The distribution of cavities or hollows in living trees varies among forest types and under different site conditions within any given forest stand [4–9]. A tree is classified as hollow- or cavity-bearing if it contains at least one hollow in its trunk or branches [6]. Furthermore, studies have shown that there are significantly more or fewer decay hollows in tree trunks in certain specific environments (humidity, diseases, drought, etc.), and that the formation and development of hollow decay in tree trunks are greatly affected by the specific site conditions [10–12]. Therefore, understanding the relationship between hollow formation or cavity occurrence in living trees and site conditions is necessary to inform forest conservation and sustainable management.

As the main body of the natural riparian vegetation in the inland river basins of Northwestern China, Euphrates poplar (*Populus euphratica*, Salicaceae) stands constitute an ecological corridor that constitutes the regional biodiversity hotspot and has the highest bioproductivity in the region [13,14]. These forests play a key role in maintaining the structure and function of riparian ecosystems in arid areas, acting as natural barriers to protect oasis agriculture and livestock production from the adverse impacts of desertification. However, in the particularly arid region of Tarim River Basin, owing to the excessive utilization of water and soil resources and anthropogenic disturbances such as deforestation over the past 50 years, natural vegetation habitats dominated by *P. euphratica* have been continuously lost. Specifically, in this region, natural oases have been shrinking, lakes have been drying up or disappearing, and desertification has increased. The natural *P. euphratica* desert forest ecosystems distributed on both banks of the lower reaches of the Tarim River have experienced dry flow for nearly 30 years, which has caused serious damage to the ecosystem [15,16].

As an indicator species of regional environmental change, *P. euphratica* riparian forests adapt their growth and development in response to biotic and abiotic stresses in different ways, including certain unique strategies that developed over evolutionary time to respond to extreme environments [17]. For example, *P. euphratica* trees grow a well-developed horizontal root system and heteromorphic leaves in response to drought stress [18]; additionally, the species responds to salt stress through selective absorption and the storage of salt ions in the body [19]. In disconnected rivers and areas with a low soil moisture content, self-renewal occurs through clonal reproduction by root suckering [20]. Moreover, *P. euphratica* exhibits a phenomenon that is easily overlooked, namely cavity or hollow formation in tree trunks.

Study have shown that, as the age of trees increases, the probability of hollow formation in the trunk also increases [21]. In addition, the degree or rate of tree hollowness is allegedly affected by its specific environment; thus, trees have a defense system against internal decay in the trunk [22]. Further, tree growth is affected by changes in site conditions, resulting in different extents of tree cavity formation [23]. Therefore, from the perspective of site conditions, researching the characteristics of hollow-bearing tree trunks would facilitate an understanding of the reasons for hollow formation and provide scientific evidence to support developing protective measures for existing natural *P. euphratica* forest resources. To date, studies in this field remain scarce, and the reasons for hollow formation and their influencing factors remain unclear [24,25]. In particular, in extremely arid habitats with strong wind erosion, *P. euphratica* forests are exposed to strong winds, rendering hollow-bearing trees susceptible to breakage and collapse [26]. A clearer understanding of the patterns of tree cavities/hollow in *P. euphratica* forests can provide basic data for urgently needed protection measures. In this study, we investigated the quantitative characteristics and distribution patterns of cavities in *P. euphratica* trees and attempted to answer the following scientific questions: What is the relationship between hollows and tree architectural characteristics? What is the influence of groundwater depth on the formation of cavities in *P. euphratica* trees? Our findings provide a useful update on the ecology of hollow *P. euphratica* and a scientific reference for maintaining and managing desert riparian forest ecosystems in arid regions.

2. Materials and Methods

2.1. Study Site

The study area is located along the Arghan transect (40°08′50″ N, 88°21′28″ E), between the Taklamakan and Kuruk Tag Deserts, in the lower reaches of the Tarim River,

Xinjiang Uyghur Autonomous Region, Northwest China (Figure 1). This area is located in an extremely arid climatic zone, with an average annual precipitation of <15 mm [13,14] and a potential annual evaporation of 2500–3000 mm [15,16]. Sparse vegetation comprising trees, shrubs, and herbs is predominantly distributed in river floodplain ecosystems. *Populus euphratica* is the dominant tree species in the study area, where nearly 70% of the existing plant species are *P. euphratica* trees [14]. Shrubs include *Tamarix ramosissima* Ledeb., *Tamarix hispida* Willd., *Tamarix elongata* Ledeb., *Lycium ruthenicum* Murr., *Halimodendron halodendron* (Pall.) Voss., *Halostachys caspica* (M.B.) C.A. Mey, *Poacynum hendersonii* (Hook. F.) Woodson., *Alhagi sparsifolia* (B. Keller et Shap.) Shap., *Glycyrrhiza inflata* Bat., *Karelinia caspica* (Pall.) Less, *Inula salsoloides* (Turcz.) Ostrnf., and *Hexinia polydichotoma* (Ostenf.) H.L. Besides *Tamarix*, most trees, shrubs, and herbs are distributed within the buffer zone at a distance of 100 m from the river. In particular, owing to the scarcity of precipitation, groundwater is the main source of the water required to maintain the structure and function of riparian ecosystems in this hyper-arid region.

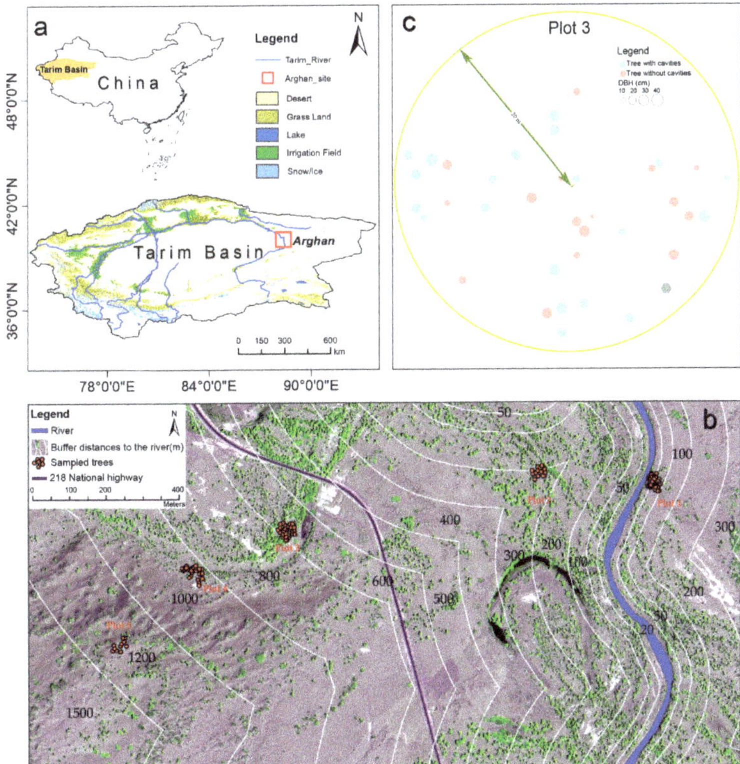

Figure 1. Map showing the location of the study site in the Tarim Basin, northwest China. (**a**) The study area in the lower reaches of the Tarim River is highlighted in the red rectangle; (**b**) the distribution of five sampled plots (plots 1–5) in the study area; (**c**) the distribution of trees with and without cavities in plot 3.

2.2. Data Collection and Processing

2.2.1. Plot Design and Measurements

Given the spatial heterogeneity of water conditions and the principle of representativeness, in August 2020 and July 2021, five circular plots with radii of 20 m at different distances from the river (≤20, 20–200, 200–500, 500–750, and 750–1050 m) were established at the Argan Transect in the lower reaches of the Tarim River. The smallest diameter at breast height (DBH) of hollow trees was used as a selection criterion: a tree-hollow survey was conducted for all *P. euphratica* trees with a DBH $\geq$ 5 cm in each plot (Figure 2). *P. euphratica* trees with one or more tree cavities on the trunk and branches were defined as hollow trees, and their related tree architectural characteristics, including DBH, tree height (TH), south–north crown width (SNCW), east–west crown width (EWCW), average crown width (ACW), height under branch (HUB), and crown loss (CL), as well as tree cavity characteristics, including the number of tree hollows, cavity types, cavity orientation, cavity height, and cavity width, were measured and recorded. The DBH of each tree was measured using a DBH meter (accuracy < 0.1 cm) and TH and HUB were determined using a laser distance meter. The SNCW and EWCW of the tree canopy were measured with a measuring tape, and the average was calculated as the ACW. The CL, with the values for an ideal tree and a completely withered tree taken as 0% and 100%, respectively, was estimated according to Aishan et al. [13]. Referring to survey methods [6] used for hollow trees, tree hollows with good visibility were counted and their parameters were measured directly from ground level, while tree hollows distributed on the higher parts of trunks or canopy branches were observed using binoculars (10 × 25). Cavity orientation was determined using a compass. The tree geographical location and the number of trees in the sample plot were recorded. Tree cavities were divided into seven types according to their location and shape: butt hollow, trunk main, trunk top, fissure, branch middle, branch end, and bayonet. The hollow size was grouped into one of four size classes based on cavity width: small (<5 cm), medium (5–15 cm), large (15–30 cm), and very large (>30 cm). In addition, partial sapwood/heartwood samples were taken from healthy and decayed tree trunks using an increment borer (Haglof, Langsele, Sweden). The presence or absence of animal utilization of tree holes was also recorded. The proportion of hollow trees (%) was the proportion of trees with hollows among all trees sampled within a plot. Hollow tree density was the total number of hollow trees per unit area of forest land (trees/ha). Groundwater depth data were obtained from the long-term monitoring wells installed by the Tarim River Basin Administration and BARO-Diver (DI800, vanEssen, Delft, The Netherlands) and set up by our research team.

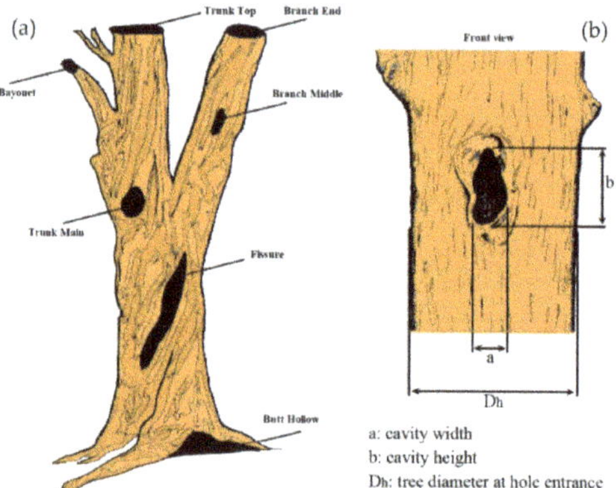

Figure 2. Cavity types and measurements; (**a**) location and type of cavities in trees, modified from [26]; (**b**) measured parameters of cavities, modified from [27].

2.2.2. Data Processing and Analysis

A map showing the location of the research area and the distribution pattern of trees with plots was created using ArcGIS 10.0 software (Esri, Inc., Redlands, CA, USA) based on a GIS database established by integrating QuickBird images with terrestrial field survey data. A schematic illustration of the cavity types was produced using Photoshop 2017 (Adobe Inc., San Jose, CA, USA). Differences in architectural characteristics among the three groups (hollow, non-hollow, and all) were tested using one-way ANOVA (significance level of $p = 0.001$) in SPSS 19.0 (IBM Corp., Armonk, NY, USA). Box–whisker plots and pie charts were produced using Origin9.4 software (OriginLab, Northampton, MA, USA). The Mantel test and random forest analyses were performed to evaluate the relationship between the cavity characteristics and tree attributes and establish the importance of factors influencing *P. euphratica* tree cavities using the "linkET" and "random forest" packages, respectively, and the results were visualized using the "ggplot2" package in R 4.2.3 (https://cran.r-project.org/web/packages/ggplot2/index.html, accessed on 13 May 2023).

3. Results

3.1. Cavity Types and Distribution Patterns

The frequency of hollow trees and number of holes in the selected sample plots indicated the characteristics of hollow trees and the abundance of holes in the study area (Table 1 and Figure 3). A total of 175 trees were investigated in the Arghan section, of which 98 (i.e., 56% of the total number of trees surveyed) were found to contain different types of hollows. A total of 352 tree holes were found in 175 trees, with densities ranging from 16 to 130 per sample plot and an average of 2.01 holes per tree. The proportions of each hollow type relative to the total number of holes were as follows: trunk main (31.3%), trunk top (20.7%), branch end (19.5%), branch middle (19.5%), butt hollow (5.1%), fissure (2.8%), and bayonet (1.1%). Holes classified as trunk main (31.3%) were significantly more abundant than the other six types of holes; the numbers of holes categorized as trunk top, branch end, and branch middle were not significantly different from each other ($p > 0.05$); and the proportions of holes categorized as butt hollow, fissure, and bayonet were smaller. These results indicate that holes are easily formed in the middle and top of the trunk and that more holes occur in the lateral branches of the middle and top sections.

Table 1. Descriptive statistics for tree attributes and tree cavity types in sample plots (mean $\pm$ SE).

Attributes at Tree and Cavity Levels	Sample Plot				
	Plot 1	Plot 2	Plot 3	Plot 4	Plot 5
Number of trees (n)	42	26	41	55	11
Distance from the river (DR) (m)	DR $\leq$ 20 m	20 < DR $\leq$ 200 m	200 < DR $\leq$ 500 m	500 < DR $\leq$ 750 m	750 < DR $\leq$ 1050 m
Number of trees with cavities (n)	11	14	21	41	11
Proportion of hollow trees (%)	26.20	53.80	51.20	74.50	100
Tree density (n/ha)	334	207	326	438	88
Mean tree height (m)	8.20 $\pm$ 2.0	7.10 $\pm$ 1.41	7.52 $\pm$ 2.13	5.54 $\pm$ 2.12	4.30 $\pm$ 1.33
Mean DBH (cm)	28.50 $\pm$ 8.90	35.42 $\pm$ 18.80	25.93 $\pm$ 8.51	29.91 $\pm$ 14.93	74.90 $\pm$ 30.20
Mean crown width (m)	7.20 $\pm$ 1.90	8.50 $\pm$ 2.00	7.13 $\pm$ 1.81	6.60 $\pm$ 1.73	6.30 $\pm$ 2.31
Mean crown loss (%)	12.3 $\pm$ 12.0	29.2 $\pm$ 0.2	26.0 $\pm$ 0.1	33.0 $\pm$ 0.2	67.0 $\pm$ 0.1
Number of butt hollows (n)	0	2	3	8	5
Number of fissures (n)	0	3	0	3	4
Number of trunk main holes (n)	4	14	11	42	36
Number of branch middle holes (n)	6	9	12	8	34
Number of branch end holes (n)	1	13	3	20	33
Number of bayonet holes (n)	0	1	1	1	1
Number of trunk top holes (n)	5	6	6	39	17

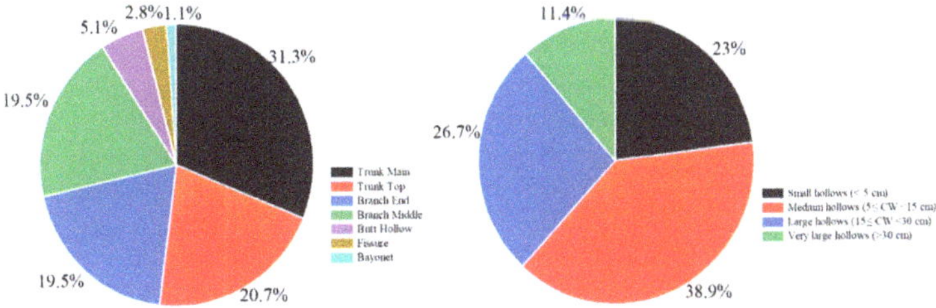

Figure 3. Percentages of hollow types and sizes found at all sites (CW: cavity width).

3.2. Architectural Traits of Trees with and without Hollows

The *P. euphratica* trees surveyed in the study area were divided into two groups according to the presence of hollows: hollow-bearing and non-hollow-bearing trees (Figure 4).

The results showed that the DBH, TH, EWCW, HUB, and CL of hollow-bearing trees were significantly different between the groups ($p < 0.001$), whereas other indices did not differ significantly. Similarly, significant differences ($p < 0.001$) were also observed for the DBH, TH, EWCW, HUB, and CL of non-hollow-bearing versus hollow-bearing trees, indicating that hollowing in *P. euphratica* trees is also related to the tree architectural characteristics. Furthermore, the DBH, EWCW, and CL of trees with hollows were larger than those of trees without hollows. For most of the trees, particularly those classified as having very large hollows (>30 cm), the height and width of the cavities increased with increasing DBH. The DBH reflects the age of *P. euphratica* [20], and the cavities formed in older trees experienced more disturbance and took longer to form, suggesting that older *P. euphratica* forests are more susceptible to cavities. However, both the TH and the HUB of living *P. euphratica* with hollows were lower than those of trees without hollows, suggesting that the taller the tree, the less likely it is to form new cavities.

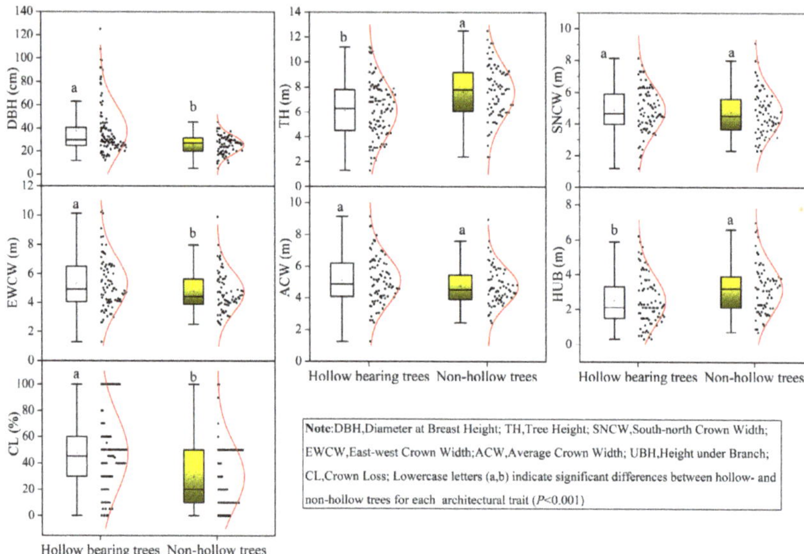

Figure 4. Architectural characteristics of hollow-bearing and non-hollow-bearing trees.

3.3. Relationship between Tree Attributes and Cavity Parameters

Figure 5 shows the results of the Mantel tests for the relationships between cavity characteristics, tree attributes, and water availability, with the orange color indicating highly significant ($p < 0.01$), green indicating significant ($p \leq 0.05$), and gray indicating non-significant ($p \geq 0.05$) relationships. As shown in Figure 5, tree cavity height showed a highly significant positive correlation ($p < 0.01$) with DBH and CL. Cavity width showed a highly significant positive correlation ($p < 0.01$) with DBH and CL and a significant negative correlation ($p \leq 0.05$) with TH. Among these variables, DBH and CL showed a highly significant positive correlation with tree cavity length and width ($p < 0.01$), indicating that DBH and CL were the main factors affecting the cavity height and width of *P. euphratica* trees, while DBH had a greater influence on tree cavity width. Random forest analysis was conducted to further investigate the correlation between the occurrence of cavities in *P. euphratica* and various tree architectural characteristics (Figure 6a,b). The results in Figure 6 show that DBH, average crown width, CL, and EWCW were more important in determining the height and width of cavities than any other variable. Random forest analysis thus showed that DBH was the most significant factor influencing the variation in both the height and width of *P. euphratica* cavities ($p < 0.001$).

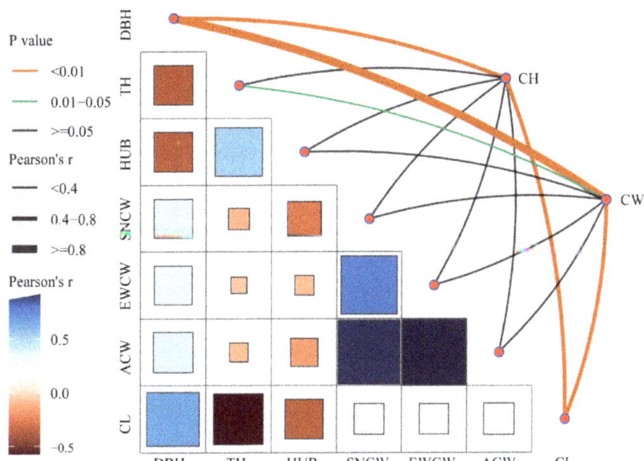

Figure 5. Mantel test results for the relationship between cavity characteristics and tree attributes (CH, cavity height; CW, cavity width; DBH, diameter at breast height; TH, tree height; HUB, height under branch; SNCW, south–north crown width; EWCW, east–west crown width; ACW, average crown width; CL, crown loss).

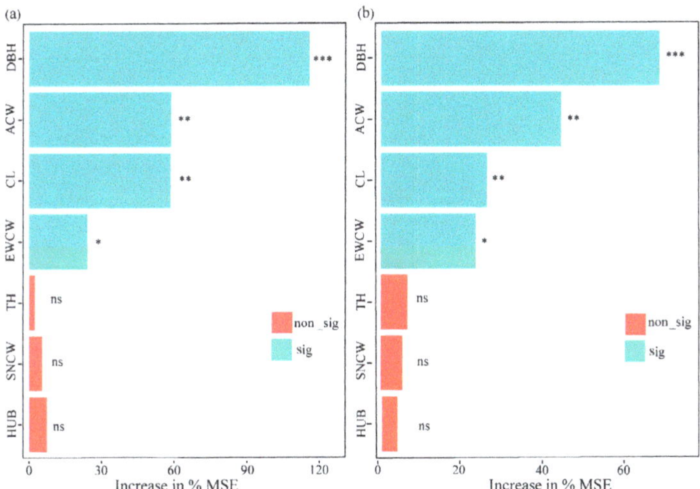

Figure 6. Importance of factors influencing *P. euphratica* tree cavities, as measured by the percentage increase in mean square error (MSE) in the random forest analysis; (**a**) presents the results of the random forest model for cavity height (CH); (**b**) results of the random forest model for cavity width (CW), * $p < 0.05$, ** $p < 0.01$, *** $p < 0.001$.

3.4. Relationship between Water Availability and Cavity Parameters

As *P. euphratica* desert riparian forests survive in extremely arid environments, water accessibility (distance to the river and groundwater) is one of the most important environmental factors affecting plant growth and development. As a result, this study hypothesized that water conditions were likely to be the main factors influencing the formation of cavities in *P. euphratica* trees. Therefore, this paper seeks to elucidate the relationship between water conditions and the characteristics of hollow-bearing *P. euphratica* trees by studying and comparing the characteristics of hollow-bearing trees among *P. euphratica* populations at different locations (distance from the river) in relation to water resources.

Figure 7 shows the responses of the cavity parameters (CW and CH) of trees to variations in distance from the river. As shown in Figure 7, as the distance from the river increased, the cavity width and cavity height of *P. euphratica* showed significant variations ($p < 0.001$); near the river (<600 m), changes in cavity width and cavity height of *P. euphratica* were not apparent, but farther from the river (>600 m), the cavity width and cavity height of *P. euphratica* tended to increase. The overall trend indicated that the development of tree cavities was influenced, to some extent, by the distance of the trees from water resources. Figure 8 shows that there was a significantly positive correlation between the proportion of trees with cavities and groundwater depth ($r = 0.98$, $p < 0.01$).

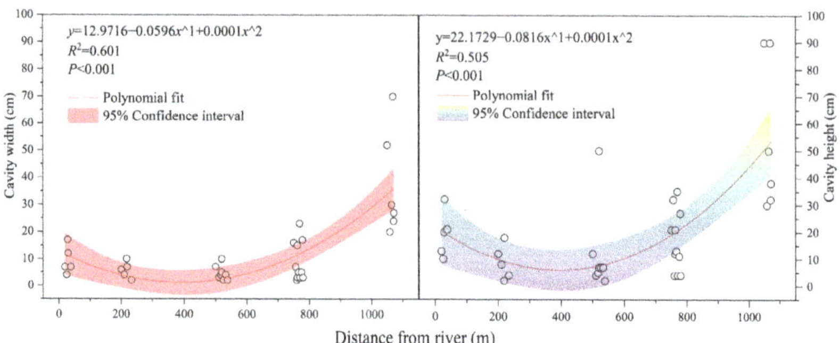

Figure 7. Variations in cavity parameters with distance from the river.

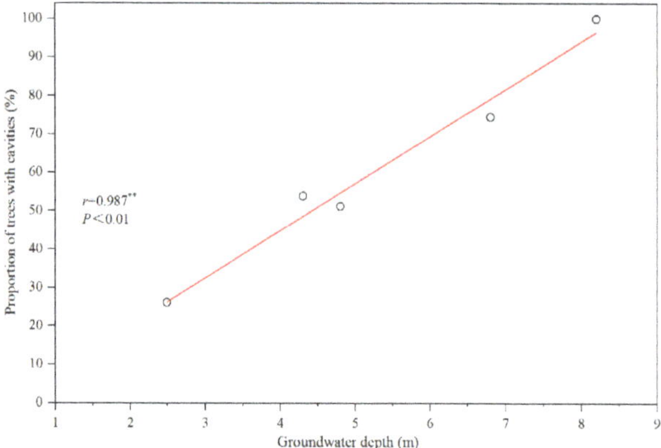

Figure 8. Relationship between proportion of hollow-bearing trees and groundwater depth (note: annual average groundwater depth data were used for this plot; **: significant correlation).

4. Discussion

Populus euphratica forests are likely to contain a high proportion of trees with hollows, with a hollow-bearing-tree density of 159 tree/ha, which is much higher than those of tropical seasonal rain forests (87 trees/ha), tropical mountain evergreen broad-leaved forests (approx. 86 trees/ha), subtropical humid evergreen broad-leaved forests (94.3 trees/ha), and even desert riparian forest in the middle reaches of the Tarim River (78 trees/ha) in China [25,29–31]. *Populus. euphratica* forests in this area are primeval, and most are over-matured [14] and contain numerous trees with hollows. External factors, such as sandstorms, strong winds, and long-term river desiccation, may cause hollows to form more easily in *P. euphratica* forests, resulting in different densities of hollow-bearing trees in different locations. Further, factors such as water conditions (mainly groundwater depth), desertification, and human interference also significantly influence the growth and distribution pattern of natural vegetation in the Tarim River Basin [13,14]. Therefore, water availability is likely the key environmental factor affecting the formation of tree cavities. In this study, the presence of cavities in *P. euphratica* increased significantly with increasing DBH and decreased significantly with increasing TH. Owing to the extremely arid conditions in the lower Tarim River, the vertical growth of most *P. euphratica* forests is limited by water; thus, lateral growth is dominant [14]. Therefore, taller trees are likely to be distributed in areas with favorable water conditions, as well as be less exposed to water stress and maintain normal growth. Temperature is the main limiting factor in tropical rainforests, and the specific orientation of tree hollows is related to thermal conditions (such as the angle of the sun) [32]. In this study, 82 (23.4%) of the hollow-bearing trees sampled exhibited clear signs of broken stems or branches. In desert riparian forests, factors such as water availability and strong winds may lead to differences in the distribution patterns of tree cavities. Hollow-bearing trees provide critical microhabitat resources for forest fauna and play an important role in maintaining biodiversity [33,34]. Of the 352 tree holes sampled, only 3 (0.85%) exhibited signs of animal use; hence, almost no tree holes were used as animal habitats. Compared with other forest types, *P. euphratica* desert riparian forests are characterized by a simple community structure and relatively poor biodiversity. Therefore, the relationship between *P. euphratica* tree cavities and biodiversity was not quantified or further analyzed in the present study.

In the study area, most of the hollows were observed on *P. euphratica* trunks that had been blown down by wind or had broken branches. The formation of tree hollows may also be an indicator of *P. euphratica* senescence. In addition, we observed that *P. euphratica* stored large amounts of water in the trunk; it can be assumed that this storage would make the middle of the trunk susceptible to porosity and eventually lead to cavity formation. The more severe drought conditions become, the more likely a *P. euphratica* trunk will develop empty spaces by storing water. Therefore, the formation of tree holes in the trunk may also be a strategy of *P. euphratica* to adapt to extreme drought conditions. In addition, our study reveals that the height and width of *P. euphratica* cavities are significantly correlated with the distance from the river. As groundwater depth increased, the proportion of hollow-bearing trees in *P. euphratica* stands increased from 26.2% to 100%. *P. euphratica* trees, as phreatophytes (i.e., plant species that have evolved the capacity to access groundwater), mainly depend on groundwater to survive [14]. As shallow soil water sources are gradually depleted with increasing distance from the river, the depth and proportion of water uptake by phreatophytes from groundwater typically increases [35–37]. Therefore, phreatophytes develop deeper roots to track the capillary fringe and/or saturated zone of aquifers. Thus, groundwater depth is considered a key limiting factor that regulates stand structure and function in desert riparian forests. The habitat conditions of the lower Tarim River are more challenging than those of the upper and middle reaches, with frequent river-flow disconnections and deeper groundwater tables; all together, these conditions cause continuous water stress and also cause *P. euphratica* forests to develop an association between increasing tree age, declining tree function, and greater hollowing. Therefore, habitat quality, particularly with respect to water availability, is an important factor in the hollowing of *P. euphratica* forests.

The formation of hollow trees is a slow and complex process that is influenced by a combination of factors, including the habitat conditions of the forest and the characteristics of the trees themselves. Hollowing is the end result of decay in living trees, with some trees forming hollows gradually from the inside out and exhibiting large external holes that are visible to the naked eye, and others starting directly from the outside, with various decay fungi invading the sapwood exterior. This study identified several issues that need to be investigated and addressed in future research. For example, drilling samples of heartwood and sapwood revealed that some of the trunks were extensively decayed internally and that the heartwood was decayed or in a crumbly state. Consequently, quantitative studies of internal decay in living trees are needed, and characteristics such as the decay ratio, decayed area, volume, and decay orientation need to be quantified using high-precision non-destructive instrumentation, such as Arbotom stress-wave detection and TreeRadar [2,38–40]. In addition, this study was mainly concerned with the physical parameters of the trees and thus can be improved by the addition of chemical (biochemical) parameters at a later stage. Therefore, it is expected that this aspect will be thoroughly investigated in depth in future studies.

5. Conclusions

We investigated and analyzed the characteristics of hollow trees in natural *P. euphratica* stands at different distances (different areas) from the lower Tarim River and concluded that the *P. euphratica* population in the study area had a high degree of hollowing, with living trees with hollows accounting for 56% of the total number of trees surveyed and large trees with very large hollows in turn accounting for nearly 66% of hollow-bearing trees sampled. In addition, hollowing in *P. euphratica* trees was closely related to tree architectural characteristics, indicating that hollowing also affects the growth pattern of *P. euphratica* to some extent. The frequency of hollowing increased with increasing DBH. Furthermore, both the cavity width and the cavity height of *P. euphratica* showed a tendency to increase with increasing distance from the river, and. the proportion of trees with cavities showed an increasing trend with increasing groundwater depth. Based on the results of our study, we propose that the elevation of groundwater at a distance from the river and optimization of stand age structure through ecological water delivery can decrease the occurrence of hollows in *P. euphratica* forests.

Author Contributions: All authors contributed to the design and development of this manuscript. T.A. and R.M. conducted the research and prepared the first draft of the manuscript; Ü.H. conceived and designed the overall concept of the research, supervised, and participated in the fieldwork; W.J. and Y.S. provided important advice and technical support on the methodology; T.A., R.M., A.Y., T.C. collected and processed the data. All authors have read and agreed to the published version of the manuscript.

Funding: This research was supported by the National Natural Science Foundation of China (Grant Nos. 32160367 and 32260285) and the Third Xinjiang Scientific Expedition and Research Program (Grant No. 2022xjkk0301).

Data Availability Statement: The datasets used in this study are available upon request. Please contact the first author if you are interested in this study.

Acknowledgments: We thank the Tarim River Basin Administration Bureau for providing hydrological data and the Forestry Department of Qarkilik (Ruoqiang) for logistical support during our field work in Arghan. The authors are grateful to the anonymous reviewers for their constructive comments. The authors thank colleagues from Xinjiang University for helping with collecting and processing the data.

Conflicts of Interest: The authors declare no conflicts of interest.

References

1. Tsutomu, U.; Masato, S.; Akio, K.; Hiroyuki, T.; JooYoung, C. Both stem and crown mass affect tree resistance to uprooting. *J. For. Res.* **2012**, *17*, 65–71.
2. Wei, Z.C.; Halik, U.; Aishan, T.; Abliz, A.; Welp, M. Spatial distribution patterns of trunk internal decay of Euphrates poplar riparian forest along the Tarim River, northwest China. *For. Ecol. Manag.* **2022**, *522*, 120434. [CrossRef]
3. Marra, R.E.; Brazee, N.J.; Fraver, S. Estimating carbon loss due to internal decay in living trees using tomography: Implications for forest carbon budgets. *Environ. Res. Lett.* **2018**, *13*, 105004. [CrossRef]
4. Niringiyimana, A.; Nzarora, A.; Twahirwa, J.C.; van der Hoek, Y. Density and characteristics of tree cavities inside and outside Volcanoes National Park, Rwanda. *Ecol. Evol.* **2022**, *12*, e9461. [CrossRef] [PubMed]
5. LaMontagne, J.M.; Kilgour, R.J.; Anderson, E.C.; Magle, S. Tree cavity availability across forest, park, and residential habitats in a highly urban area. *Urban Ecosyst.* **2015**, *18*, 151–167. [CrossRef]
6. Harper, M.J.; McCarthy, M.A.; van der Ree, R. The abundance of hollow-bearing trees in urban dry sclerophyll forest and the effect of wind on hollow development. *Biol. Conserv.* **2005**, *122*, 181–192. [CrossRef]
7. Fan, Z.F.; Shifley, S.R.; Spetich, M.A.; Thompson, F.R.; Larsen, D.R. Distribution of cavity trees in midwestern old-growth and second-growth forests. *Can. J. For. Res.* **2003**, *33*, 1481–1494. [CrossRef]

8. Liu, J.Y.; Zheng, Z.; Xu, X.; Dong, T.F.; Chen, S.C. Abundance and distribution of cavity trees and the effect of topography on cavity presence in a tropical rainforest, southwestern China. *Can. J. For. Res.* **2018**, *48*, 1058–1066. [CrossRef]
9. Takashima, A.; Nakanishi, A.; Morishita, M.; Abe, S.; Saito, K.; Kotaka, N. Tree-cavity formation in the mature subtropical forests of Yambaru, Okinawa Island. *J. For. Res.-Jpn.* **2021**, *26*, 410–418. [CrossRef]
10. Soge, A.O.; Popoola, O.I.; Adetoyinbo, A.A. Detection of wood decay and cavities in living trees: A review. *Can. J. For. Res.* **2021**, *51*, 937–947. [CrossRef]
11. Jauregui, A.; Rodriguez, S.A.; Garcia, L.N.G.; Gonzalez, E.; Segura, L.N. Wood density and tree size used as cues to locate and excavate cavities in two Colaptes woodpeckers inhabiting a threatened southern temperate forest of Argentina. *For. Ecol. Manag.* **2021**, *502*, 119723. [CrossRef]
12. Dudinszky, N.; Ippi, S.; Kitzberger, T.; Ceron, G.; Ojeda, V. Tree size and crown structure explain the presence of cavities required by wildlife in cool-temperate forests of South America. *For. Ecol. Manag.* **2021**, *494*, 119295. [CrossRef]
13. Peng, Y.; He, G.J.; Wang, G.Z. Spatial-temporal analysis of the changes in Populus euphratica distribution in the Tarim National Nature Reserve over the past 60 years. *Int. J. Appl. Earth Obs. Geoinf.* **2022**, *113*, 103000.
14. Aishan, T. Degraded Tugai Forests under Rehabilitation in the Tarim Riparian Ecosystem, Northwest China: Monitoring, Assessing and Modelling. Ph.D. Thesis, Katholische Universitat Eichstatt-Ingolstadt, Eichstätt, Germany, 2016; 152p.
15. Ling, H.B.; Guo, B.; Zhang, G.P.; Xu, H.L.; Deng, X.Y. Evaluation of the ecological protective effect of the "large basin" comprehensive management system in the Tarim River basin, China. *Sci. Total Environ.* **2019**, *650*, 1696–1706. [CrossRef] [PubMed]
16. Ling, H.B.; Zhang, P.; Xu, H.L.; Zhao, X.F. How to Regenerate and Protect Desert Riparian *Populus euphratica* Forest in Arid Areas. *Sci. Rep.* **2015**, *5*, 15418. [CrossRef] [PubMed]
17. Zhuang, L.; Li, W.H.; Yuan, F.; Gong, W.C.; Tian, Z.P. Ecological adaptation characteristics of *Populus euphratica* and *Tamarix ramosissima* leaf microstructures in the lower reaches of Tarim River. *Acta Ecol. Sin.* **2010**, *30*, 62–66. [CrossRef]
18. Li, D.; Si, J.H.; Zhang, X.Y.; Gao, Y.Y.; Luo, H.; Qin, J.; Ren, L.X. Ecological adaptation of *Populus euphratica* to drought stress. *J. Desert Res.* **2020**, *40*, 17–23. (In Chinese)
19. Zhao, R.; Chen, S.L. The salt-stress signaling network involved in the regulation of ionic and ROS homeostasis in poplar. *Sci. Sinica Vitae* **2020**, *50*, 167–175. (In Chinese)
20. Eusemann, P.; Petzold, A.; Thevs, N.; Schnittler, M. Growth patterns and genetic structure of *Populus euphratica* Oliv. (Salicaceae) forests in NW China–Implications for conservation and management. *Forest Ecol. Manag.* **2013**, *297*, 27–36. [CrossRef]
21. Renton, K.; Salinas-Melgoza, A.; Rueda-Hernandez, R.; Vazquez-Reyes, L.D. Differential resilience to extreme climate events of tree phenology and cavity resources in tropical dry forest: Cascading effects on a threatened species. *Forest Ecol. Manag.* **2018**, *426*, 164–175. [CrossRef]
22. Vazquez, L.; Renton, K. High density of tree-cavities and snags in tropical dry forest of western Mexico raises questions for a latitudinal gradient. *PLoS ONE* **2015**, *10*, e0116745. [CrossRef] [PubMed]
23. Onodera, K.; Tokuda, S.; Abe, T.; Nagasaka, A. Occurrence probabilities of tree cavities classified by entrance width and internal dimensions in hardwood forests in Hokkaido, Japan. *J. Forest Res.-Jpn.* **2013**, *18*, 101–110. [CrossRef]
24. Mumin, R.; Aishan, T.; Halik, U. Hollow-bearing characteristics of *Populus euphratica* in the lower reaches of Tarim River, China. *Chin. J. Appl. Ecol.* **2020**, *31*, 1933–1940. (In Chinese)
25. Cheng, Q.; Aishan, T.; Halik, U.; Wang, X.Y. Hollow tree characteristics of different aged *Populus euphratica* forests in the middle reaches of the Tarim River. *Arid Zone Res.* **2023**, *40*, 247–256. (In Chinese)
26. Li, Z.W.; Yu, G.A.; Brierley, G.J.; Wang, Z.Y.; Jia, Y.H. Migration and cutoff of meanders in the hyperarid environment of the middle Tarim River, northwestern China. *Geomorphology* **2017**, *276*, 116–124. [CrossRef]
27. Lindenmayer, D.B. Cavity sizes and types in Australian eucalypts from wet and dry forest types—A simple of rule of thumb for estimating size and number of cavities. *Forest Ecol. Manag.* **2000**, *137*, 139–150. [CrossRef]
28. Wang, S.J.; Chen, B.H.; Li, H.Q. *Populus euphratica Forest*; China Environmental Science Press: Beijing, China, 1995; pp. 57–59. (In Chinese)
29. Remm, J.; Lõhmus, A.; Remm, K. Tree cavities in riverine forests: What determines their occurrence and use by hole-nesting passerines? *Forest Ecol. Manag.* **2006**, *221*, 267–277. [CrossRef]
30. Wang, Y.T.; Xu, H.D.; Wang, L.H.; Li, F.R.; Sun, H. Field investigation of decay rate of Korean pine standing trees in natural forests in Lesser Xing'an Mountains. *J. Beijing For. Univ.* **2015**, *37*, 97–104. (In Chinese)
31. Brandeis, T.J.; Newton, M.; Filip, G.M. Cavity-nester habitat development in artificially made Douglas-fir snags. *J. Wildl. Manag.* **2002**, *66*, 625–633. [CrossRef]
32. Boyle, W.A.; Ganong, C.N.; Clark, D.B. Density, distribution, and attributes of tree cavities in an old-growth tropical rain forest. *Biotropica* **2008**, *40*, 241–245. [CrossRef]
33. Gruebler, M.U.; Schaller, S.; Keil, H.; Naef-Daenzer, B. The occurrence of cavities in fruit trees: Effects of tree age and management on biodiversity in traditional European orchards. *Biodivers. Conserv.* **2013**, *22*, 3233–3246. [CrossRef]
34. Ibarra, J.T.; Novoa, F.J.; Jaillard, H.; Altamirano, T.A. Large trees and decay: Suppliers of a keystone resource for cavity-using wildlife in old-growth and secondary Andean temperate forests. *Austral Ecol.* **2020**, *45*, 1135–1144. [CrossRef]
35. Thomas, F.M. Ecology of phreatophytes. *Prog. Bot.* **2014**, *75*, 335–375.
36. Thomas, F.M.; Jeschke, M.; Zhang, X.; Lang, P. Stand structure and productivity of *Populus euphratica* along a gradient of groundwater distances at the Tarim River (NW China). *J. Plant Ecol.* **2017**, *10*, 753–764.

37. Si, J.H.; Feng, Q.; Cao, S.K.; Yu, T.F.; Zhao, C.Y. Water use sources of desert riparian Populus euphratica forests. *Environ. Monit. Assess* **2014**, *186*, 5469–5477. [CrossRef] [PubMed]
38. Goh, C.L.; Rahim, R.A.; Rahiman, M.H.F.; Talib, M.T.M.; Tee, Z.C. Sensing wood decay in standing trees: A review. *Sensor Actuat. A Phys.* **2018**, *269*, 276–282. [CrossRef]
39. Li, H.B.; Zhang, X.W.; Li, Z.Q.; Wen, J.; Tan, X. A Review of Research on Tree Risk Assessment Methods. *Forests* **2022**, *13*, 1556. [CrossRef]
40. Okun, A.; Brazee, N.J.; Clark, J.R.; Cunningham-Minnick, M.J.; Burcham, D.C.; Kane, B. Assessing the Likelihood of Failure Due to Stem Decay Using Different Assessment Techniques. *Forests* **2023**, *14*, 1043. [CrossRef]

Disclaimer/Publisher's Note: The statements, opinions and data contained in all publications are solely those of the individual author(s) and contributor(s) and not of MDPI and/or the editor(s). MDPI and/or the editor(s) disclaim responsibility for any injury to people or property resulting from any ideas, methods, instructions or products referred to in the content.

Article

Effects of Water Control and Nitrogen Addition on Functional Traits and Rhizosphere Microbial Community Diversity of *Haloxylon ammodendron* Seedlings

Menghao Zhu [1,2], Lamei Jiang [1,2], Deyan Wu [1,2], Wenjing Li [1,2], Huifang Yang [1,2] and Xuemin He [1,2,*]

[1] College of Ecology and Environment, Xinjiang University, Urumqi 830046, China; zhumenghao1999@163.com (M.Z.); jianglam0108@126.com (L.J.); bingyeshifu@163.com (D.W.); liwenjing0624@stu.xju.edu.cn (W.L.); hfyang1005@163.com (H.Y.)

[2] Key Laboratory of Oasis Ecology of Education Ministry, Xinjiang University, Urumqi 830046, China

* Correspondence: hxm@xju.edu.cn

Abstract: Water and nitrogen sources have always been the primary limiting factors for vegetation growth in arid and semi-arid regions and play an important role in the physiological ecology of vegetation. In this work, we studied the effects of water deficit and nitrogen addition on the physiological traits and rhizosphere bacterial microbial community of *Haloxylon ammodendron* seedlings in sterilized and non-sterilized soil habitats. A pot experiment was conducted to control the water and nitrogen sources of *H. ammodendron* seedlings. The water deficit treatment was divided into two groups based on gradient: a normal water group (CK, 70% field water holding capacity) and water deficit group (D, 30% field water holding capacity). The nitrogen addition treatment was divided into a no addition group (CK, 2.8 mg·kg^{-1}) and addition group (N, 22.4 mg·kg^{-1}). At the end of the growing season, the biochemical indexes of *H. ammodendron* seedlings were measured, and the rhizosphere soil was subjected to 16S rDNA-high-throughput sequencing to determine the rhizosphere bacterial community composition of *H. ammodendron* seedlings under different treatments. The results showed that the root-to-crown ratio of *H. ammodendron* seedlings increased significantly ($p < 0.05$) under the water deficit treatment compared to the control and nitrogen addition treatments, indicating that *H. ammodendron* seedlings preferred to allocate biological carbon to the lower part of the ground. In contrast, plant height and root length were significantly lower ($p < 0.05$) under water deficit treatment compared to the control, and no significant change was observed under water deficit and nitrogen addition compared to the control, indicating that water deficit inhibited the growth of *H. ammodendron* seedlings and nitrogen addition mitigated the effect of water deficit on the growth of *H. ammodendron* seedlings. Under sterilized soil conditions, both water deficit and nitrogen addition significantly increased the abundance and diversity of bacterial communities in *H. ammodendron* seedlings ($p < 0.05$). Conversely, under non-sterilized conditions, both inhibited the diversity of microbial bacterial communities, and the microbial characteristic species under different controls were different. Therefore, in the short-term experiment, *H. ammodendron* seedlings were affected by water deficit and allocated greater quantities of biomass to the underground part, especially in the non-sterile microbial environment; different initial soil conditions resulted in divergent responses of rhizosphere bacterial communities to water deficit and nitrogen addition. Under different initial soil conditions, the same water deficit and nitrogen addition treatment will lead to the development of distinct differences in rhizosphere bacterial community composition.

Keywords: water deficit; nitrogen addition; functional traits; *Haloxylon ammodendron*; microbial community diversity

Citation: Zhu, M.; Jiang, L.; Wu, D.; Li, W.; Yang, H.; He, X. Effects of Water Control and Nitrogen Addition on Functional Traits and Rhizosphere Microbial Community Diversity of *Haloxylon ammodendron* Seedlings. *Forests* **2023**, *14*, 1879. https://doi.org/10.3390/f14091879

Academic Editor: Luca Belelli Marchesini

Received: 8 June 2023
Revised: 16 July 2023
Accepted: 18 July 2023
Published: 15 September 2023

Copyright: © 2023 by the authors. Licensee MDPI, Basel, Switzerland. This article is an open access article distributed under the terms and conditions of the Creative Commons Attribution (CC BY) license (https://creativecommons.org/licenses/by/4.0/).

1. Introduction

The physiological and ecological processes of plants are susceptible to external habitats, inducing a variety of different changes. Soil physicochemical properties are the most important external factors affecting plant physiology and ecology [1,2]. Plant functional traits such as plant height, root length, aboveground and underground biomass can measure plant productivity. As such, they are often used to reflect the growth status of plants and the structural and functional statuses of plant ecosystems [3]. Water deficit is one of the main limiting factors in plant growth and has an important impact on plant growth, especially in arid and semi-arid areas [4]. Researchers have found that water deficit not only changes the physiological status of plants, but also indirectly affects the rhizosphere microbial community of vegetation through water control. At the same time, plants and their rhizosphere microorganisms also interact with each other. They can use the changes in rhizosphere microbial communities to change the rhizosphere soil environment of plants and improve the adaptability of plants to extreme environments, such as drought, salt, and heavy metal deficit [5–8]. Nutrient limitation is another major limiting factor in arid and semi-arid areas, including nitrogen, phosphorus, iron, and other elements. Nitrogen source is one of the primary nutrient elements and plays an indispensable role in the growth of desert vegetation. For example, plants have higher competitiveness in high-nutrient environments [9].

H. ammodendron is the constructive species or dominant species of desert plant communities. It plays a significant role in afforestation and water conservation in desert areas. It has the characteristics of plants from extreme habitats such as drought tolerance, salinity, and high light intensity [10,11]. In the growth process of *H. ammodendron*, it is easily affected by the external environment, such as water, soil, climate and other abiotic environments, as well animal, insect, microbial, and other biological factors. Changes in the physiological and ecological status of *H. ammodendron* will also affect the status of its rhizosphere microbial community, resulting in changes in its composition.

Soil bacteria are the most abundant soil microorganisms, and they are indispensable biological groups in important processes such as soil organic matter decomposition, humus formation, and nutrient transformation [12]. Developing an in-depth understanding of the deficit adaptation mechanism of *H. ammodendron* rhizosphere soil bacterial community diversity in desert area is of great significance for afforestation and vegetation restoration in desert areas [13]. Plants can change the nutrient status of rhizosphere soil and other soil physical and chemical properties through root activity, and then change the rhizosphere microbial community structure [14]. Studies have found that rhizosphere microorganisms can improve the adaptability of sorghum to water deficit [15]. Nitrogen addition may lead to changes in rhizosphere microbial community composition and affect plant growth and health [16]. Nitrogen addition also reduces soil fungal community abundance and arbuscular fungal biomass, but significantly increases bacterial biomass or has no significant effect on it [17]. Most studies only consider the effect of single factor on the rhizosphere bacterial community, while the effects of multi-factor interaction on the functional traits of *H. ammodendron* seedlings and the diversity of rhizosphere microbial communities, areas in urgent need of study, are typically ignored. Through the utilization of water deficit and nitrogen addition control experiments, 16S rDNA was deployed to determine the rhizosphere bacterial community of plant seedlings, and we analyzed the differences in physiological traits and rhizosphere microbial community composition and diversity caused by operating in different treatment conditions in a variety of habitats.

2. Materials and Methods

2.1. Materials

In situ soil was collected from the field scientific observation station of the temperate desert ecosystem in Jinghe, Xinjiang, the Ebinur Lake Wetland National Nature Reserve. The Ebinur Lake Wetland National Nature Reserve is located in Jinghe County, Bortala Mongol Autonomous Prefecture, Xinjiang (44°30′–45°09′ N, 82°36′–83°50′ E), and has a total area of about 2670.8 km^2. It is a typical temperate continental climate. The annual sunshine duration is approximately 2800 h, and the average annual temperature is about 6–8 °C. Historically, the highest temperature under extreme conditions can reach

44 °C, while the lowest temperature is −33 °C. The main dominant plant species in this reserve are *Populus euphratica*, *H. ammodendron*, *Nitraria tangutorum*, *Phragmites australis*, and *Apocynum venetum*.

The seeds of *H. ammodendron* were obtained from Luze Ecological Seeds Co., Ltd. (Guyuan, China), and are soaked in 1% sodium hypochlorite solution (diluted with water) for 20~30 min, then washed with water and used. Soil was collected from 0 to 60 cm between the roots of pike in the study area, mixed well, and used as in situ soil. The ratio of the soil matrix was peat soil: in situ soil: perlite = 3:1:1. The treatment method of soil sterilization involved high-pressure steam sterilization, which was then repeated three times. The duration of each sterilization time was 1.5 h. The pot experiment was conducted with a pot height of 30 cm and a diameter of 10 cm. The weight of each basin was 1.9 kg. We used ordinary urea (containing N 46%) as the source for nitrogen addition and dissolved it in deionized water in the form of a solution with a concentration of 22.4 mg·kg^{-1}.

2.2. Methods

2.2.1. Experimental Design

Under the precondition of sterilized (M) and non-sterilized (S) soil, four treatments were set up with four replicates in each group: control group (CK), nitrogen addition group (N), water deficit group (D), and water deficit nitrogen addition group (ND), for a total of 32 samples.

The water treatment was divided into normal water group (70% of field capacity) and water deficit group (30% of field capacity). During the experimental water control process, with sterile water used for irrigation, daily weighing was carried out from 18:00 to 20:00 Beijing time, combined with the use of EM50 soil moisture data collector measurement to observe the dynamics of the relative water content of the soil; the water consumed on that day was replenished, so that the soil field water holding capacity was controlled within the set gradient. The nitrogen addition group was divided into a no addition group and a nitrogen addition group, and was added once a month for a total of three additions from July to September. (Nitrogen addition level was based on the nitrogen content per kg of soil in *H. ammodendron* habitat [18]).

The experimental environment for potted *H. ammodendron* seedlings was conducted outdoors under the same temperature and light conditions as in the study area for a total of 5 months from planting to sampling.

2.2.2. Sample Collection and Processing

The control pot plants were excavated from the pot, the surrounding soil was carefully stripped, and the rhizosphere soil was collected via the PBS root washing method [19]. The rhizosphere soil of each plant was put into sterile tubes and marked and stored at −80 °C for the determination of soil microbial community composition and diversity.

Plant samples were collected and plant height, root length, above-ground biomass, and below-ground biomass were measured, where plant fresh weight was used to calculate the root-to-crown ratio of *H. ammodendron* seedlings with the following equation:

$$RSR = \frac{BB}{AB}$$

RSR: Root-shoot ratio; *BB*: Below-ground biomass; *AB*: Above-ground biomass.

Plant organic carbon was measured using the potassium dichromate dilution heat method, and then the organic matter content was obtained by converting the formula as follows:

$$SOM = SOC * 1.724$$

SOM: Soil organic matter; *SOC*: Soil organic carbon.

2.2.3. 16s rDNA Sequencing

Microbial Diversity is based on the Illumina NovaSeq sequencing platform and uses a two-end sequencing (Paired-End) approach to build small fragment libraries for sequencing. Reads are filtered by splicing, clustering, or denoising, and species annotation and abundance analysis are performed. Primers using 16s rDNA universal primers.

2.2.4. Calculation of Rhizosphere Soil Microbial Community Diversity

The raw data obtained by sequencing were spliced after the removal of the primer adapter sequence and the low quality bases (Phred quality score = 20). The sequences with a length of less than 200 bp were discarded, and the non-specific amplification sequences and chimeras were removed to obtain the effective sequence data of each sample. The 16 S sequence was divided into operational taxonomic units (OTUs), with 97% set as the threshold. RDP classifier 2.12 was used to analyze the representative sequences of OTUs at 97% similarity level, and the species classification information corresponding to each OTU was obtained. The bacterial community composition and relative abundance of each sample were counted, and the α diversity index (Chao1, Shannon index, Simpson index and coverage) was calculated.

2.3. Data Analysis

The differences of functional traits and rhizosphere microbial diversity of *H. ammodendron* seedlings under water control and nitrogen addition were analyzed using one-way ANOVA and an independent sample *t* test. Multivariate permutation analysis of variance (PERMANOVA) and principal components were used. Analysis (PCA) was used to analyze and compare the effects of water deficit and nitrogen addition on microbial community composition. Linear discriminant analysis (LEfSe) was used to detect components with significant differences in microbial community composition under conditions of water deficit and nitrogen addition. One-way ANOVA was completed in SPSS26.0. Multivariate permutation analysis of variance and PCA were completed in R, and data visualization was performed in Excel 2019 and Origin 2023b.

3. Results

3.1. Differences of Aboveground and Underground Functional Traits of H. ammodendron Seedlings

3.1.1. Differences in Aboveground and Underground Morphological Indexes of *H. ammodendron* Seedlings

Under different soil habitat conditions, the morphological structure of *H. ammodendron* seedlings changed significantly (Figure 1). Under sterilized and non-sterilized conditions, the root length of *H. ammodendron* seedlings was affected by water deficit (MD), which was significantly lower than that of the nitrogen addition treatment (MN) ($p < 0.05$). Similarly, the plant height of *H. ammodendron* seedlings was also significantly affected ($p < 0.05$). Under the conditions of sterilization, the plant height of *H. ammodendron* seedlings under water deficit (MD) treatment was significantly lower than those of the control group (MCK), nitrogen addition group (MN), and water deficit nitrogen addition group (MND) ($p < 0.05$). Under non-sterile conditions, the plant height of *H. ammodendron* seedlings was also affected by water deficit (SD), the plant height was significantly lower than that of nitrogen addition (SN), and water deficit nitrogen addition (SND) ($p < 0.05$). When compared with identical treatments in the sterilization and non-sterilization groups, we found that although the plant height and root length of *H. ammodendron* seedlings were larger under non-sterilized conditions, this difference was non-significant ($p > 0.05$).

Under sterilized conditions, the root–shoot ratio of water deficit group was significantly different from that of the nitrogen addition group and water deficit nitrogen addition group ($p < 0.05$). Under non-sterile conditions, the root–shoot ratio of water deficit group differed significantly from that of the nitrogen addition group ($p < 0.05$). There was no significant difference in terms of root–shoot ratio between sterilized and non-sterilized conditions ($p > 0.05$) (Figure 2).

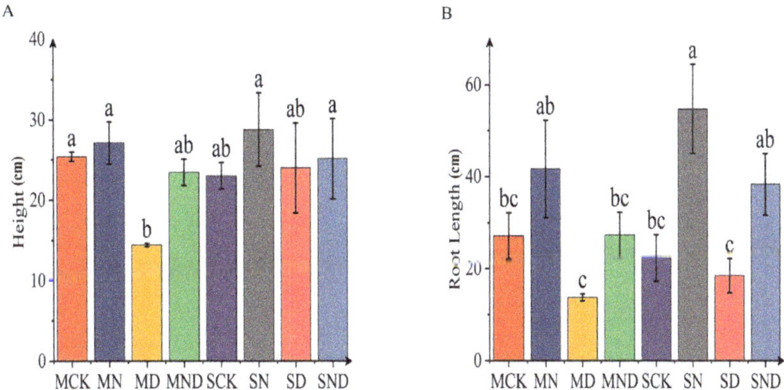

Figure 1. Plant physiological and ecological indicators. The bar above the column is Standard Error. (**A**) Plant height; (**B**) root length. Note: Different letters indicate significant differences between different treatment groups ($p < 0.05$). M: sterilization treatment; S: non-sterile treatment; CK: control group; N: nitrogen addition group; D: water deficit group; ND: water deficit, nitrogen addition group.

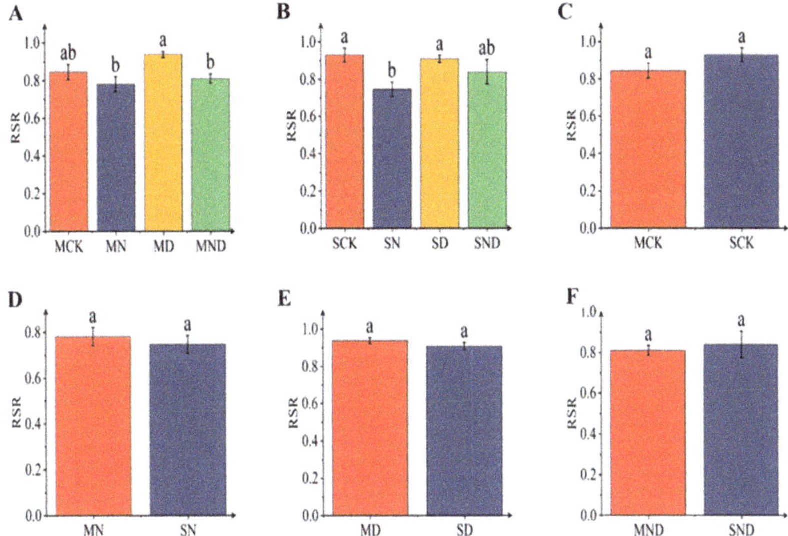

Figure 2. Root–shoot ratio of *H. ammodendron* seedlings. The bar above the column is Standard Error. (**A**) Root–shoot ratio of sterilization group; (**B**) Root–shoot ratio of non-sterile group; (**C**) Root–shoot ratio of sterilized non-sterilized control group; (**D**) Root–shoot ratio of sterilized non-sterilized nitrogen addition group; (**E**) Root–shoot ratio of sterilized non-sterilized water deficit group; (**F**) Root–shoot ratio of sterilized non-sterilized water deficit nitrogen addition group. Note: Different letters indicate significant differences between different treatment groups ($p < 0.05$). M: sterilization treatment; S: non-sterile treatment; CK: control group; N: nitrogen addition group; D: water deficit group; ND: water deficit, nitrogen addition group.

There were significant differences in aboveground and underground biomass of *H. ammodendron* seedlings under different treatment regimens ($p < 0.05$). Specifically, under sterile and non-sterile conditions, the aboveground biomass of *H. ammodendron* seedlings under nitrogen addition treatment (MN) was significantly higher than that under water deficit treatment (MD) ($p < 0.05$). We observed no significant difference between the nitrogen addition group and the water deficit and nitrogen addition group ($p > 0.05$). Under the sterilization conditions, the underground biomass of water deficit group was significantly lower than that of nitrogen addition group ($p < 0.05$). Under non-sterile conditions, with the underground biomass of *H. ammodendron* seedlings in water deficit, nitrogen addition treatment groups were significantly higher than that in the control group (SCK) and water deficit group (SD) ($p < 0.05$). The comparison of aboveground and underground biomass between sterilized and non-sterilized groups showed that the aboveground and underground biomasses were higher under sterilized conditions, but that there was no significant difference ($p > 0.05$) (Figure 3).

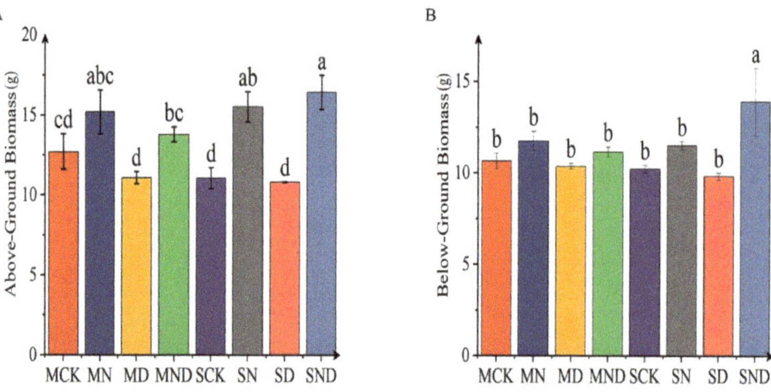

Figure 3. Displays the aboveground and belowground biomass of *H. ammodendron seedlings*. The bar above the column is Standard Error. (**A**) Above-ground biomass; (**B**) Below-ground biomass. Note: Different letters indicate significant differences between different treatment groups ($p < 0.05$). M: sterilization treatment; S: non-sterile treatment; CK: control group; N: nitrogen addition group; D: water deficit group; ND: water deficit, nitrogen addition group.

3.1.2. Correlation Analysis between Plant Aboveground and Underground Traits and Environmental Factors

Under sterilization conditions, plant root length (RL) was strongly correlated with soil organic carbon (SOC) and total phosphorus (TP) and displayed a positive correlation with RDA1. Aboveground biomass (AB) and belowground biomass (BB) were negatively correlated with ammonium nitrogen (SAN) in RDA1 and RDA2. On the RDA1 axis, plant height (HT) was positively correlated with available phosphorus (SAP), electrical conductivity (EC), total nitrogen (TN), and nitrate nitrogen (SNN). Conversely, it displayed a negative correlation with ammonium nitrogen (Figure 4).

Under non-sterile conditions, root length (RL) was positively correlated with total phosphorus (P) in terms of RDA1, negatively correlated with pH, and positively correlated with ammonium nitrogen, nitrate nitrogen, total nitrogen, and available phosphorus on the RDA2 axis. The aboveground biomass and belowground biomass were positively correlated with pH and EC, and negatively correlated with ammonium nitrogen, nitrate nitrogen, total nitrogen, and available phosphorus. Plant height was positively correlated with conductivity, ammonium nitrogen, nitrate nitrogen, available phosphorus, and total nitrogen on the RDA2 axis (Figure 5).

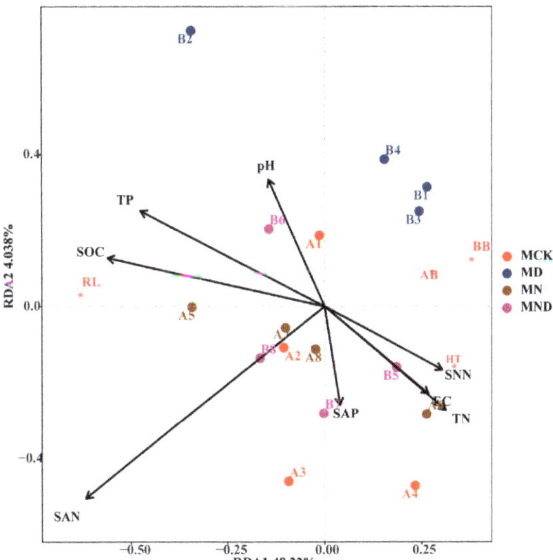

Figure 4. RDA analysis of aboveground and underground morphological characteristics of plants and environmental factors under sterilization conditions. Note: M: sterilization treatment; CK: control group; N: nitrogen addition group; D: water deficit group; ND: water deficit, nitrogen addition group; AB*: Above-ground biomass; BB*: Below-ground biomass; RL*: Root Length; HT*: Height.

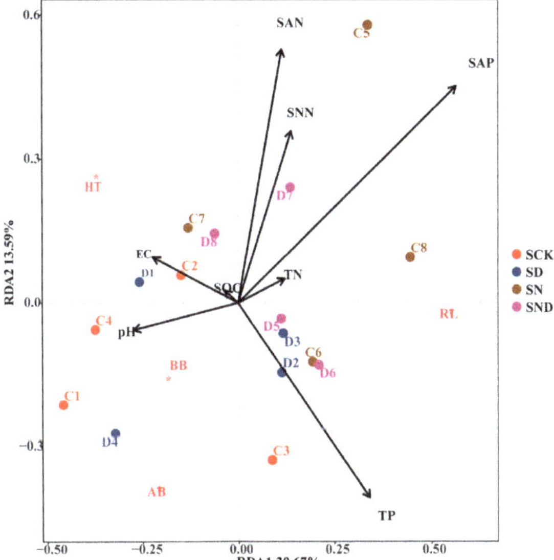

Figure 5. RDA Analysis of Aboveground and Belowground Morphological Characteristics of Plants and Environmental Factors under Non-Sterile Conditions. Note: S: non-sterile treatment; CK: control group; N: nitrogen addition group; D: water deficit group; ND: water deficit, nitrogen addition group. AB*: Above-ground biomass; BB*: aboveground Below-ground biomass; RL*: Root Length; HT*: Height.

3.2. *Diversity of Rhizosphere Bacterial Community in H. ammodendron Seedlings*

3.2.1. Differences in Rhizosphere Bacterial Community Abundance of *H. ammodendron* Seedlings

The number of OTUs can directly reflect the difference in bacterial community abundance between groups of samples (Figure 6). Under sterilization treatment, the bacterial abundance of the control group (MCK) was lower than that of the nitrogen addition group (MN), water deficit group (MD) and water deficit nitrogen addition group (MND), and there was a significant difference ($p < 0.05$) (Figure 7).

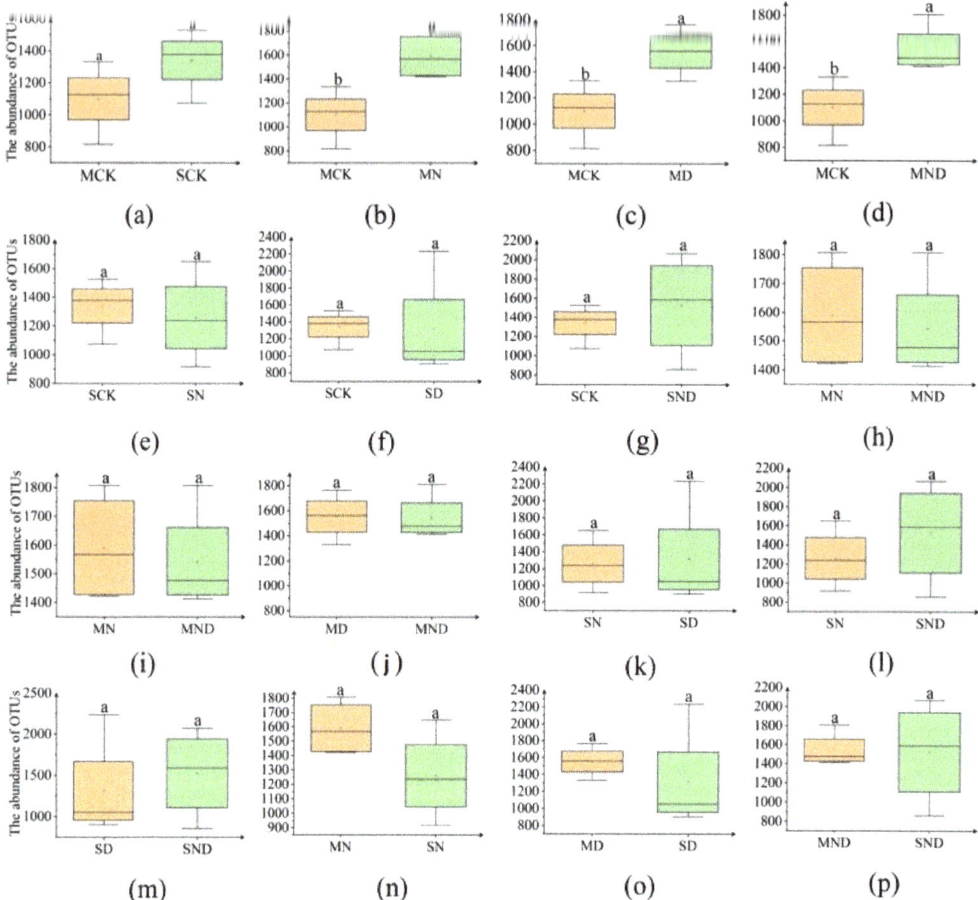

Figure 6. Bar plot of bacterial OTU numbers for different treatment groups. The bar above the column is Standard Error. (**a**) Sterilization control group vs. non-sterilization control group; (**b**) sterilization control group vs. sterilization nitrogen addition group; (**c**) sterilization control group vs. sterilization water deficit group; (**d**) sterilization control group vs. sterilization water deficit nitrogen addition group; (**e**) non-sterile control group vs. non-sterile nitrogen addition group; (**f**) non-sterile control group vs. non-sterile water deficit group; (**g**) non-sterile control group vs. non-sterile water deficit nitrogen addition group; (**h**) sterilized nitrogen addition group vs. sterilized water deficit nitrogen addition group; (**i**) sterilized nitrogen addition group vs. sterilized water deficit nitrogen addition group; (**j**) sterilization water deficit group vs. sterilization water deficit nitrogen addition group; (**k**) non-sterile nitrogen addition group vs. non-sterile water deficit group; (**l**) non-sterile nitrogen

addition group vs. non-sterile water deficit nitrogen addition group; (**m**) non-sterile water deficit group vs. non-sterile water deficit nitrogen addition group; (**n**) sterilized nitrogen addition group vs. non-sterilized nitrogen addition group; (**o**) sterilized water deficit group vs. non-sterilized water deficit group; (**p**) sterilized water deficit nitrogen addition group vs. non-sterilized water deficit nitrogen addition group. Note: M: sterilization treatment; S: non-sterile treatment; CK: control group; n: nitrogen addition group; d: water deficit group; ND: water deficit, nitrogen addition group; different lowercase letters indicate significant differences between groups ($p < 0.05$).

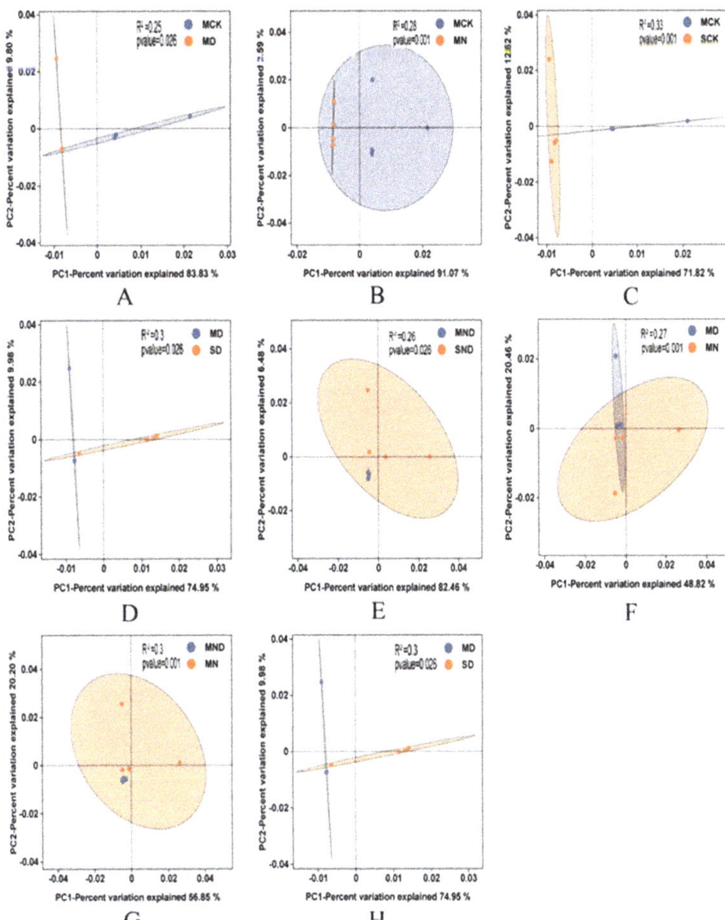

Figure 7. PCA analysis of different treatment groups vs. control groups. (**A**) Sterilization control group vs. sterilization water deficit group; (**B**) Sterilization control group vs. sterilization nitrogen addition group; (**C**) Sterilization control group vs. non-sterilization control group; (**D**) Sterilized water deficit group vs. non-sterilized water deficit group; (**E**) Sterilized water deficit nitrogen addition group vs. non-sterilized water deficit nitrogen addition group; (**F**) sterilized water deficit group vs. sterilized nitrogen addition group; (**G**) Sterilized water deficit nitrogen addition group vs. sterilized nitrogen addition group; (**H**) Sterilized water deficit group vs. non-sterilized water deficit group. Note: M: sterilization treatment; S: non-sterile treatment; CK: control group; N: nitrogen addition group; D: water deficit group; ND: water deficit, nitrogen addition group.

3.2.2. Alpha Diversity of Bacterial Community in *H. ammodendron* Seedlings

Under sterilization conditions, the control group (MCK) and water deficit (MD), nitrogen addition group (MND) had significant differences in terms of Simpson index value ($p < 0.05$); there were significant differences between water deficit group (MD), nitrogen addition group (MN), and the water deficit nitrogen addition group (ND) ($p < 0.05$). The nitrogen addition group (N) and water deficit nitrogen addition group (MND) displayed significant differences ($p < 0.05$). Under non-sterile conditions, there was no significant difference in Simpson index between the control group (MCK), the water deficit group (MD), nitrogen addition group (MN), and water deficit nitrogen addition group (MND) ($p > 0.05$). The ACE, Chao1, and Shannon index values of soil bacteria under non-sterile nitrogen addition treatment conditions were the largest, and the soil bacterial diversity index value of non-sterile control group was the smallest. There was a significant difference between the non-sterile nitrogen addition treatment and the non-sterile water deficit nitrogen addition treatment and the control group ($p < 0.05$) (Table 1).

Table 1. The diversity of bacteria in different soil samples.

Sample ID	ACE Mean ± Se	Chao1 Mean ± Se	Simpson Mean ± Se	Shannon Mean ± Se
MCK	1360.69 ± 97 a	1361.36 ± 97 a	0.992 ± 0.002 a	8.893 ± 0.193 abc
MN	1276.92 ± 152 a	1280.19 ± 153 a	0.989 ± 0.005 a ·	8.649 ± 0.211 abc
MD	1334.75 ± 317 a	1338.82 ± 321 a	0.975 ± 0.008 a	8.264 ± 0.662 c
MND	1571.13 ± 278 a	1565.63 ± 276 a	0.974 ± 0.02 a	8.412 ± 0.708 bc
SCK	1118.13 ± 111 a	1119.48 ± 112 a	0.971 ± 0.015 a	8.234 ± 0.468 c
SN	1636.32 ± 105 a	1635.75 ± 106 a	0.998 ± 0.001 a	9.737 ± 0.07 a
SD	1585.50 ± 91 a	1593.35 ± 94 a	0.996 ± 0.002 a	9.542 ± 0.266 ab
SND	1572.96 ± 94 a	1576.39 ± 96 a	0.998 ± 0.001 a	9.688 ± 0.093 a

Note: M: sterilization treatment; S: non-sterile treatment; CK: control group; N: nitrogen addition group; D: water deficit group; ND: water deficit, nitrogen addition group; different lowercase letters indicated significant differences between groups ($p < 0.05$).

3.2.3. Changes in Soil Bacterial Community Composition under Different Treatments

Through PCA analysis, we uncovered that, under sterilization conditions, nitrogen addition and water deficit groups were significantly different from the control group ($p < 0.05$). Indeed, nitrogen addition, water deficit, and water deficit nitrogen addition treatment also had significant differences ($p < 0.05$). There were further significant differences between the results of the control group, nitrogen addition, water deficit, and water deficit nitrogen addition ($p < 0.05$).

The results showed that, at the gate level, the 8 measured treatment groups belonged to 45 gates. Among them, under sterilization treatment conditions, the top 10 groups included Proteobacteria, Acidobacteria, Gemmatimonadetes, Bacteroides, Chloroflexi, Actinobacteria, Firmicutes, Myxococcota, unclassified_Bacteria, Patescibacteria, and a number of others. In non-sterilized treatment conditions, the top 10 groups included Proteobacteria, Firmicutes, Bacteroides, Gemmatimonadetes, Acidobacteria, Actinobacteria, Myxococcota, Chloroflexi, Cyanobacteria, unclassified_Bacteria. The rhizosphere bacterial community composition of each treatment was the same, and the dominant species were similar, but the abundance was slightly different (Figure 8).

To analyze the specific taxonomic groups of bacteria in different treatment groups, linear discriminant analysis (LDA) effect size (LEfSe) difference analysis was performed to select biomarkers from the phylum to the genus level. The analysis showed that in the sterilization and non-sterilization treatment, each treatment group had characteristic biological species. These tended to be concentrated in Proteobacteria, Acidobacteria and Bacteroidetes. Under sterilized conditions, Gemmatimonadales and SBR1031 were significantly enriched in the nitrogen addition group (MN). In the nitrogen addition group (MND) under water deficit, sphingomonas, sulphurococcus, and thermoaerophilic bacteria were significantly enriched. Under non-sterile conditions, the enrichment of Pseudomonas in nitrogen addition group (N) was significant; in the water deficit nitrogen addition group (ND), cyanobacteria were significantly enriched (Figure 9).

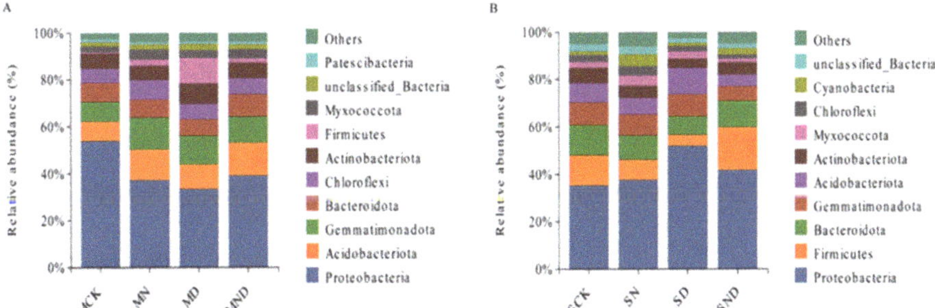

Figure 8. Classification of root-associated microbial communities in different treatment groups at the phylum level. (**A**) The sterilization treatment group ranks among the top 10 microbial taxa at the phylum level; (**B**) The non-sterile treatment group ranks among the top 10 microbial taxa at the phylum level. Note: MCK, MN, MD, MND, SCK, SN, SD, SND represent different treatment groups.

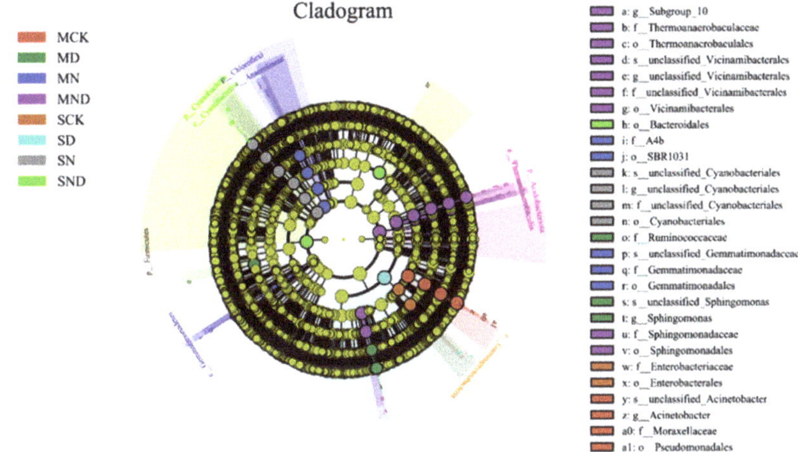

Figure 9. LEfSe analysis of rhizosphere bacterial communities under different treatments. Note: MCK, MN, MD, MND, SCK, SN, SD, SND represent different treatment groups.

4. Discussion

4.1. Physiological and Ecological Changes in H. ammodendron Seedlings under Nitrogen and Water Coupling

In the present study, the growth status of *H. ammodendron* seedlings was affected by water deficit and nitrogen addition. Under non-sterilized conditions, root length and plant height of *H. ammodendron* seedlings were significantly increased in the nitrogen-added group compared to the control group ($p < 0.05$). Similarly, plant height and root length in the water-deficient nitrogen addition group also increased significantly ($p < 0.05$) compared to the water deficit group and the control group, a result consistent with previous studies indicating that nitrogen addition promotes the growth of *H. ammodendron* seedlings and water deficit inhibits the growth of *H. ammodendron* seedlings [20,21]. However, in the sterilization treatment, there was no significant difference in root length and plant height between the nitrogen addition group and the control group. The root length and plant height of the water deficit group were significantly lower than those of the control group ($p < 0.05$), indicating that the growth of the plant was affected under conditions water deficit, and that nitrogen addition slowed down the effect of water deficit on the physiological state of the plant [22]. In addition, sterilization and non-sterilization treatments affect the root length of plants, and plants under sterilization treatment show a significant increase in root length and plant height compared to non-sterilization treatment ($p < 0.05$), with the maximum plant height in the non-sterilized group being significantly greater than that in the sterilized group [23,24]. Some microorganisms in the soil under non-sterilized conditions can effectively promote plant growth [25].

The experimental results show that water deficit and nitrogen addition can significantly affect the aboveground and underground biomass of plants, and previous studies have also demonstrated the validity of the relevant conclusions [26]. The aboveground and underground biomass of the nitrogen addition treatment group under sterilized and non-sterilized conditions was significantly higher than that of the water deficit group ($p < 0.05$), but the root length and plant height of the plants under sterilized treatment were significantly different from those under non-sterilized treatment ($p < 0.05$). In this study, we found that the root–shoot ratio of plants under water deficit treatment was higher compared to other treatments under both sterilized and non-sterilized conditions, especially significantly different from the nitrogen addition treatment group ($p < 0.05$) [27], which indicates that *H. ammodendron* seedlings under water deficit conditions allocated more biomass to the lower part of the ground to ensure the growth of the lower part of the ground, while the root–shoot ratio of *H. ammodendron* seedlings under the combined effect of water deficit and nitrogen addition was not significantly different relative to the control ($p > 0.05$), indicating that nitrogen addition would alleviate the water deficit to which the plants were subjected.

4.2. Changes in Rhizosphere Microbial Community in H. ammodendron Seedlings

Studies have shown that the composition and richness of plant rhizosphere microbial communities vary among different external habitats [28]. Water deficit will have a negative impact on the microbial community, resulting in a decrease in rhizosphere bacterial community diversity [29]. The same study found that excessive nitrogen addition can also lead to decreases in rhizosphere bacterial community diversity [30]. However, scholars have drawn multiple different conclusions. Some studies have found that drought disturbance can sometimes promote the diversity of rhizosphere bacterial communities. Similarly, nitrogen addition treatment has also been found to promote significant increases in rhizosphere bacterial community abundance [31].

This study found that under non-sterile conditions, water deficit reduced the abundance of rhizosphere bacterial communities and reduced microbial community diversity, which was consistent with previous studies [32]. At the same time, it was also found that the diversity of rhizosphere bacterial communities treated with nitrogen addition was also reduced, that is, both water deficit and nitrogen addition lead to decreases in the abundance of bacterial microbial communities. This result mirrored those of previous studies [33]. However, under the combined treatment of the two, the abundance of rhizosphere bacterial communities increased, and the diversity of the bacterial communities increased [34]. Additionally, because the initial soil is barren, nitrogen addition has a greater advantage in promoting the diversity of bacterial microbial communities, resulting in an increase in bacterial microbial diversity [35]. Under the condition of sterilization, water deficit and nitrogen addition alone can promote the abundance of the rhizosphere soil microbial community. Compared with the effects of the control treatment, the increase in bacterial community abundance is significant. This may be because the rhizosphere bacterial community is in the process of community restoration and reconstruction after sterilization treatment, which indicates that the diversity of rhizosphere microbial community is far lower than that of the normal community level, meaning that the application of external environmental factors cannot inhibit the growth of rhizosphere bacteria. When the two work together, water deficit and nitrogen addition will reduce the diversity of rhizosphere bacterial microbial communities. This may be due to the combination of the two, resulting in a great increase in the amount of bacterial microorganisms and resource competition, which in turn decreases the abundance of rhizosphere bacterial microbial communities. It is speculated that the initial soil habitat conditions of sterilized or non-sterilized soil will affect the diversity of rhizosphere bacterial communities [36].

The bacterial community in rhizosphere soil is affected by water deficit and nitrogen addition, leading to differences in community composition [30,33]. Under sterilized conditions, the composition of rhizosphere bacterial microbial communities at the phylum level was different under different treatments. Acidobacteria, Gemmatimonadetes, and Chloroflexi were more abundant in the nitrogen addition treatment group than in the control group. Acidobacteria lived in an oligotrophic manner in the ecosystem, and their high content indicated poor soil nutrition. Among them, Gemmatimonadetes lived more in nitrogen-rich soil, and Chloroflexi participated in nitrogen cycle. Their higher abundance indicated that nitrogen addition influenced soil bacterial community composition [37–39]. In water deficit treatment, the abundance of Actinobacteria and Firmicutes was higher than that of the control group, and the proportion of bacterial community increased. This showed that water deficit had a greater impact on Actinomycetes and Firmicutes. Under the combined action of water deficit and nitrogen addition, the abundance of Acidobacteria and Bacteroidetes was higher, and Bacteroidetes played a greater role in degrading polysaccharides [40–42]. Under non-sterile conditions, the abundance of Myxobacteria, Chloroflexi, and Cyanobacteria in the nitrogen addition group was higher than that in the control group. In the water deficit treatment group, the abundance of Proteobacteria and Myxococcus was higher than that of the control group and the water deficit nitrogen addition group, indicating that some genera of Proteobacteria participated in plant physiological processes and played a significant role in plant deficit resistance [43]. The above viewpoint was also confirmed by LEfSe analysis. Due to different treatment conditions, the specific flora of the rhizosphere bacterial community was enriched, giving nitrogen fixation in Cyanobacteria and strong drought tolerance in Gemmatimonadetes [44].

This study also has some limitations. In the future, root exudates and microorganisms should be combined to study the response of plants to soil environmental changes and determine how root exudates mediate rhizosphere microbial community changes by changing the soil habitats of plants.

5. Conclusions

In short-term control trials, water deficiency inhibited the growth of pokeweed seedlings and was more pronounced under non-sterilized conditions. Nitrogen addition was effective in alleviating water deficit in plants. Water deficiency changed the original carbon allocation ratio of the plants, transporting more carbon underground thus promoting the growth of the below-ground organs of the plants.

The effects of water deficit and nitrogen addition on the diversity of inter-rooted bacterial communities of *P. sylvestris* seedlings under different initial soil conditions were significantly different. Under soil sterilization conditions, both water deficit and N addition alone promoted an increase in bacterial community abundance and diversity and altered the composition of the inter-root microbial community. However, the diversity of the inter-rooted microbial community was suppressed by the combined effect of both. Under non-sterilized conditions, water deficiency and nitrogen addition alone suppressed the diversity of the inter-rooted bacterial community of *P.* spp. seedlings, while the combination of both promoted bacterial diversity.

Author Contributions: Conceptualization, M.Z.; methodology, M.Z.; software, M.Z.; validation, L.J., D.W., H.Y. and W.L.; formal analysis, M.Z.; investigation, M.Z.; resources, M.Z.; data curation, M.Z.; writing—original draft preparation, M.Z.; writing—review and editing, M.Z.; visualization, M.Z.; supervision, X.H.; project administration, X.H.; funding acquisition, X.H. All authors have read and agreed to the published version of the manuscript.

Funding: This research was financially supported by the National Natural Science Foundation of China (32260266): National Natural Science Foundation of China (32101360) and Xinjiang Uygur Autonomous Region innovation environment Construction special project & Science and technology innovation base construction project (PT2107).

Data Availability Statement: Data will be made available on request.

Conflicts of Interest: The authors declare no conflict of interest.

References

1. Wang, J.; Duan, B.; Zhang, Y. Effects of Experimental Warming on Growth, Biomass Allocation, and Needle Chemistry of Abies Faxoniana in Even-Aged Monospecific Stands. *Plant Ecol.* **2012**, *213*, 47–55. [CrossRef]
2. Yang, X.; Long, Y.; Sarkar, B.; Li, Y.; Lü, G.; Ali, A.; Yang, J.; Cao, Y.-E. Influence of Soil Microorganisms and Physicochemical Properties on Plant Diversity in an Arid Desert of Western China. *J. For. Res.* **2021**, *32*, 2645–2659. [CrossRef]
3. Mokany, K.; Raison, R.J.; Prokushkin, A.S. Critical Analysis of Root: Shoot Ratios in Terrestrial Biomes. *Glob. Chang. Biol.* **2005**, *12*, 84–96. [CrossRef]
4. Singla-Pareek, S.L. Transcription Factors and Plants Response to Drought Deficit: Current Understanding and Future Directions. *Front. Plant Sci.* **2016**, *7*, 1029. [CrossRef]
5. Xu, L.; Naylor, D.; Dong, Z.; Simmons, T.; Pierroz, G.; Hixson, K.K.; Kim, Y.-M.; Zink, E.M.; Engbrecht, K.M.; Wang, Y.; et al. Drought Delays Development of the Sorghum Root Microbiome and Enriches for Monoderm Bacteria. *Proc. Natl. Acad. Sci. USA* **2018**, *115*, E4284–E4293. [CrossRef]
6. Fang, S.; Liang, X. Response Mechanisms of Plants Under Saline-Alkali Deficit. *Front. Plant Sci.* **2021**, *72*, 673–689. [CrossRef]
7. Fang, Y.; Xiong, L. General Mechanisms of Drought Response and Their Application in Drought Resistance Improvement in Plants. *Front. Plant Sci.* **2021**, *12*, 667458. [CrossRef]
8. Etesami, H. Bacterial Mediated Alleviation of Heavy Metal Deficit and Decreased Accumulation of Metals in Plant Tissues_ Mechanisms and Future Prospects. *Ecotoxicol. Environ. Saf.* **2018**, *147*, 175–191. [CrossRef]
9. Li, W.J.; Wang, J.L.; Jiang, L.M.; Lv, G.H.; Hu, D.; Wu, D.Y.; Yang, X.D. Rhizosphere effect and water constraint jointly determined the roles of microorganism in soil phosphorus cycling in arid desert regions. *CATENA* **2023**, *222*, 106809. [CrossRef]
10. Huang, Z.; Zhang, X.; Zheng, G.; Gutterman, Y. Influence of Light, Temperature, Salinity and Storage on Seed Germination of *Haloxylon ammodendron*. *J. Arid Environ.* **2003**, *55*, 453–464. [CrossRef]
11. Sheng, Y.A.N.; Zheng, W.H.; Pei, K.Q.; Ma, K.P. Genetic Variation within and Among Populations of a Dominant Desert Tree *H. ammodendron* (Amaranthaceae) in China. *Ann. Bot.* **2005**, *96*, 245–252. [CrossRef] [PubMed]
12. Kour, D.; Yadav, A.N. Bacterial Mitigation of Drought Deficit in Plants: Current Perspectives and Future Challenges. *Curr. Microbiol.* **2022**, *79*, 248. [CrossRef] [PubMed]
13. Hassan, F.A.S.; Ali, E.F.; Mahfouz, S.A. Comparison between different fertilization sources, irrigation frequency and their combinations on the growth and yield of coriander plant. *Aust. J. Basic Appl. Sci.* **2012**, *6*, 600–615.
14. Boyle, S.A.; Yarwood, R.R.; Bottomley, P.J.; Myrold, D.D. Bacterial and Fungal Contributions to Soil Nitrogen Cycling under Douglas FIr and Red Alder at Two Sites in Oregon. *Soil Biol.* **2008**, *40*, 443–450. [CrossRef]
15. He, K.N.; Wang, H.; Wang, W.L.; Zhang, T. Salinity Effects on Germination and Plant Growth of H. ammodendron at Qaidam Basin. *J. Phys. Conf. Ser.* **2020**, *1578*, 012234. [CrossRef]
16. Vos, M.; Wolf, A.B.; Jennings, S.J.; Kowalchuk, G.A. Micro-Scale Determinants of Bacterial Diversity in Soil. *FEMS Microbiol. Rev.* **2013**, *37*, 936–954. [CrossRef]
17. Treseder, K.K. Nitrogen Additions and Microbial Biomass: A Meta-analysis of Ecosystem Studies. *Ecol. Lett.* **2008**, *11*, 1111–1120. [CrossRef]
18. Gutknecht, J.M.; Field, C.B.; Balser, T.C. Microbial Communities and Their Responses to Simulated Global Change Fluctuate Greatly Over Multiple Years. *Glob. Chang. Biol.* **2012**, *18*, 2256–2269. [CrossRef]
19. Edwards, J.; Johnson, C.; Santos-Medellín, C.; Lurie, E.; Podishetty, N.K.; Bhatnagar, S.; Eisen, J.A.; Sundaresan, V. Structure, Variation, and Assembly of the Root-Associated Microbiomes of Rice. *Proc. Natl. Acad. Sci. USA* **2015**, *112*, E911–E920. [CrossRef]
20. Turner, B.L.; Driessen, J.P.; Haygarth, P.M.; Mckelvie, I.D. Potential Contribution of Lysed Bacterial Cells to Phosphorus Solubilisation in Two Rewetted Australian Pasture Soils. *Soil Biol. Biochem.* **2003**, *35*, 187–189. [CrossRef]
21. Etzold, S.; Ferretti, M.; Reinds, G.J.; Solberg, S.; Gessler, A.; Waldner, P.; Schaub, M.; Simpson, D.; Benham, S.; Hansen, K.; et al. Nitrogen Deposition Is the Most Important Environmental Driver of Growth of Pure, Even-Aged and Managed European Forests. *For. Ecol. Manag.* **2020**, *458*, 117762. [CrossRef]
22. Valliere, J.M.; Allen, E.B. Interactive Effects of Nitrogen Deposition and Drought-Deficit on Plant-Soil Feedbacks of Artemisia Californica Seedlings. *Plant Soil.* **2016**, *403*, 277–290. [CrossRef]
23. Li, Y.; Hu, W.; Zou, J.; He, J.; Wang, Y.; Chen, B.; Meng, Y.; Wang, S.; Zhou, Z. Effects of Soil Drought on Cottonseed Kernel Carbohydrate Metabolism and Kernel Biomass Accumulation. *Plant Physiol. Biochem.* **2023**, *195*, 170–181. [CrossRef]
24. Li, W.; Jin, C.; Guan, D.; Wang, Q.; Wang, A.; Yuan, F.; Wu, J. The Effects of Simulated Nitrogen Deposition on Plant Root Traits: A Meta-Analysis. *Soil Biol. Biochem.* **2015**, *82*, 112–118. [CrossRef]

25. Zhang, X.; Lei, L.; Lai, J.; Zhao, H.; Song, W. Effects of Drought Deficit and Water Recovery on Physiological Responses and Gene Expression in Maize Seedlings. *BMC Plant Biol.* **2018**, *18*, 68. [CrossRef]
26. Ravelo-Ortega, G.; Raya-González, J.; López-Bucio, J. Compounds from Rhizosphere Microbes That Promote Plant Growth. *Curr. Opin. Plant Biol.* **2023**, *73*, 102336. [CrossRef]
27. Li, W.; Lei, X.; Zhang, R.; Cao, Q.; Yang, H.; Zhang, N.; Liu, S.; Wang, Y. Shifts in Rhizosphere Microbial Communities in Oplopanax Elatus Nakai Are Related to Soil Chemical Properties under Different Growth Conditions. *Sci. Rep.* **2022**, *12*, 11485. [CrossRef]
28. Bach, E.M.; Williams, R.J.; Hargreaves, S.K.; Yang, F.; Hofmockel, K.S. Greatest Soil Microbial Diversity Found in Micro-Habitats. *Soil Biol. Biochem.* **2018**, *118*, 217–226. [CrossRef]
29. Fuchslueger, L.; Bahn, M.; Fritz, K.; Hasibeder, R.; Richter, A. Experimental Drought Reduces the Transfer of Recently Fixed Plant Carbon to Soil Microbes and Alters the Bacterial Community Composition in a Mountain Meadow. *New Phytol.* **2014**, *201*, 916–927. [CrossRef]
30. Kaurin, A.; Mihelič, R.; Kastelec, D.; Grčman, H.; Bru, D.; Philippot, L.; Suhadolc, M. Resilience of Bacteria, Archaea, Fungi and N-Cycling Microbial Guilds under Plough and Conservation Tillage, to Agricultural Drought. *Soil Biol. Biochem.* **2018**, *120*, 233–245. [CrossRef]
31. Van, D.L.T.A.; Lilleskov, E.A.; Pregitzer, K.S.; Miller, R.M. Simulated Nitrogen Deposition Causes a Decline of Intra- and Extraradical Abundance of Arbuscular Mycorrhizal Fungi and Changes in Microbial Community Structure in Northern Hardwood Forests. *Ecosystems* **2010**, *13*, 683–695. [CrossRef]
32. Singh, P.; Singh, R.K.; Zhou, Y.; Wang, J.; Jiang, Y.; Shen, N.; Wang, Y.; Yang, L.; Jiang, M. Unlocking the Strength of Plant Growth Promoting *Pseudomonas* in Improving Crop Productivity in Normal and Challenging Environments: A Review. *J. Plant Interact.* **2022**, *17*, 220–238. [CrossRef]
33. Fitzpatrick, C.R.; Copeland, J.; Wang, P.W.; Guttman, D.S.; Kotanen, P.M.; Johnson, M.T.J. Assembly and Ecological Function of the Root Microbiome across Angiosperm Plant Species. *Proc. Natl. Acad. Sci. USA* **2018**, *115*, E1157–E1165. [CrossRef] [PubMed]
34. Hillebrand, H. Meta-analysis on Pulse Disturbances Reveals Differences in Functional and Compositional Recovery across Ecosystems. *Ecol. Lett.* **2020**, *23*, 575–585. [CrossRef]
35. Zhou, Z.; Wang, C.; Zheng, M.; Jiang, L.; Luo, Y. Patterns and Mechanisms of Responses by Soil Microbial Communities to Nitrogen Addition. *Soil Biol. Biochem.* **2017**, *115*, 433–441. [CrossRef]
36. Zhang, T.; Chen, H.Y.H.; Ruan, H. Global Negative Effects of Nitrogen Deposition on Soil Microbes. *ISME J.* **2018**, *12*, 1817–1825. [CrossRef]
37. Arnett, A.E.; Louda, S.M. Re-Test of Rhinocyllus Conicus Host Specificity, and the Prediction of Ecological Risk in Biological Control. *Biol. Conserv.* **2002**, *106*, 251–257. [CrossRef]
38. Moreira, X.; Abdala-Roberts, L. Specificity and Context-Dependency of Plant–Plant Communication in Response to Insect Herbivory. *Curr. Opin. Insect Sci.* **2019**, *32*, 15–21. [CrossRef]
39. Dowie, N.J.; Grubisha, L.C.; Trowbridge, S.M.; Klooster, M.R.; Miller, S.L. Variability of Ecological and Autotrophic Host Specificity in a Mycoheterotrophic System: *Pterospora andromedea* and Associated Fungal and Conifer Hosts. *Fungal Ecol.* **2016**, *20*, 97–107. [CrossRef]
40. Zhou, Z.; Zheng, M.; Xia, J.; Wang, C. Nitrogen Addition Promotes Soil Microbial Beta Diversity, and the Stochastic Assembly. *Sci. Total Environ.* **2022**, *806*, 150569. [CrossRef]
41. Cheng, J. Bacteroides Utilization for Dietary Polysaccharides and Their Beneficial Effects on Gut Health. *Food Sci. Hum. Wellness* **2022**, *11*, 1101–1110. [CrossRef]
42. Hu, Y.; Chen, M.; Yang, Z.; Cong, M.; Zhu, X.; Jia, H. Soil Microbial Community Response to Nitrogen Application on a Swamp Meadow in the Arid Region of Central Asia. *Front. Microbiol.* **2022**, *12*, 797306. [CrossRef] [PubMed]
43. Genderjahn, S.; Mashal, A.; Kai, M.; Fabian, H.; Dirk, W. Desiccation- and Saline-Tolerant Bacteria and Archaea in Kalahari Pan Sediments. *Front. Microbiol.* **2018**, *9*, 2082. [CrossRef] [PubMed]
44. Xie, J.B.; Peng, L.Z.; Li, Y.M.; Li, Y. Competitive interactions between two desert shrub seedings towards variation in soil nitrogen and phosphorus content. *Arid. Land Geog. Raphy.* **2018**, *41*, 83–91. [CrossRef]

Disclaimer/Publisher's Note: The statements, opinions and data contained in all publications are solely those of the individual author(s) and contributor(s) and not of MDPI and/or the editor(s). MDPI and/or the editor(s) disclaim responsibility for any injury to people or property resulting from any ideas, methods, instructions or products referred to in the content.

Article

Soil Bacterial Community Structure and Physicochemical Influencing Factors of Artificial *Haloxylon ammodendron* Forest in the Sand Blocking and Fixing Belt of Minqin, China

Anlin Wang, Rui Ma *, Yanjun Ma, Danni Niu, Teng Liu, Yongsheng Tian, Zhenghu Dong and Qiaodi Chai

College of Forestry, Gansu Agricultural University, Lanzhou 730070, China; wangal@st.gsau.edu.cn (A.W.); mayanjun@gsau.edu.cn (Y.M.)

* Correspondence: mar@gsau.edu.cn

Abstract: Microbial activity plays a crucial role in upholding the functional stability of vegetation–soil ecosystems. Nevertheless, there exists a paucity of studies concerning the impact of sand-fixing vegetation (*Haloxylon ammodendron*) on the structure and functional attributes of soil microbial communities. We employed Illumina high-throughput sequencing and PICRUSt2 functional prediction technology to investigate the characteristics of soil bacterial community structure, diversity, and metabolic functions in an artificial *H. ammodendron* forest, and RDA analysis and the Mantel test were used to reveal the main environmental factors affecting the structure and ecological functions of soil bacterial communities. The findings revealed a significant increase in the principal nutrient contents (organic matter, total nitrogen, total phosphorus) in the *H. ammodendron* forest soil compared to the mobile dune soil, while a reduction of 17.17% in the surface soil water content was observed. The *H. ammodendron* forest exhibited a significant enhancement in the diversity and richness index of soil bacteria. Specifically, Actinobacteria (24.94% ± 11.85%), Proteobacteria (29.99% ± 11.56%), and Chloroflexi (11.14% ± 4.55%) emerged as the dominant bacterial phyla, with Actinobacteria displaying significantly higher abundance compared to the mobile dune soil. PICRUSt2 analyses revealed that the predominant secondary metabolic functions of soil bacteria were carbohydrate metabolism, amino acid metabolism, and the metabolism of cofactors and vitamins. Additionally, the tertiary metabolic pathways exhibited greater activity in relation to enzyme function, nucleotide metabolism, energy metabolism, and antibiotics. The RDA results demonstrated that SOM, AK, and pH collectively accounted for 82.4% of the cumulative contribution, significantly influencing the bacterial community. Moreover, the Mantel test revealed that the metabolic function of soil bacteria primarily relied on five environmental factors, namely SOM, TN, AK, pH, and EC. This study significantly advances our understanding of the structural and functional changes in soil bacterial communities during the reclamation of sandy land through the establishment of artificial *H. ammodendron* forests.

Keywords: soil bacterial community; PICRUSt2 function prediction; soil physicochemical properties; arid desert area; sandy land restoration

Citation: Wang, A.; Ma, R.; Ma, Y.; Niu, D.; Liu, T.; Tian, Y.; Dong, Z.; Chai, Q. Soil Bacterial Community Structure and Physicochemical Influencing Factors of Artificial *Haloxylon ammodendron* Forest in the Sand Blocking and Fixing Belt of Minqin, China. *Forests* **2023**, *14*, 2244. https://doi.org/10.3390/f14112244

Academic Editor: Anna Zavarzina

Received: 3 October 2023
Revised: 3 November 2023
Accepted: 8 November 2023
Published: 14 November 2023

Copyright: © 2023 by the authors. Licensee MDPI, Basel, Switzerland. This article is an open access article distributed under the terms and conditions of the Creative Commons Attribution (CC BY) license (https://creativecommons.org/licenses/by/4.0/).

1. Introduction

The desert–oasis transition zone represents a pivotal ecological juncture wherein desert and oasis ecosystems converge under a shared influence [1]. The sand blocking and fixing belt at the edge of the oasis, which consists of the protective forest belt and the peripheral sealing and protection belt, is the core barrier belt of the artificial oasis, playing a key role in preventing the invasion of wind and sand, maintaining the stability of the transition zone ecosystem, and guaranteeing the oasis's survival and development. At the same time, it is also an important part of the fragile ecosystem of the desert–oasis transition zone. However, the advent of global warming and irrational human activities has exacerbated the desertification process in this transition zone, rendering its ecological environment increasingly fragile and susceptible [2,3]. Therefore, the region's ecological stability and sustainable economic development have been severely impeded [4]. Nevertheless, the introduction of artificial vegetation specifically designed to stabilize the sand in the transitional zone has proven to be an efficient and expeditious means of mitigating desertification [5–9]. This approach not only ensures the preservation of oasis ecosystem stability but also enhances land productivity. Over the course of approximately fifty years, experts and scholars have widely embraced this perspective as an effective technique for combating desertification.

Considering its structurally intricate and non-homogeneous nature, soil plays a pivotal role in sustaining terrestrial ecosystem functions and services [10]. Soil nutrients are limited within desert ecosystems, necessitating the implementation of ecological mechanisms that expedite the soil element cycling process, regulate soil fertility transformation, and enhance the absorption and utilization of plant nutrients [11]. These mechanisms are crucial for ensuring the sustainable restoration of desert vegetation productivity and effectively mitigating land desertification [12]. Soil microorganisms, often referred to as the "engines" of biogeochemical cycles, assume a pivotal position in facilitating the circulation of materials, the flow of energy, and the preservation of stable ecological functions within soil ecosystems [13–15]. On one hand, the participation of soil microorganisms in ecological processes comprises the decomposition of apoplastic material and humus, the breakdown of organic matter, and the facilitation of symbiotic nitrogen fixation through plant–microbe systems [16–18]. On the other hand, the soil microbial communities exhibit remarkable sensitivity to changes in environmental factors and harbor a wealth of information pertaining to regional soil environmental changes [19]. This information is frequently employed to evaluate the stability and vitality of soil ecological functions. Notably, bacteria represent the predominant constituents of the soil microbial community, owing to their abundant population, extensive diversity, and intricate complexity. In addition, they possess the ability to adapt to extreme habitats, rendering them valuable as indicator species for assessing the stability, quality, and functionality of soil ecosystems [20].

The burgeoning popularity of the high-throughput sequencing of 16S rRNA and the utilization of PICRUSt2 macrogenomic function prediction technologies present innovative methodologies for unraveling the intricate microbial community structure that underlies the functional equilibrium of diverse soil ecosystems [21,22]. The study determined that the introduction of desert flora exerts a momentous influence on the micro-ecological rehabilitation of arid and semi-arid desert soils [9,11,17,19]. For instance, in the Horqin sandy land, the rehabilitation process involving *Caragana microphylla* vegetation not only contributed to enhancements in vegetation coverage and diversity but also entailed a significant enrichment of the soil microbial community structure [23]. Following two decades of growth and development in the Mu Us Sandy Land, *Pinus sylvestris* var. *mongolica* has demonstrably impacted the microbial diversity in the rhizosphere soil, thereby liberating an array of ecologically functional metabolites, thereby improving the metabolic functionality of soil bacteria [24]. The cultivation of *Artemisia ordosica* has not only ameliorated soil microbial enzyme activity and biomass but has also enhanced the abundance of Actinobacteria and Proteobacteria [25]. *Haloxylon ammodendron*, a diminutive tree belonging to the Chenopodiaceae family, has emerged as the favored and extensively implemented windbreak and sand-fixing vegetation in arid desert regions [26]. This phenomenon can be primarily attributed to its inherent characteristics of drought resistance, minimal water requirements, and significant capacity for carbon sequestration [27]. Therefore, it has garnered significant scholarly attention in sand control. The analysis conducted on the soil of the Urad natural *H. ammodendron* forest revealed a marked improvement in bacterial diversity, with surface vegetation and pH exerting a significant influence on said diversity [28]. In addition, the introduction of *H. ammodendron* vegetation induced changes in the spatial heterogeneity of microbial communities within desert ecosystems [29]. A study into the long-term impact of *H. ammodendron* forests on soil bacteria in sandy regions unveiled a significant enhancement in the bacterial richness index and the abundance of dominant bacteria, such as Actinomycetes, during a finite period of ecological succession [30]. Additionally, it has been demonstrated that vegetation not only impacts soil bacterial communities and their diversity but also shapes the functional composition of soil bacteria [15,17].

An exploration of the soil bacterial community and its associated influencing factors in *H. ammodendron* forests offers valuable insights into the underlying mechanistic principles governing the adaptation of *H. ammodendron* to arid environments. However, the existing literature on the impact of artificially planted *H. ammodendron* forests on the bacterial composition and functional attributes of sandy soils, as well as their correlation with physicochemical properties, remains limited and inconclusive. Therefore, this study is centered on the soil composition of an artificial *H. ammodendron* forest situated in the sand blocking and fixing belts on the periphery of the Minqin oasis. The purpose of this study was to investigate the effects of *H. ammodendron* forest planting on the structure and function of desert soil bacterial communities and to reveal the main environmental variables affecting the structure and function of soil bacterial communities in *H. ammodendron* forest. We hypothesized that (1) the planting of artificial *H. ammodendron* forest significantly affects the structure, diversity, and metabolic function of soil bacterial communities, and (2) soil physical and chemical properties are the key driving factors affecting soil bacterial community structure and potential metabolic functions. This research seeks to enhance the existing body of knowledge pertaining to the impact of planting artificial sand-fixing vegetation (*H. ammodendron*) on the micro-ecological functions of sandy land.

2. Materials and Methods

2.1. Study Site Description

This study was carried out in the sand blocking and fixing belt (Laohukou Desert Control and Prevention Demonstration Area), situated on the periphery of the Minqin Oasis in Northwest China (38°39′14″ N–38°55′23″ N, 103°02′86″ E–103°12′37″ E). The geographical location of this region places it in the transitional zone between two major deserts, namely the Badain Jaran Desert to the west and the Tengger Desert to the east. These natural characteristics render the area an optimal field research site for the study of desertification prevention and control [31]. The average annual precipitation in this area is approximately 113.5 mm, while the average annual evaporation exceeds 2383.5 mm. The region benefits from ample solar radiation and heat resources; however, it is also subject to frequent and intense wind and sand activities. The prevailing wind direction is northwest, with an annual average wind speed of approximately 3.1 m·s^{-1} and a maximum wind speed of 21 m·s^{-1}. The predominant soil types consist predominantly of sandy soil. The topography is primarily characterized by three distinct types of sand dunes: mobile, semi-stable, and stable sand dunes. The vegetation in the region can be classified into two categories: natural and artificial. The primary species of vegetation include *Haloxylon ammodendron*, *Tamarix ramosissima*, *Nitraria tangutorum*, *Elaeagnus angustifolia*, *Calligonum mongolicum*, and other desert vegetation [32–34].

2.2. Experimental Design and Sample Collection

In August 2022, during the zenith of vegetative growth, field surveys were undertaken to evaluate the flora and procure soil specimens in the Laohukou Desert Control and Prevention Demonstration Area. Following a three-year effort toward artificial *H. ammodendron* afforestation, the sand pressure afforestation and beach afforestation areas had expanded to 6666.67 hm^2 and 3066.68 hm^2, respectively, by 2008. Presently, the *H. ammodendron* forest belt, serving as a windbreak and sand stabilizer, has attained a state of rudimentary establishment. For the purpose of this study, the soil derived from an *H. ammodendron* forest (HAS) that had been cultivated for a span of 15 years was designated as the experimental group, while the soil originating from the mobile dune (MDS) in the upwind direction was selected as the control group. In the *H. ammodendron* forest, single standardized sampling plots measuring 20 m × 20 m were randomly established at intervals of 100 m along a direction perpendicular to the principal wind direction (northwestward). A cumulative 15 standardized quadrats were established in the experimental group, wherein the stature and extent of the *H. ammodendron* forest in each plot were analyzed (Table 1). Correspondingly, a total of 15 (20 m × 20 m) control groups were instituted in the vicinity of the upwind region of the mobile dune soil area, perpendicular to the standardized quadrats of each experimental group (Figure 1). The 5-point technique was employed to demarcate five diminutive sample squares (1 m × 1 m) uniformly in each standardized sample plot. Surface weeds and litter were removed, and soil samples were procured from a depth of 0–20 cm in the confines of the five small sample squares. These five portions of soil were subsequently integrated thoroughly to yield a singular standardized soil sample. In the control group, 15 standardized soil samples were collected utilizing the same methodology, resulting in a total of 30 soil samples. Each standardized soil sample was promptly partitioned into two segments. Approximately 10 mL of soil sample was transferred into a sterile centrifuge tube and preserved in a refrigerator at −80 °C for subsequent soil microbial analysis. The remaining segment was subjected to natural desiccation and subsequently sieved using a 2 mm mesh to evaluate the physicochemical properties of the soil [24].

Table 1. Height and coverage characteristics of *Haloxylon ammodendron* forest.

Sampling Area	Latitude	Longitude	Altitude (m)	Dominant Tree	Height (cm)	Coverage (%)
MDS	38°45′ N	103°05′ E	1311.56	—	—	—
HAS	38°50′ N	103°08′ E	1304.95	*Haloxylon ammodendron*	163.22 ± 44.05	35.30 ± 7.25

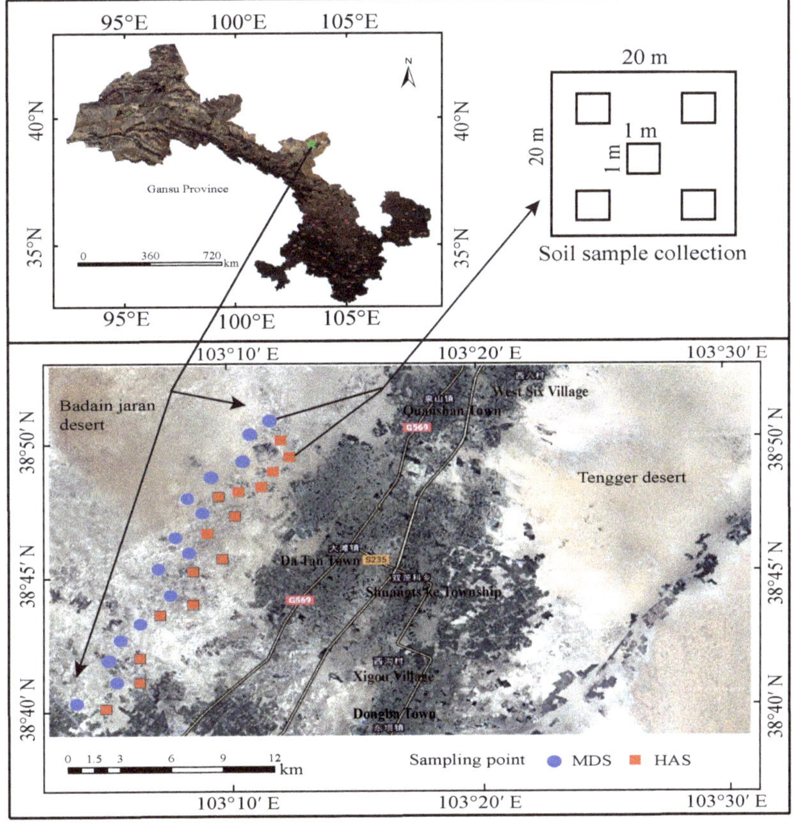

Figure 1. Geographical location of study area. MDS and HAS represent mobile dune and *H. ammodendron* forest soil, respectively.

2.3. Soil Physicochemical Measurements

Soil water content (SWC) was identified through the utilization of the drying–weighing method. The determination of soil pH and electrical conductivity (EC) was conducted employing the potentiometric method (utilizing water-to-soil ratios of 2.5:1 and 5:1, respectively) (MP-551 pH/Conductivity Meter, Shanghai, China.) [23,24]. The quantification of soil organic matter (SOM) was accomplished utilizing the potassium dichromate sulfuric acid volumetric method [24]. The determination of soil total nitrogen (TN) was executed using the Kjeldahl method (K9840 Automatic Kjeldahl Nitrogen Analyzer, Shandong, China). Soil total phosphorus (TP) was determined employing the acid digestion–molybdenum antimony anti-colorimetric method. Soil total potassium (TK) was quantified utilizing flame spectrophotometry (Flame photometer FP6410, Shanghai, China). The determination of soil available phosphorus (AP) was carried out through the sodium hydrogen hydrochloride extraction–molybdenum antimony colorimetric method. Soil available potassium (AK) was determined using ammonium acetate extraction–flame spectrophotometry (Flame photometer FP6410, Shanghai, China) [35]. The detailed methodologies for each index determination can be found in the relevant literature [36].

2.4. Illumina Sequencing Analysis of 16S rRNA Gene Amplicons

The total DNA of soil bacteria was extracted using the Magabi soil genome 50 ng DNA purification kit (Maga Bio Soil Genomic DNA Purification Kit, Carlsbad, CA, USA). Genomic DNA served as the template for amplifying the V3-V4 variable region of the bacterial 16S rRNA gene. This amplification was carried out with barcode-specific primers and TaKaRa Premix Taq® Version 2.0 (TaKaRa Biotechnology Co, Dalian, China) [30]. The universal primers used were 338F (5′-ACTCCTACGGGGAGGCAGCA-3′) and 806R (5′-GGACTACHVGGGTWTCTAAT-3′) [37]. The polymerase chain reaction (PCR) amplification reaction system consisted of 25 µL of 2 x Premix Taq, 1 µL of Primer-F and Primer-R, 50 ng of template DNA, and nuclease-free water to a final volume of 50 µL. The PCR procedure involved an initial denaturation step at 94 °C for 5 min, followed by 30 cycles of denaturation at 94 °C for 30 s, annealing at 52 °C for 30 s, and extension at 72 °C for 30 s. A final extension step was performed at 72 °C for 10 min [30,38]. The resulting PCR products were analyzed by 1% agarose gel electrophoresis to determine fragment length and concentration. The amplicon library was constructed using the Illumina Nova 6000 platform standard process and subjected to PE250 high-throughput sequencing.

2.5. Bioinformatic Analysis

The RawReads data at both ends were subjected to individual quality clipping using the Fastp software (version 0.14.4), while the primers at both ends were eliminated through the employment of the Cutadapt software (version 1.14), thereby yielding valid sequences [39]. Subsequently, the Usearch software's Uparse (version 10.0.240) algorithm was employed to cluster the sample quality sequences into OTUs, and the Silva 16S database was utilized to annotate the species, thereby obtaining taxonomic information for the OTUs. Statistical information on the taxonomic level sequences of phylum and genus in OTUs was collected and the calculation of the relative abundance of each taxon was performed. The bacterial Alpha diversity index was calculated using the QIIME2 software (version 2020.11). To elucidate the Beta diversity of soil bacteria in the artificial *H. ammodendron* forest, principal coordinate analysis (PCoA) was employed. The functional information of the samples in terms of KEGG Orthology was obtained by comparing the KEGG database information using the PICRUSt2 function prediction software (version 2.3.0-b).

2.6. Data Analysis

The standardized soil microbiological and physicochemical data were subjected to statistical analysis using SPSS 26.0 software (SPSS Inc., Chicago, IL, USA), and the differences in soil physicochemical properties, soil bacterial community structure, diversity, and the relative abundance of metabolic function between HAS and MDS were analyzed by independent sample *T*-tests ($p < 0.05$). In order to explore the key physicochemical drivers affecting bacterial structure, redundancy analysis (RDA) was conducted using the Canoco 5 software. The Monte Carlo replacement test with 499 test replicates was utilized in this analysis (Microcomputer Power, Ithaca, NY, USA). In addition, the environmental factors influencing bacterial function were assessed using the "ggcor" package Mantel test correlation tests in R software (version 4.2.1).

3. Results

3.1. Soil Physical and Chemical Properties

The variation in soil factors subsequent to the establishment of *H. ammodendron* forests is illustrated in Figure 2. Planting *H. ammodendron* forests consumed the surface SWC, which decreased by 17.17% in HAS compared to MDS. However, it resulted in significant improvements in various soil nutrient factors. The pH of the study area ranged between 8.11 and 9.01, and EC ranged between 149 and 364 $\mu S \cdot cm^{-1}$, and both pH and EC in HAS exhibited significant differences compared to MDS. The contents of SOM, TN, TP, and AK in HAS soil increased ($p < 0.05$). SOM increased by 35.59%, TN by 36.36%, TP by 32.26%, and AK by 20.18% compared to MDS. However, there were no major differences in TK and AP content.

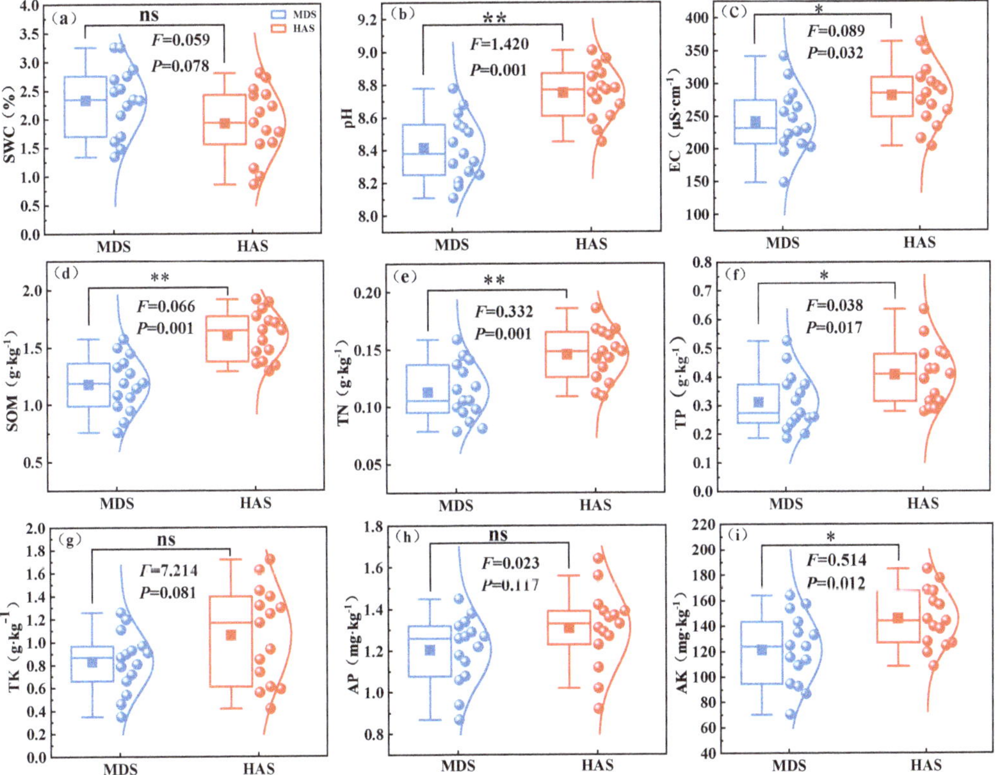

Figure 2. Soil physicochemical properties of *H. ammodendron* forest and mobile dune. Note: MDS and HAS represent mobile dune and *H. ammodendron* forest soil, respectively. (**a**): SWC, soil water content; (**b**): pH, soil pH; (**c**): EC, soil electrical conductivity; (**d**): SOM, soil organic matter; (**e**): TN, total nitrogen; (**f**): TP, total phosphorus; (**g**): TK, total potassium; (**h**): AP, available phosphorus; (**i**): AK, available potassium. "**", "*" and "ns" indicate $p < 0.01$, $p < 0.05$ and $p > 0.05$, respectively. The same as below.

3.2. Soil Bacterial Diversity

The total number of original sequences obtained from all samples of MDS and HAS was 1,714,258 and 1,796,338, respectively. The effective sequences obtained after undergoing quality control filtration and chimera removal were 1,690,381 and 1,771,937, respectively. Subsequently, the effective sequences were subjected to random sampling in order to derive representative bacterial dilution curves (Figure 3d). These curves were constructed by plotting the extracted sequences alongside the corresponding number of OTUs they represented. As the number of sequences progressively increased, the dilution curve gradually reached a plateau, thereby indicating that the inclusion of additional sequences yielded reducing returns in terms of the discovery of novel OTUs. This observation strongly suggests that the OTU coverage of the sample approached saturation, and thus, the sequencing results effectively encapsulated the entirety of pertinent microbiological information. The standardization process involved utilizing the minimum number of sample sequences, which were uniformly drawn and subsequently clustered based on a similarity threshold of greater than 97%. This procedure yielded a total of 46,187 OTUs across all samples. Notably, there was a significant difference between the mean number of OTUs observed in the HAS group (2110) and the MDS group (969) ($p < 0.01$). Moreover, the mean Chao1 and Shannon index values for the HAS group were 2158 and 9.34, respectively, while the corresponding values for the MDS group were 979 and 7.27, respectively. These differences were also found to be statistically significant between the two groups.

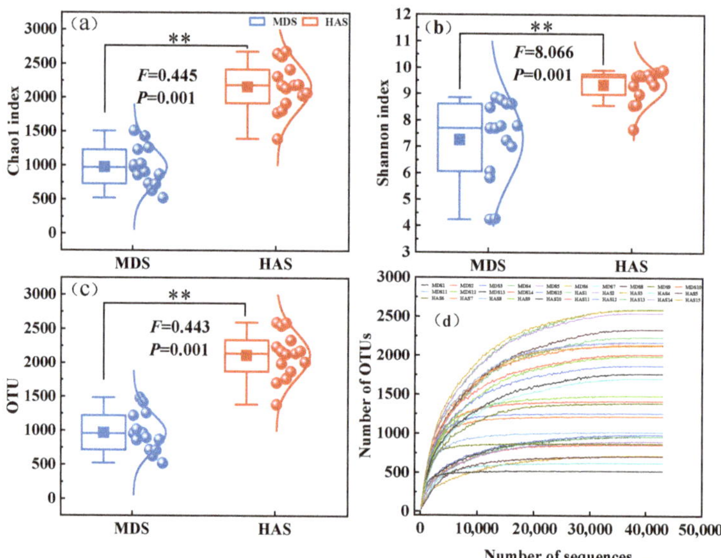

Figure 3. Soil bacterial α diversity characteristics of artificial *H. ammodendron* forest. MDS and HAS represent mobile dune and *H. ammodendron* forest soil, respectively. (**a–c**) Soil bacterial Chao1 index, Shannon index, and OTU numbers, respectively; (**d**) soil bacterial dilution curve, MDS 1–15 and HAS 1–15 represent mobile dune and *H. ammodendron* forest soil samples. "**" indicate $p < 0.01$.

Principal coordinate analysis (PCoA) was employed to assess the β diversity of soil bacterial communities at the OTU level. The results of the analysis revealed distinct spatial distributions between the HAS and MDS groups, it indicates that the structure of soil bacterial community structures changed significantly after the planting of *H. ammodendron* forest ($p < 0.01$) (Figure 4).

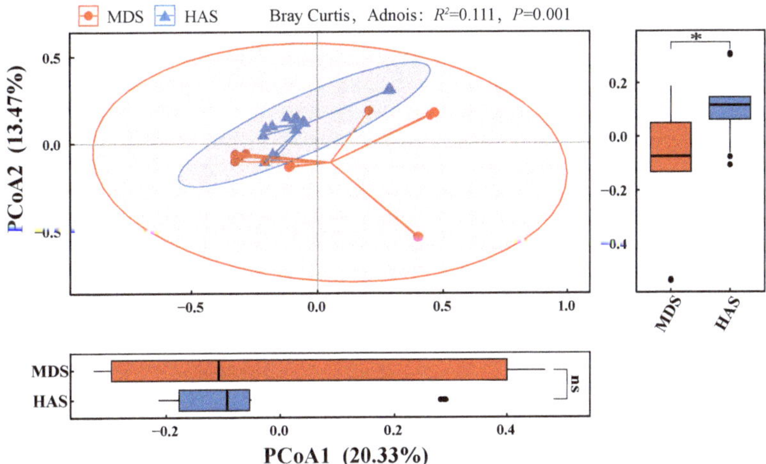

Figure 4. Principal coordinate analysis (PCoA) of soil bacterial structure in artificial *H. ammodendron* forest. MDS and HAS represent mobile dune and *H. ammodendron* forest soil, respectively. "*" and "ns": indicate $p < 0.05$ and $p > 0.05$.

3.3. Soil Bacterial Community Structure and Differences

All 16S rRNA sequences were subjected to comprehensive annotation, resulting in the identification of a total of 27 bacterial phyla. Among these, there were seven bacterial phyla exhibiting a relative abundance of >1% (Figure 5a). The bacterial community structure observed in the MDS and HAS plots exhibited similarities, albeit with notable differences in relative abundance. Notably, Actinobacteria, Proteobacteria, and Chloroflexi emerged as the top three bacterial phyla in terms of their relative abundance in the soil of the study area. Their respective average abundances were 24.94% ± 11.85%, 29.99% ± 11.56%, and 11.14% ± 4.55%. Additionally, the cumulative abundance of Bacteroidetes, Firmicutes, Acidobacteria, and Gemmatimonadetes reached 21.12%, thereby establishing their prominence as major bacterial phyla in the soil of the *H. ammodendron* forest. Notably, the abundances of Actinobacteria, Acidobacteria, and Gemmatimonadetes exhibited a significant increase in the HAS plot ($p < 0.05$). Conversely, the abundances of Proteobacteria and Bacteroidetes experienced respective decreases of 11.95% and 12.69%. At the genus level, a significant proportion of bacterial genera, amounting to 82.37%, eluded classification and exhibited a relative abundance of less than 1%, thereby presenting certain impediments to the comprehensive analysis of bacterial ecological functions. Moreover, a total of eight distinct categories of bacterial genera were identified (Figure 5b). The relative abundance of these aforementioned genera displayed an evident pattern across the two groups of plots. Specifically, in comparison to the MDS group, the relative abundance of *Rubrobacter*, *Sphingomonas*, *Bacillus*, and *Microvirga* in the HAS group experienced a significant increase of 53.53%, 34.52%, 124.7%, and 13.19%, respectively. Conversely, the relative abundance of *Ralstonia*, *Nocardioides*, *Pontibacter*, and *Pseudomonas* witnessed a decline of 57.29%, 19.35%, 55.47%, and 51.47%, respectively, in the HAS group when compared to the MDS group.

3.4. Characteristics of the Metabolic Function of Soil Bacteria

The functional prediction conducted by PICRUSt2 revealed the acquisition of six distinct classes of primary biometabolic pathways across all samples, namely Metabolism, Genetic Information Processing, Cellular Processes, Environmental Information Processing, Organismal Systems, and Human Diseases. Notably, the functional abundance of Genetic Information Processing and Human Diseases exhibited a higher prevalence in the HAS group, whereas the functional abundance of Metabolism demonstrated a greater predominance in the MDS group ($p < 0.05$) (Table 2).

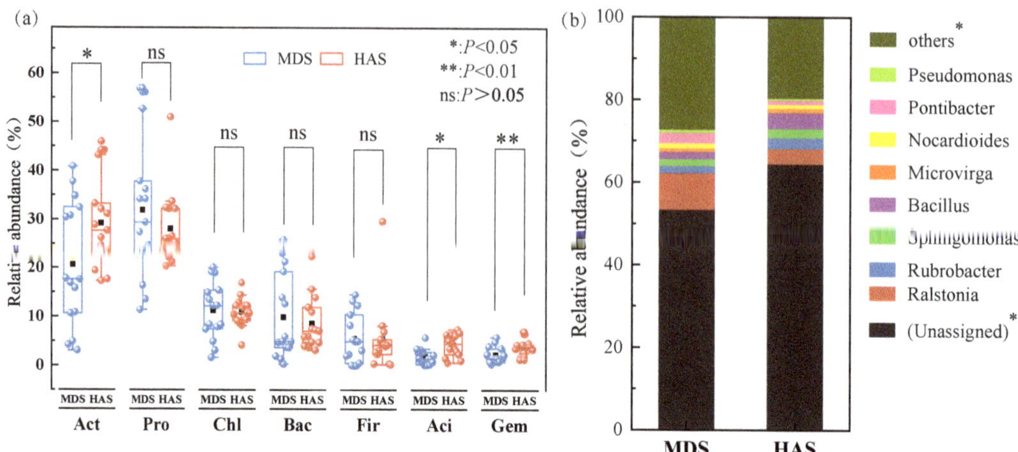

Figure 5. Relative abundance and difference in soil bacterial community at phylum (**a**) and genus (**b**) levels. (**a**), Act: Actinobacteria; Pro: Proteobacteria; Chl: Chloroflexi; Bac: Bacteroidetes; Fir: Firmicutes; Aci: Acidobacteria; Gem: Gemmatimonadetes. MDS and HAS represent mobile dune and *H. ammodendron* forest soil, respectively. (**b**): "*", $p < 0.05$.

Table 2. Relative abundance and difference in the primary function of soil bacteria.

Primary Function of Bacteria	Relative Abundance (%)		T-Test		
	MDS	HAS	F	P	
Metabolism	82.10 ± 0.23	81.79 ± 0.44	5.397	0.023	<0.05
Genetic Information Processing	11.11 ± 0.65	11.70 ± 0.36	3.267	0.005	<0.01
Cellular Processes	4.03 ± 0.44	3.90 ± 0.34	2.633	0.368	>0.05
Environmental Information Processing	2.28 ± 0.33	2.09 ± 0.17	4.998	0.059	>0.05
Organismal Systems	0.27 ± 0.04	0.27 ± 0.10	0.641	0.915	>0.05
Human Diseases	0.19 ± 0.05	0.23 ± 0.06	3.616	0.041	<0.05

A total of 15 secondary functions, exhibiting an abundance exceeding 1%, were identified in the bacterial communities dispersed across six metabolic pathways. The principal secondary metabolic functions of soil bacteria in the study area comprised carbohydrate metabolism (13.52% ± 0.24%), amino acid metabolism (13.49% ± 0.20%), and the metabolism of cofactors and vitamins (11.85% ± 0.56%) (Figure 6a). In comparison to MDS, the relative abundance of carbohydrate metabolism, the metabolism of terpenoids and polyketides, replication and repair, folding, sorting, and degradation, and translation exhibited a significant increase in HAS ($p < 0.05$). Conversely, the relative abundance of the metabolism of other amino acids and xenobiotics biodegradation and metabolism experienced a significant decrease ($p < 0.05$). However, the relative abundance of the remaining eight secondary functions displayed no significant differences.

To appraise the impact of *H. ammodendron* plantation on soil functional traits, an analysis was conducted on the tertiary metabolic functional pathways of soil bacteria. A total of 15 categories exhibited a relative abundance exceeding 1% (Figure 6b). While most tertiary metabolic pathways demonstrated similarity, the relative abundance of each pathway varied. Notably, crucial functions pertaining to energy metabolism, nucleotide metabolism, antibiotics, and enzymes, such as the biosynthesis of ansamycins and valine, were observed. The primary functional characteristics of soil bacteria included leucine and isoleucine biosynthesis, the biosynthesis of vancomycin group antibiotics, the synthesis and degradation of ketone bodies, C5-branched dibasic acid metabolism, and fatty acid biosynthesis. In summary, the relative abundances of metabolic pathways, including the biosynthesis of ansamycins, valine, leucine, and isoleucine, the biosynthesis of vancomycin group antibiotics, D-Alanine metabolism, and citrate cycle_TCA cycle_, exhibited a significant increase in HAS ($p < 0.05$) compared to MDS. This suggests that the plantation of *H. ammodendron* forest contributes to the enhancement of ecological functions, such as soil bacterial metabolism.

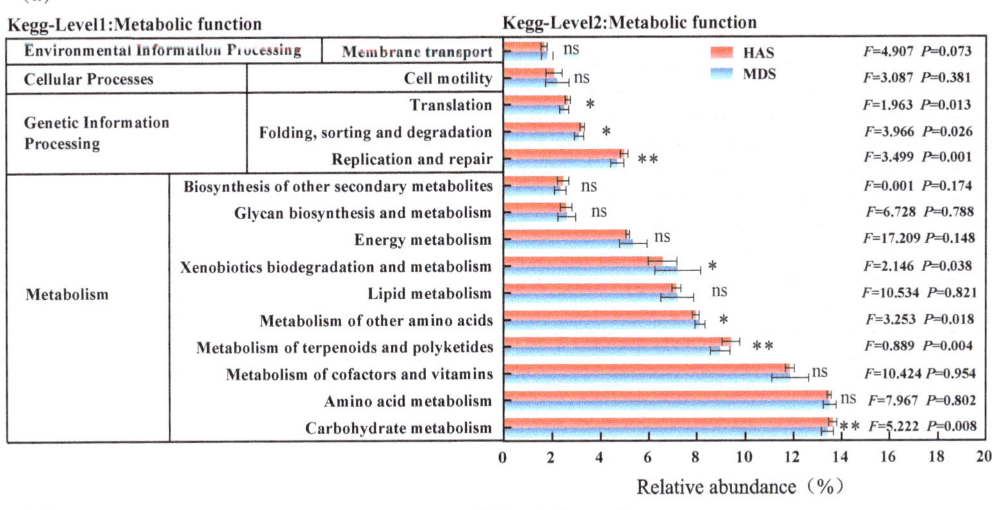

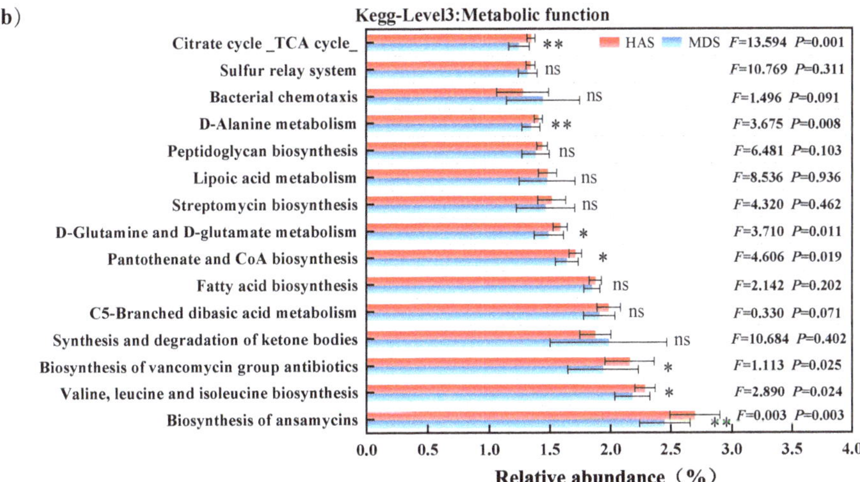

Figure 6. Prediction of the metabolic functions of soil bacteria. Metabolic functional pathways in HAS and MDS soil bacteria; (**a**,**b**) denote secondary and tertiary metabolic pathways. "**", "*" and "ns" indicate $p < 0.01$ and $p < 0.05$ and $p > 0.05$, respectively.

3.5. The Main Physicochemical Factors Affecting the Structure and Function of Soil Bacterial Communities

Redundancy analysis (RDA) refers to a linear model ranking variables by combining correspondence analysis and multiple regression analysis. It provides an intuitive measure of the influence of environmental factors on the impact of soil bacteria. The cumulative interpretation rate of the first and second axes in RDA was found to be 65.36%. The correlation coefficient between soil bacterial properties and physicochemical attributes on the first axis was 0.8086, explaining 99.96% of the cumulative variance. This suggests that the first axis plays a decisive role in determining the relationship between these variables (Table 3).

Table 3. Soil bacterial community structure RDA ordination characteristic value and interpretation.

Item	Axis 1	Axis 2	Axis 3	Axis 4
Characteristic values of soil bacterial community structure	0.6534	0.0002	0.0001	0.0000
Correlation between soil bacterial community structure and environmental factor	0.0006	0.7403	0.4797	0.3822
Cumulative interpretation of soil bacterial community structure	65.34	65.36	65.37	65.37
Cumulative interpretation of soil bacterial community and environmental factors	99.96	99.99	100.00	100.00
Canonical eigenvalue		0.6537		
Total eigenvalue		1.0000		

The RDA ordination model diagram reveals that amongst the three dominant phyla, Actinobacteria and the SOM, TN, TP, AK, and AP are situated in the second quadrant. This positioning suggests a positive correlation between Actinobacteria and these physicochemical factors. On the other hand, Proteobacteria exhibit a positive correlation with SWC and AP, while displaying a negative correlation with soil pH, EC, and TK. Conversely, Chloroflexi demonstrate a positive correlation with pH, EC, and TK, but exhibit a negative correlation with SOM, AP, and SWC. The Chao1 and Shannon indices were significantly positively correlated with SOM, pH, AK, and EC (Figure 7a). Monte Carlo tests (Figure 7b) demonstrate that SOM exerts the greatest influence on the soil bacterial community, accounting for 51.7% of the variation, and this effect is highly significant ($p = 0.002$). Additionally, AK ($p = 0.032$) and pH ($p = 0.048$) are also found to significantly impact the community composition.

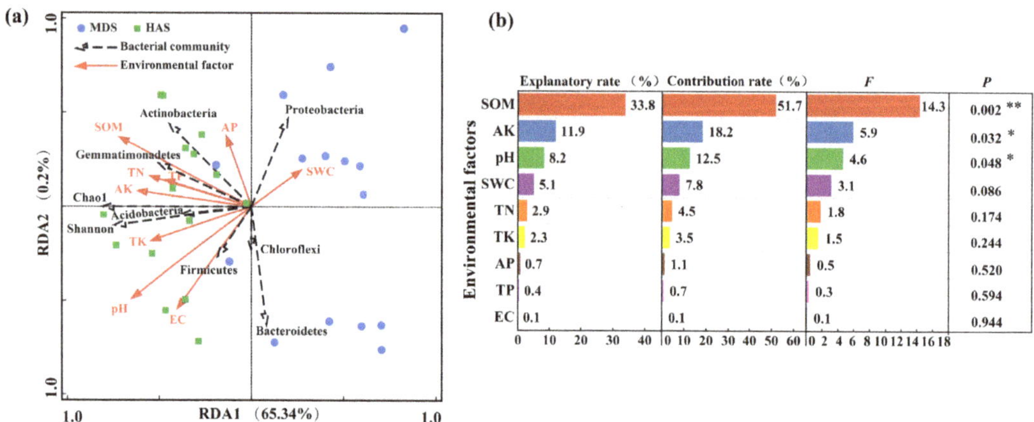

Figure 7. (**a**) RDA analysis of the effect of soil physicochemical factors on soil bacterial community structure; (**b**) explanation and significance test of soil physicochemical factors. ** $p < 0.01$, * $p < 0.05$.

The Bray–Curtis distance matrix of secondary metabolic function abundance and the Euclidean distance matrix of soil physicochemical parameters were utilized for the Mantel test correlation analysis. The findings revealed significant correlations between SOM, TN, AK, pH, and EC with the secondary metabolic functions of specific soil bacteria (Figure 8).

Notably, SOM exhibited significant positive correlations with amino acid metabolism, lipid metabolism, energy metabolism, and glycan biosynthesis and metabolism ($p < 0.05$). TN demonstrated significant positive associations with carbohydrate metabolism ($p < 0.01$) and energy metabolism ($p < 0.05$). AK displayed a positive correlation with carbohydrate metabolism ($p < 0.05$). pH exhibited a significant positive relationship with xenobiotics biodegradation and metabolism, as well as replication and repair ($p < 0.01$), and significant positive associations with folding, sorting, and degradation and translation ($p < 0.05$). Finally, EC exhibited significant positive correlations with xenobiotics biodegradation and metabolism, as well as folding, sorting, and degradation ($p < 0.05$) (Table 4). The results verify the significant role played by the aforementioned soil physicochemical factors in both the stability of the soil bacterial community structure and its function.

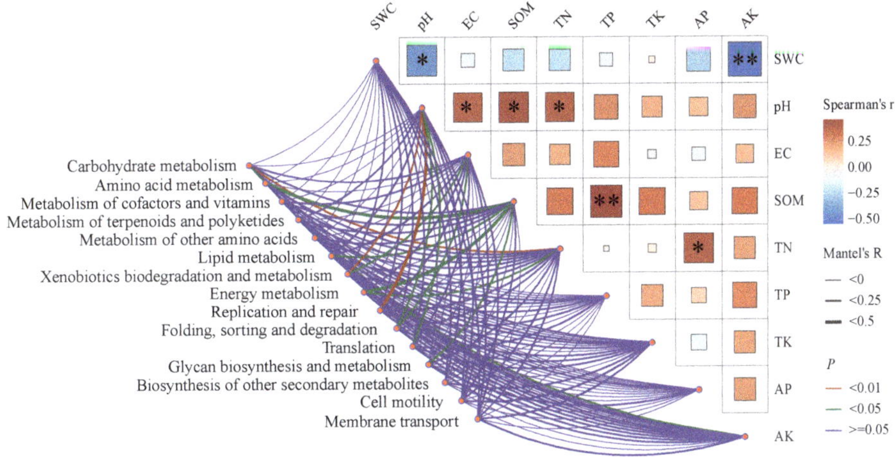

Figure 8. Mantel test of the effect of soil physicochemical factors on soil bacterial functions. ** $p < 0.01$, * $p < 0.05$.

Table 4. The significance of bacterial metabolic function and main physicocshemical factors in the Mantel test.

The Second-Level Function of Bacteria	Physicochemical Factors									
	pH		EC		SOM		TN		AK	
	r	*p*	*r*	*p*	*r*	*p*	*r*	*p*	*r*	*p*
Carbohydrate metabolism							0.209	**	0.173	*
Amino acid metabolism					0.290	*				
Lipid metabolism					0.192	*				
Xenobiotics biodegradation and metabolism	0.225	**	0.202	*						
Energy metabolism					0.275	*	0.194	*		
Replication and repair	0.259	**								
Folding, sorting, and degradation	0.207	*	0.240	*						
Translation	0.199	*								
Glycan biosynthesis and metabolism					0.171	*				

"**" and "*" indicate $p < 0.01$ and $p < 0.05$, respectively. Abbreviations for soil physicochemical factors are shown in Figure 2.

4. Discussion

4.1. Effects of Vegetation Reconstruction on Soil Physicochemical Properties

Afforestation is widely recognized as a potent strategy for mitigating desertification in arid regions across the globe [9,40]. The restoration of vegetation assumes a pivotal role in the broader context of ecological restoration, as it leads to a reciprocal relationship between the soil and plants. This symbiotic interaction between vegetation and soil results in a complex system that facilitates the circulation of ecosystem materials and the flow of energy [41]. The study observed a decrease of 17.17% in the surface soil water content in the *H. ammodendron* forest. This observed change can be ascribed to the utilization of surface soil water during the growth phase of the vegetation. The enhancement of evapotranspiration, which arises from the expansion and maturation of the root, stem, and leaf structures in the *H. ammodendron* forest, significantly contributes to this phenomenon [42,43]. However, it is noteworthy that the soil pH and EC exhibited a significant increase, predominantly attributable to the cumulative impact of *H. ammodendron* on salt ions under conditions devoid of irrigation. The accumulation of salts in the soil resulted in a reduction in tissue osmotic potential and water potential, while concurrently enhancing the capacity for water uptake and strengthening drought resistance through salt accumulation [44–46]. Prior studies have explained that *H. ammodendron* predominantly absorbs soil NO_3^- and subsequently releases one OH^- for each absorbed NO_3^- ion, thereby exerting an influence on soil pH [40]. The presence of vegetation drives changes in soil factors through ecological mechanisms, including the secretion of root exudates, the decomposition of deceased foliage, and the "fertile island effect" [16,47,48]. Measures such as the stabilization of sand through the implementation of artificial vegetation can yield significant enhancements in the physicochemical properties of soils, as well as other habitat qualities [40]. This is substantiated by the observed increments in the soil quality index, as well as the significant amplification of soil organic carbon, nitrogen, phosphorus, and other nutrient contents [49].

The SOM, TN, TP, and AK contents of the *H. ammodendron* forests exhibited a consistent and notable pattern of significant increases in the scope of this study. The primary rationale behind this phenomenon can be attributed to the establishment of *H. ammodendron*, whereby the towering flora effectively intercepted and amassed the minute particulate matter, as well as branch and leaf detritus, carried by the wind–sand flow [50,51]. Therefore, this interception mechanism served to mitigate wind–sand erosion while concurrently enriching the topsoil's fertility. It is worth noting that, in arid and semi-arid sandy regions, the soil parent material contributes only a negligible proportion of the diverse array of nutrients, with soil organic matter and total nitrogen constituting the principal constituents [52]. Instead, the metabolic processes responsible for nutrient accumulation are predominantly derived from particle deposition and the inherent vegetation dynamics [53].

4.2. Soil Bacterial Communities and Diversity in H. ammodendron Forests

The mechanism underlying plant community construction is primarily attributed to habitat heterogeneity under the context of environmental filtering [54]. Soil factors, being pivotal constituents that exert a profound influence on the growth and development of vegetation, also exert a strong impact on the differentiation of ecological niches. Simultaneously, the decomposition function of soil microorganisms exerts a significant influence on the transformation of soil nutrients and the growth of aboveground vegetation [55,56]. Research findings suggest that a higher soil bacterial diversity index corresponds to greater resistance to environmental stress, thereby fostering a more stable micro-ecological function in the soil [57]. The introduction of sand-fixing vegetation not only expedites soil formation but also offers essential nutrients for the growth and development of surface soil microorganisms through the decomposition of its organic matter and root exudates [16–18,23–25,40,47,48]. Therefore, this process significantly modifies the structure and diversity of the soil microbial community. This study has demonstrated a significant improvement in both the Chao1 and Shannon indices of soil bacteria in *H. ammodendron* forests. In addition, significant differences in the composition of soil bacterial communities were also observed (Figure 3). These findings serve to indicate that the growth process and root exudation of *H. ammodendron* forests lead to a propitious habitat for the growth of soil bacteria [30].

The subtle shifts occurring in the soil habitat possess the capacity to directly or indirectly exert a profound influence on the configuration of soil bacteria [16,45,46]. The assemblage of soil bacterial structure in arid and desert regions exhibits a striking resemblance, primarily comprising approximately ten groups, including *Actinobacteria*, *Proteobacteria*, *Chloroflexi*, *Bacteroidetes*, *Acidobacteria*, and *Cyanobacteria* [23–25,30,46,58–60]. This consensus, which has been established by scholars, was derived from an extensive analysis of diverse representative artificial vegetation sand-fixing areas. The overall characteristics of bacterial communities in the soil of both the *H. ammodendron* forest and mobile dunes were found to be congruous. However, notable variations were observed in their respective relative abundances. Specifically, *Actinobacteria*, *Proteobacteria*, and *Chloroflexi* were identified as the preeminent phyla, a finding that aligns with the results reported by Bi et al. [24] and Sun et al. [60]. The exacerbation of soil salinization in the transitional zone of the Minqin desert oasis can be attributed to the adverse climatic factors of drought and intense radiation, which have contributed to the debilitation of elemental cycling processes, particularly those pertaining to soil carbon and nitrogen [61,62]. *Actinobacteria*, a member of the Gram-positive bacterial group, exhibits remarkable competitiveness in harsh saline–alkali and arid environments owing to its diverse physiological traits. These include spore reproduction, multiple mechanisms for repairing UV-induced damage, heat and drought resistance, and active involvement in the intricate processes of organic matter decomposition [59,63]. In addition, the growth of *H. ammodendron*, a notable carbon sink, creates favorable conditions for the growth and reproduction of actinomycetes [27]. This phenomenon may account for the greater abundance of soil actinomycetes in the *H. ammodendron* forest. *Proteobacteria*, on the other hand, are widely distributed and comprise a vast number of species. They possess significant morphological and physiological characteristics, endowing them with a competitive advantage in various ecological niches [64]. These bacteria demonstrate a robust capacity to adapt to their surroundings, making them highly adaptable organisms. In desert soil, the primary pathways governing the carbon and nitrogen cycles involve dissimilatory nitrate reduction, nitrification, and denitrification. *Actinobacteria* and *Proteobacteria* assume a pivotal role in the aforementioned cycles as they serve as the primary contributors [15,18,65]. Moreover, *Chloroflexi*, a light-energy-trophic microorganism, exerts dominance in CO_2 fixation and microbial photosynthesis through the reductive tricarboxylic acid cycle and the C4-dicarboxylic acid cycle [66]. This study was conducted in an arid desert region that exhibits abundant light resources, thereby providing optimal conditions for the growth and maturation of these bacteria. A staggering 82.37% of the bacterial genera at the taxonomic level pose impediments to the analysis of ecological functions, while a mere 17.63% of the known bacterial genus structure may selectively shape the soil bacterial type suitable for the habitat due to distinct site conditions and root exudates [67]. Therefore, this difference may lead to alienation characteristics between the *H. ammodendron* forest and the mobile dune soil. As the biomass of *H. ammodendron* gradually increases, certain nutrients accumulate in the soil, primarily from the decomposition of apoplastic material and root secretions [17,28–30]. This accumulation significantly contributes to the decomposition of organic matter by the bacterial community, as well as to their own growth and development. The influence of carbon and nitrogen sources, derived from residual vegetation roots and litter decomposition, on soil bacteria was explained by Urbanova et al. [68]. The prevalence of microorganisms engaged in carbon fixation and nitrogen fixation sustains the capacity for carbon and nitrogen cycling within desert ecosystems, thereby enhancing material transformation and energy flow in the soil. Moreover, this process facilitates an amplified provision of chemical elements essential for vegetation growth, thus enhancing the ecological carrying capacity and stability of the soil [65]. The observed changes in soil bacteria and the greater diversity in the *H. ammodendron* forest, as documented in this study, signify an improvement in the desert soil environment and the nutritional status of the plants.

4.3. Functional Characteristics of Soil Bacteria in H. ammodendron Forest

The establishment of artificial sand-fixing vegetation has elicited profound shifts in the bacterial community in desert soil, thus exerting an impact on the functional attributes of soil bacteria [30,58]. Soil bacteria, through their metabolic activities, actively participate in nearly all physiological and biochemical reactions, comprising the decomposition of organic matter and the transformation of nutrients [58,69]. Our findings have unequivocally demonstrated that metabolic function constitutes the primary role of soil bacteria, with carbohydrate metabolism, amino acid metabolism, and the metabolism of cofactors and vitamins emerging as the principal metabolic functions in the study area. Studies conducted in the Horqin and Tengger deserts have consistently underscored the paramount importance of amino acid and carbohydrate metabolism as the predominant metabolic pathways [30,58,59]. Similarly, investigations exploring the functional response of two salt-tolerant plants to soil bacteria in saline regions have revealed a greater abundance of amino acid metabolism and carbohydrate metabolism [46]. Remarkably, even in *Picea asperata* forest soil, which exhibits significant differences in physicochemical properties when compared to desert soil, carbohydrate metabolism and other metabolic functions have been identified as the primary metabolic functions of soil bacteria [70]. Recent studies have brought to light the existence of partial incongruity in the gene sequences pertaining to specific metabolic pathways among microbial populations that share similar overall metabolic functions. This divergence can be ascribed to variations in the microbial constituents responsible for these specific metabolic functions, which arise due to differences in environmental conditions. Therefore, such diversity in microbial composition plays a pivotal role in the effective regulation and preservation of ecosystem function stability [71]. Notably, the soil carbohydrate metabolism function in the *H. ammodendron* forest exhibited a substantial relative abundance in this study. These findings highlight a robust correlation between carbohydrate metabolism and nitrogen fixation, as well as phosphorus solubilization, thereby facilitating the assimilation of nitrogen and phosphorus by the roots of *H. ammodendron* [72]. In addition, amino acid metabolism, as the second most prominent metabolic function, contributes to the nitrogen cycle through processes such as deamination and transamination, thereby providing indispensable nutritional conditions for bacterial survival and reproduction [73]. The third tier of metabolism exhibited a greater relative abundance of significant functions associated with enzymes, nucleotide metabolism, energy metabolism, and antibiotics. These metabolic pathways represent a crucial aspect of the metabolic function exhibited by soil bacteria and the survival strategy adopted by bacterial communities in nutrient-depleted desert soil environments [30,58–60,63,74]. Notably, the biosynthesis of ansamycins, valine, leucine, and isoleucine and the biosynthesis of vancomycin-group antibiotics exhibited a significant upregulation in the *H. ammodendron* forest soil in comparison to the mobile dune soil. These metabolic functions actively facilitated plant growth and development through the production of antibiotics, growth-promoting hormones, and antibacterial proteins, thereby resulting in an enhanced relative abundance of these metabolic functions [75]. The findings of this study unequivocally demonstrate that the cultivation of *H. ammodendron* can expedite bacterial metabolic activity and enhance the efficiency of carbon, nitrogen, and other elemental circulation in the soil.

4.4. Effects of Soil Physicochemical Factors on the Structure and Metabolic Function of Bacterial Communities

The functional structure of microbial communities is intricately linked to the environmental factors present in their respective habitats [56,60,69]. Particularly, the physicochemical properties of soil exert a profound influence on the composition and diversity of microbial communities, particularly in the face of the rapid and far-reaching effects of global climate change. These effects, in turn, have the potential to significantly impact delicate and vulnerable desert ecosystems [76,77]. Through the RDA analysis, it was determined that soil physicochemical properties accounted for a significant 65.36% of the observed variations in bacterial community structure. Notably, non-biological factors such as SOM, AK, and pH emerged as significant contributors to the overall explanatory changes. These findings align with the results reported by Bi et al. [23] in their research on rhizosphere soil bacteria associated with *Pinus sylvestris* var. *mongolica* in the Mu Us Desert, as well as the research conducted by Xu et al. [57] on soil bacteria inhabiting artificial *Robinia pseudoacacia* ecosystems. Simultaneously, the SOM, TN, AK, pH, and EC exhibited significant impacts on the primary metabolic functions of soil bacteria. This phenomenon can be readily explained. Firstly, in desert ecosystems characterized by sparse vegetation and limited soil nutrient availability, essential nutrients such as carbon and nitrogen, which serve as pivotal energy sources shaping soil bacterial communities, assume the role of key limiting environmental factors [78,79]. Secondly, Gemmatimonadetes, known for its remarkable capacity to solubilize potassium, phosphorus, and other elements, facilitates the uptake of available nutrients by plants from the soil. The growth of Gemmatimonadetes in *H. ammodendron* forests enhances the absorption and utilization of available soil nutrients by *H. ammodendron*, thereby potentially accounting for the significant influence of available potassium content on the soil bacterial community structure in *H. ammodendron* forests [56,77]. Moreover, the study of global geographical patterns necessitates the consideration of soil pH as a pivotal determinant, as it exerts a profound influence on the characteristics of bacterial communities. This influence is primarily reflected through its direct impact on cell membrane permeability and stability [76,80]. It is widely acknowledged that EC exerts a detrimental effect on bacterial growth. This is attributable to the strong association between EC and soil salinity, which can disrupt cell permeability, induce nutrient imbalances, reduce enzyme activity, and even elicit toxic effects on microorganisms [81]. The varying degrees of salt tolerance and sensitivity exhibited by diverse microorganisms contribute to the divergence observed in soil microbial communities. Remarkably, the findings of this study reveal a significant and positive correlation between EC content and the majority of bacterial species, thereby providing further evidence that bacteria inhabiting desert soils have adeptly adapted to high-salt environments. The findings of this study explain that the impact of SOM, TN, AK, pH, and EC on bacterial community composition surpasses that of other nutrients. Nevertheless, it is important to note that this does not negate the influence of other nutrient characteristics on bacterial communities [14,19]. The complex relationship between soil physicochemical factors and bacterial communities is not unidimensional, as there may exist inherent correlations among the environmental factors themselves, which in turn affect the structure of microbial communities through interconnected effects [56,76,77]. Therefore, future research should focus on employing a combination of macrogenome sequencing and functional gene chip technology to delve into the specific mechanisms by which individual environmental factors, such as nitrogen, phosphorus, and other elements, exert their influence on the structure and functional potential of soil bacterial communities in *H. ammodendron* forests.

5. Conclusions

This study assessed the impact of an *H. ammodendron* plantation on the structural composition and functional dynamics of soil bacterial communities in the sand blocking and fixing belts situated at the periphery of oases. The prevailing bacterial taxa identified in the *H. ammodendron* forest soil were Actinobacteria, Proteobacteria, and Chloroflexi. Notably, Actinobacteria exhibited a significantly greater relative abundance in comparison to the mobile dune soil, thereby contributing to an overall enhancement in bacterial diversity. Moreover, the metabolic activities of the bacterial communities were primarily characterized by carbohydrate metabolism, amino acid metabolism, and cofactor and vitamin metabolism. These findings highlight the pivotal role played by the *H. ammodendron* plantation in shaping the structure and function of soil bacterial communities in sandy land ecosystems. The introduction of *H. ammodendron* cultivation resulted in a significant improvement in soil nutrient levels, comprising SOM, TN, and TP. The structure and function of the soil bacterial community were found to be significantly influenced by physicochemical factors such as soil SOM, TN, AK, pH, and EC. However, it is important to acknowledge that this intervention also involves specific adverse consequences, primarily characterized by the reduction in surface soil moisture and the accumulation of salt content. This study contributes to a deeper understanding of the ecological transformations experienced by soil microorganisms and their intricate complexity with physical and chemical factors in sandy land environments through the implementation of artificial vegetation-based sand-fixing measures.

Author Contributions: Methodology, software, and writing—original draft preparation, A.W.; resources, conceptualization, and writing—review and editing, R.M. and Y.M.; investigation and sampling, A.W., D.N., T.L., Y.T., Z.D., Q.C.; formal analysis and validation, T.L. and Y.T.; data curation, Z.D. and Q.C. All authors have read and agreed to the published version of the manuscript.

Funding: This research was funded by the National Natural Science Foundation of China Joint Fund for Regional Innovation and Development (U21A2001) and the Youth Tutor Support Fund of Gansu Agricultural University (GAU-QDFC-2020-09).

Data Availability Statement: Relevant data for this study can be obtained by contacting the corresponding author in a reasonable manner.

Acknowledgments: We are grateful to the editors and reviewers for their valuable suggestions.

Conflicts of Interest: The authors declare no conflict of interest.

References

1. Zheng, Y.; Yang, Q.; Ren, H.; Wang, D.; Zhao, C.; Zhao, W. Spatial pattern variation of artificial sand-binding vegetation based on UAV imagery and its influencing factors in an oasis–desert transitional zone. *Ecol. Indic.* **2022**, *141*, 109068. [CrossRef]
2. Huang, J.; Yu, H.; Han, D.; Zhang, G.; Wei, Y.; Huang, J.; An, L.; Liu, X.; Ren, Y. Declines in global ecological security under climate change. *Ecol. Indic.* **2020**, *117*, 106651. [CrossRef]
3. Zhang, D.; Deng, H. Historical human activities accelerated climate-driven desertification in China's Mu Us Desert. *Sci. Total Environ.* **2020**, *708*, 134771. [CrossRef]
4. Shao, W.; Wang, Q.; Guan, Q.; Zhang, J.; Yang, X.; Liu, Z. Environmental sensitivity assessment of land desertification in the Hexi Corridor, China. *Catena* **2023**, *220*, 106728. [CrossRef]
5. Zhao, W.; Hu, G.; Zhang, Z.; He, Z. Shielding effect of oasis-protection systems composed of various forms of wind break on sand fixation in an arid region: A case study in the Hexi Corridor, northwest China. *Ecol. Eng.* **2008**, *33*, 119–125. [CrossRef]
6. Su, Y.Z.; Zhao, W.Z.; Su, P.X.; Zhang, Z.H.; Wang, T.; Ram, R. Ecological effects of desertification control and desertified land reclamation in an oasis–desert ecotone in an arid region: A case study in Hexi Corridor, northwest China. *Ecol. Eng.* **2005**, *29*, 117–124. [CrossRef]
7. Li, X.R.; Zhang, Z.S.; Tan, H.J.; Gao, Y.H.; Liu, G.L.; Wang, X.P. Ecological restoration and recovery in the wind-blown sand hazard areas of northern China: Relationship between soil water and carrying capacity for vegetation in the Tengger Desert. *Sci. China Life Sci.* **2014**, *57*, 539–548. [CrossRef]
8. Zhang, C.; Wang, Y.; Jia, X.; Shao, M.; An, Z. Impacts of shrub introduction on soil properties and implications for dryland revegetation. *Sci. Total Environ.* **2020**, *742*, 140498. [CrossRef]
9. Zhou, W.; Li, C.; Wang, S.; Ren, Z.; Stringer, L.C. Effects of vegetation restoration on soil properties and vegetation attributes in the arid and semi-arid regions of China. *JEM* **2023**, *343*, 118186. [CrossRef]
10. Aksoy, E.; Louwagie, G.; Gardi, C.; Gregor, M.; Schröder, C.; Löhnertz, M. Assessing soil biodiversity potentials in Europe. *Sci. Total Environ.* **2017**, *589*, 236–249. [CrossRef]
11. Song, S.; Xiong, K.; Chi, Y. Ecological Stoichiometric Characteristics of Plant–Soil–Microorganism of Grassland Ecosystems under Different Restoration Modes in the Karst Desertification Area. *Agronomy* **2016**, *13*, 2016. [CrossRef]
12. Kimmell, L.B.; Fagan, J.M.; Havrilla, C.A. Soil restoration increases soil health across global drylands: A meta-analysis. *J. Appl. Ecol.* **2023**, *60*, 1939–1951. [CrossRef]
13. Falkowski, G.P.; Fenchel, T.; Delong, F.E. The Microbial Engines That Drive Earth's Biogeochemical Cycles. *Science* **2008**, *320*, 1034–1039. [CrossRef] [PubMed]
14. Delgado-Baquerizo, M.; Grinyer, J.; Reich, P.B.; Singh, B.K. Relative importance of soil properties and microbial community for soil functionality: Insights from a microbial swap experiment. *Funct. Ecol.* **2016**, *30*, 1862–1873. [CrossRef]
15. Maestre, F.T.; Reich, P.B.; Jeffries, T.C.; Gaitan, J.J.; Encinar, D.; Berdugo, M.; Campbell, C.D.; Singh, B.K. Microbial diversity drives multifunctionality in terrestrial ecosystems. *Nat. Commun.* **2016**, *7*, 10541.
16. Bourget, M.Y.; Fanin, N.; Fromin, N.; Hättenschwiler, S.; Roumet, C.; Shihan, A.; Huys, R.; Sauvadet, M.; Freschet, G.T. Plant litter chemistry drives long-lasting changes in the catabolic capacities of soil microbial communities. *Funct. Ecol.* **2023**, *37*, 2014–2028. [CrossRef]
17. Leloup, J.; Baude, M.; Nunan, N.; Meriguet, J.; Dajoz, I.; Le Roux, X.; Raynaud, X. Unravelling the effects of plant species diversity and aboveground litter input on soil bacterial communities. *Geoderma* **2018**, *317*, 1–7. [CrossRef]
18. Waksman, S.A.; Tenney, F.G.; Stevens, K.R. The Role of Microorganisms in the Transformation of Organic Matter in Forest Soils. *Ecology* **1928**, *9*, 126–144. [CrossRef]
19. Xu, Y.; Sun, R.; Yan, W.; Zhong, Y. Divergent response of soil microbes to environmental stress change under different plant communities in the Loess Plateau. *Catena* **2023**, *230*, 107240. [CrossRef]
20. Hermans, S.M.; Buckley, H.L.; Case, B.S.; Curran-Cournane, F.; Taylor, M.; Lear, G. Bacteria as emerging indicators of soil condition. *Appl. Environ. Microbiol.* **2017**, *83*, e02826-16. [CrossRef]

21. Callahan, B.J.; Wong, J.; Heiner, C.; Oh, S.; Theriot, C.M.; Gulati, A.S.; McGill, S.K.; Dougherty, M.K. High-throughput amplicon sequencing of the full-length 16S rRNA gene with single-nucleotide resolution. *Nucleic Acids Res.* **2019**, *47*, e103. [CrossRef] [PubMed]

22. Douglas, G.M.; Maffei, V.J.; Zaneveld, J.R.; Yurgel, S.N.; Brown, J.R.; Taylor, C.M.; Huttenhower, C.; Langille, M.G. PICRUSt2 for prediction of metagenome functions. *Nat. Biotechnol.* **2020**, *38*, 685–688. [CrossRef] [PubMed]

23. Yu, J.; Liu, F.; Tripathi, B.M.; Steinberger, Y. Changes in the composition of soil bacterial and fungal communities after revegetation with *Caragana microphylla* in a desertified semiarid grassland. *J. Arid. Environ.* **2020**, *182*, 104262. [CrossRef]

24. Bi, B.; Yuan, Y.; Zhang, H.; Wu, Z.; Wang, Y.; Han, F. Rhizosphere soil metabolites mediated microbial community changes of *Pinus sylvestris* var. mongolica across stand ages in the Mu Us Desert. *Appl. Soil Ecol.* **2022**, *169*, 104222. [CrossRef]

25. Bai, Y.; She, W.; Miao, L.; Qin, S.; Zhang, Y. Soil microbial interactions modulate the effect of *Artemisia ordosica* on herbaceous species in a desert ecosystem, northern China. *Soil Biol. Biochem.* **2020**, *150*, 108013. [CrossRef]

26. Yang, F.; Lv, G. Metabolomic Analysis of the Response of *Haloxylon ammodendron* and *Haloxylon persicum* to Drought. *Int. J. Mol. Sci.* **2023**, *24*, 9099. [CrossRef]

27. Yu, T.; Han, T.; Feng, Q.; Chen, W.; Zhao, C.; Li, H.; Liu, J. Divergent response to abiotic factor determines the decoupling of water and carbon fluxes over an artificial C4 shrub in desert. *J. Environ. Manag.* **2023**, *344*, 118416. [CrossRef]

28. Chen, F.; Zhang, J.; Han, E. Soil microbial diversity and its relationship with soil physicochemical properties in Urat natural *Haloxylon ammodendron* forest. *J. Desert Res.* **2022**, *42*, 207–214.

29. Cao, Y.; Li, Y.; Li, C.; Huang, G.; Lü, G. Relationship between presence of the desert shrub *Haloxylon ammodendron* and microbial communities in two soils with contrasting textures. *Appl. Soil Ecol.* **2016**, *103*, 93–100. [CrossRef]

30. An, F.; Niu, Z.; Liu, T.; Su, Y. Succession of soil bacterial community along a 46-year choronsequence artificial revegetation in an arid oasis-desert ecotone. *Sci. Total Environ.* **2021**, *814*, 152496. [CrossRef]

31. Wu, C.; Deng, L.; Huang, C.; Chen, Y.; Peng, C. Effects of vegetation restoration on soil nutrients, plant diversity, and its spatiotemporal heterogeneity in a desert–oasis ecotone. *Land Degrad. Dev.* **2021**, *32*, 670–683. [CrossRef]

32. Wang, X.Y.; Chen, X.S.; Ding, Q.P. Vegetation and soil environmental factor characteristics, and their relationship at different desertification stages: A case study in the Minqin desert-oasis ecotone. *Acta Ecol. Sin.* **2018**, *38*, 1569–1580.

33. Hu, Y.; Wang, Z.; Zhang, Z.; Song, N.; Zhou, H.; Li, Y.; Wang, Y.; Li, C.; Hale, L. Alteration of desert soil microbial community structure in response to agricultural reclamation and abandonment. *Catena* **2021**, *207*, 105678. [CrossRef]

34. Peng, M.; He, H.; Wang, Z.; Li, G.; Lv, X.; Pu, X.; Zhuang, L. Responses and comprehensive evaluation of growth characteristics of ephemeral plants in the desert–oasis ecotone to soil types. *J. Environ. Manag.* **2022**, *316*, 115288. [CrossRef]

35. Cao, X.; Mo, Y.; Yan, W.; Zhang, Z.; Peng, Y. Evaluation of Soil Quality in Five Ages of Chinese Fir Plantations in Subtropical China Based on a Structural Equation Model. *Forests* **2023**, *14*, 1217. [CrossRef]

36. Lu, R. *Methods for Agrochemical Analysis of Soil*; China Agricultural Science and Technology Press: Beijing, China, 2000.

37. Hu, M.; Li, C.; Zhou, X.; Xue, Y.; Wang, S.; Hu, A.; Chen, S.; Mo, X.; Zhou, J. Microbial Diversity Analysis and Genome Sequencing Identify *Xanthomonas perforans* as the Pathogen of Bacterial Leaf Canker of Water Spinach (*Ipomoea aquatic*). *Front. Microbiol.* **2021**, *12*, 752760. [CrossRef]

38. Xi, J.; Yang, D.; Xue, H.; Liu, Z.; Bi, Y.; Zhang, Y.; Yang, X.; Shang, S. Isolation of the Main Pathogens Causing Postharvest Disease in Fresh *Angelica sinensis* during Different Storage Stages and Impacts of Ozone Treatment on Disease Development and Mycotoxin Production. *Toxins* **2023**, *15*, 154. [CrossRef]

39. Yu, L.; Zhang, Z.; Zhou, L.; Huang, K. Effects of Altitude and Continuous Cropping on Arbuscular Mycorrhizal Fungi Community in *Siraitia grosvenorii* Rhizosphere. *Agriculture* **2023**, *13*, 1548. [CrossRef]

40. Li, X.; Li, Y.; Xie, T.; Chang, Z.; Li, X. Recovery of soil carbon and nitrogen stocks following afforestation with xerophytic shrubs in the Tengger Desert, North China. *Catena* **2022**, *214*, 106277. [CrossRef]

41. Van der Putten, W.H.; Bardgett, R.D.; Bever, J.D. Plant–soil feedbacks: The past, the present and future challenges. *J. Ecol.* **2013**, *101*, 265–276. [CrossRef]

42. Zhang, Y.; Zhao, W. Vegetation and soil property response of short-time fencing in temperate desert of the Hexi Corridor, northwestern China. *Catena* **2015**, *133*, 43–51. [CrossRef]

43. Zuo, X.; Zhao, X.; Zhao, H.; Zhang, T.; Guo, Y.; Li, Y.; Huang, Y. Spatial heterogeneity of soil properties and vegetation–soil relationships following vegetation restoration of mobile dunes in Horqin Sandy Land, Northern China. *Plant Soil* **2009**, *318*, 153–167. [CrossRef]

44. Zhang, K.; Su, Y.; Wang, T.; Liu, T. Soil properties and herbaceous characteristics in an age sequence of *Haloxylon ammodendron* plantations in an oasis-desert ecotone of northwestern China. *J. Arid Land* **2016**, *8*, 960–972. [CrossRef]

45. Rath, K.M.; Maheshwari, A.; Rousk, J. The impact of salinity on the microbial response to drying and rewetting in soil. *Soil Biol. Biochem.* **2017**, *108*, 17–26. [CrossRef]

46. James, J.J.; Tiller, L.R.; Richards, H.J. Multiple Resources Limit Plant Growth and Function in a Saline-Alkaline Desert Community. *J. Ecol.* **2005**, *93*, 113–126. [CrossRef]

47. Lei, X.; Shen, Y.; Zhao, J.; Huang, J.; Wang, H.; Yu, Y.; Xiao, C. Root Exudates Mediate the Processes of Soil Organic Carbon Input and Efflux. *Plants* **2023**, *12*, 630. [CrossRef] [PubMed]

48. Gao, Y.; Tariq, A.; Zeng, F.; Sardans, J.; Peñuelas, J.; Zhang, Z.; Islam, W.; Xu, M. "Fertile islands" beneath three desert vegetation on soil phosphorus fractions, enzymatic activities, and microbial biomass in the desert-oasis transition zone. *Catena* **2022**, *212*, 106090. [CrossRef]

49. Dong, S.; Liu, S.; Cui, S.; Zhou, X.; Gao, Q. Responses of soil properties and bacterial community to the application of sulfur fertilizers in black and sandy soils. *Pol. J. Environ. Stud.* **2022**, *1*, 31. [CrossRef]

50. Su, Y.Z.; Wang, X.F.; Yang, R.; Lee, J. Effects of sandy desertified land rehabilitation on soil carbon sequestration and aggregation in an arid region in China. *J. Environ. Manag.* **2009**, *91*, 2109–2116. [CrossRef]

51. Wang, G.H.; Zhao, W.Z.; Liu, H.; Zhang, G.; Li, F. Changes in soil and vegetation with stabilization of dunes in a desert–oasis ecotone. *Ecol. Res.* **2015**, *30*, 639–650. [CrossRef]

52. Zhao, L.; Li, W.; Wu, Y.; Q. Y.; Li, P.; Guo, Y.; Han, L. Moisture, Temperature, and Salinity of a Typical Desert Plant (*Haloxylon ammodendron*) in an Arid Oasis of Northwest China. *Sustainability* **2020**, *13*, 1908. [CrossRef]

53. Li, Y.; Cui, J.; Zhang, T.; Okuro, T.; Drake, S. Effectiveness of sand-fixing measures on desert land restoration in Kerqin Sandy Land, northern China. *Ecol. Eng.* **2009**, *35*, 118–127. [CrossRef]

54. Murrell, D.J.; Law, R. Heteromyopia and the spatial coexistence of similar competitors. *Ecol. Lett.* **2003**, *6*, 48–59. [CrossRef]

55. Shao, W.; Wang, Q.; Guan, Q.; Luo, H.; Ma, Y.; Zhang, J. Distribution of soil available nutrients and their response to environmental factors based on path analysis model in arid and semi-arid area of northwest China. *Sci. Total Environ.* **2022**, *827*, 154254. [CrossRef] [PubMed]

56. Toledo, S.; Bondaruk, V.F.; Yahdjian, L.; Oñatibia, G.R.; Loydi, A.; Alberti, J.; Bruschetti, M.; Pascual, J.; Peter, G.; Agüero, W.D.; et al. Environmental factors regulate soil microbial attributes and their response to drought in rangeland ecosystems. *Sci. Total Environ.* **2023**, *892*, 164406. [CrossRef]

57. Xu, M.; Lu, X.; Xu, Y.; Zhong, Z.; Zhang, W.; Ren, C.; Han, X.; Yang, G.; Feng, Y. Dynamics of bacterial community in litter and soil along a chronosequence of Robinia pseudoacacia plantations. *Sci. Total Environ.* **2020**, *703*, 135613. [CrossRef]

58. Cao, H.; Du, Y.; Gao, G.; Rao, L.; Ding, G.; Zhang, Y. Afforestation with *Pinus sylvestris* var. *mongolica* remodelled soil bacterial community and potential metabolic function in the Horqin Desert. *Glob. Ecol. Conserv.* **2021**, *29*, e01716. [CrossRef]

59. Sun, Y.; Shi, Y.; Wang, H.; Zhang, T.; Yu, L.; Sun, H.; Zhang, Y. Diversity of bacteria and the characteristics of actinobacteria community structure in Badain Jaran Desert and Tengger Desert of China. *Front. Microbiol.* **2018**, *9*, 68–82. [CrossRef]

60. Sun, Y.; Zhang, Y.; Feng, W. Effects of xeric shrubs on soil microbial communities in a desert in northern China. *Plant Soil* **2017**, *414*, 281–294. [CrossRef]

61. Li, S.; Fan, W.; Xu, G.; Cao, Y.; Zhao, X.; Hao, S.; Deng, B.; Ren, S.; Hu, S. Bio-organic fertilizers improve *Dendrocalamus farinosus* growth by remolding the soil microbiome and metabolome. *Front. Microbiol.* **2023**, *14*, 1117355. [CrossRef]

62. Yu, X.J.; Zhang, L.X.; Zhou, T.; Zhang, X. Long-term changes in the effect of drought stress on ecosystems across global drylands. *Sci. China Earth Sci.* **2023**, *66*, 146–160. [CrossRef]

63. Zhang, X.; Myrold, D.D.; Shi, L.; Kuzyakov, Y.; Dai, H.; Thu Hoang, D.T.; Dippold, M.A.; Meng, X.; Song, X.; Li, Z.; et al. Resistance of microbial community and its functional sensitivity in the rhizosphere hotspots to drought. *Soil Biol. Biochem.* **2021**, *161*, 108360. [CrossRef]

64. Spain, A.M.; Krumholz, L.R.; Elshahed, M.S. Abundance, composition, diversity and novelty of soil Proteobacteria. *ISME J.* **2009**, *3*, 992–1000. [CrossRef]

65. Pajares, S.; Bohannan, B.J. Ecology of Nitrogen Fixing, Nitrifying, and Denitrifying Microorganisms in Tropical Forest Soils. *Front. Microbiol.* **2016**, *7*, 209327. [CrossRef] [PubMed]

66. Huang, Q.; Huang, Y.; Wang, B.; Dippold, M.A.; Li, H.; Li, N.; Jia, P.; Zhang, H.; An, S.; Kuzyakov, Y. Metabolic pathways of CO_2 fixing microorganisms determined C-fixation rates in grassland soils along the precipitation gradient. *Soil Biol. Biochem.* **2022**, *172*, 108764. [CrossRef]

67. Wang, R.; Wang, M.; Wang, J.; Lin, Y. Habitats Are More Important Than Seasons in Shaping Soil Bacterial Communities on the Qinghai-Tibetan Plateau. *Microorganisms* **2021**, *9*, 1595. [CrossRef]

68. Urbanová, M.; Šnajdr, J.; Baldrian, P. Composition of fungal and bacterial communities in forest litter and soil is largely determined by dominant trees. *Soil Biol. Biochem.* **2015**, *84*, 53–64. [CrossRef]

69. Cheng, H.; Wu, B.; Wei, M.; Wang, S.; Rong, X.; Du, D.; Wang, C. Changes in community structure and metabolic function of soil bacteria depending on the type restoration processing in the degraded alpine grassland ecosystems in Northern Tibet. *Sci. Total Environ.* **2021**, *755*, 142619. [CrossRef]

70. Zhu, H.; Gong, L.; Luo, Y.; Tang, J.; Ding, Z.; Li, X. Effects of litter and root manipulations on soil bacterial and fungal community structure and function in a schrenk's spruce (*Picea schrenkiana*) forest. *Front. Plant Sci.* **2022**, *13*, 849483. [CrossRef]

71. Wu, C.; Ma, Y.; Wang, D.; Shan, Y.; Song, X.; Hu, H.; Ren, X.; Ma, X.; Cui, J.; Ma, Y. Integrated microbiology and metabolomics analysis reveal plastic mulch film residue affects soil microorganisms and their metabolic functions. *J. Hazard. Mater.* **2022**, *423*, 127258. [CrossRef]

72. Li, Y.Y.; Xu, T.T.; Ai, Z. Diversity and predictive functional of caragana jubata bacterial community in rhizosphere and non-Rhizosphere soil at different elevations. *Environ. Sci.* **2023**, *44*, 2304–2314.

73. Bromke, A.M. Amino Acid Biosynthesis Pathways in Diatoms. *Metabolites* **2013**, *3*, 294–311. [CrossRef] [PubMed]

74. Chen, Y.; Neilson, J.W.; Kushwaha, P.; Maier, R.M.; Barberán, A. Life-history strategies of soil microbial communities in an arid ecosystem. *ISME J.* **2021**, *15*, 649–657. [CrossRef]

75.] Ashry, N.M.; Alaidaroos, B.A.; Mohamed, S.A.; Badr, O.A.; El-Saadony, M.T.; Esmael, A. Utilization of drought-tolerant bacterial strains isolated from harsh soils as a plant growth-promoting rhizobacteria (PGPR). *Saudi J. Biol. Sci.* **2022**, *29*, 1760–1769. [CrossRef] [PubMed]

76. Zhou, Z.; Wang, C.; Luo, Y. Meta-analysis of the impacts of global change factors on soil microbial diversity and functionality. *Nat. Commun.* **2020**, *11*, 3072. [CrossRef]

77. Hermans, S.M.; Buckley, H.L.; Case, B.S. Using soil bacterial communities to predict physico-chemical variables and soil quality. *Microbiome* **2020**, *8*, 79. [CrossRef]

78. Bastida, F.; Eldridge, D.J.; García, C.; Kenny Png, G.; Bardgett, R.D. Soil microbial diversity–biomass relationships are driven by soil carbon content across global biomes. *ISME J.* **2021**, *15*, 2081–2091. [CrossRef]

79. Zeng, J.; Liu, X.; Song, L.; Lin, X.; Zhang, H.; Shen, C.; Chu, H. Nitrogen fertilization directly affects soil bacterial diversity and indirectly affects bacterial community composition. *Soil Biol. Biochem.* **2016**, *92*, 41–49. [CrossRef]

80. Naz, M.; Dai, Z.; Hussain, S.; Tariq, M.; Danish, S.; Khan, I.U.; Qi, S.; Du, D. The soil pH and heavy metals revealed their impact on soil microbial community. *J. Environ. Manag.* **2022**, *321*, 115770. [CrossRef]

81. Kim, J.M.; Roh, A.S.; Choi, S.C. Soil pH and electrical conductivity are key edaphic factors shaping bacterial communities of greenhouse soils in Korea. *J. Microbiol.* **2016**, *54*, 838–845. [CrossRef]

Disclaimer/Publisher's Note: The statements, opinions and data contained in all publications are solely those of the individual author(s) and contributor(s) and not of MDPI and/or the editor(s). MDPI and/or the editor(s) disclaim responsibility for any injury to people or property resulting from any ideas, methods, instructions or products referred to in the content.

Article

Spatial and Temporal Variation in Vegetation Response to Runoff in the Ebinur Lake Basin

Chenglong Yao [1,2], Yuejian Wang [1,2,*], Guang Yang [3,4,*], Baofei Xia [1,2], Yongpeng Tong [1,2], Junqiang Yao [5] and Huanhuan Chen [1,2]

1. College of Science, Shihezi University, Shihezi 832000, China
2. Key Laboratory of Oasis Town and Mountain Basin System Ecology of Xinjiang Production and Construction Corps, Shihezi 832000, China
3. College of Water Conservancy & Architectural Engineering, Shihezi University, Shihezi 832000, China
4. Key Laboratory of Cold and Arid Regions Eco-Hydraulic Engineering of Xinjiang Production & Construction Corps, Shihezi 832000, China
5. Institute of Desert Meteorology, China Meteorological Administration, Urumqi 830002, China

* Correspondence: wangyuejian0808@163.com (Y.W.); mikeyork@163.com (G.Y.)

Citation: Yao, C.; Wang, Y.; Yang, G.; Xia, B.; Tong, Y.; Yao, J.; Chen, H. Spatial and Temporal Variation in Vegetation Response to Runoff in the Ebinur Lake Basin. *Forests* **2023**, *14*, 1699. https://doi.org/10.3390/f14091699

Academic Editor: Daniel L. McLaughlin

Received: 16 July 2023
Revised: 5 August 2023
Accepted: 18 August 2023
Published: 23 August 2023

Copyright: © 2023 by the authors. Licensee MDPI, Basel, Switzerland. This article is an open access article distributed under the terms and conditions of the Creative Commons Attribution (CC BY) license (https:// creativecommons.org/licenses/by/ 4.0/).

Abstract: The response of spatial and temporal vegetation changes to runoff is a complex process involving the interaction of several factors and mechanisms. Timely and accurate vegetation and runoff change information is an important reference for the water cycle and water resource security. The Ebinur Lake Basin is representative of arid areas worldwide. This basin has been affected by climate change and human activities for a long time, resulting in the destruction of the basin's ecological environment, and especially its vegetation. However, there have been few studies that have focused on watershed vegetation and runoff changes. Therefore, we combined Generalized Information System and remote sensing technology, used SWAT and InVEST models based on the Google Earth Engine platform, and used the vegetation normalization index method to calculate the spatial distribution of vegetation and water production from 2000 to 2020 in Ebinur Lake. Sen's trend analysis and the M–K test were used to calculate vegetation and runoff trends. The relationship between vegetation and runoff variation was studied using bivariate spatial autocorrelation based on sub-basins and plant types. The results showed that the Z parameter in the InVEST model spanned from 1–2. The spatial distribution of the water yield in a watershed is similar to the elevation of the watershed, showing a trend of higher altitude leading to a higher water yield. Its water yield capacity tends to saturate at elevations greater than 3500 m. The local spatial distribution of the Normalized Difference Vegetation Index(NDVI) values and water yield clustering in the watershed were consistent and reproducible. Interannual runoff based on sub-basins correlated positively with the overall NDVI, whereas interannual runoff based on plant type correlated negatively with the overall NDVI.

Keywords: Ebinur Lake watershed; bivariate spatial autocorrelation; Google Earth Engine; InVEST model

1. Introduction

Vegetation is an important component of terrestrial ecosystems and is a foundation for the survival of other organisms [1,2]. In the context of global climate change, the frequency of droughts, areas affected, and degree of damage are increasing annually [3,4]. Many rivers worldwide are experiencing substantial declines in water flow, with some completely drying. This has resulted in severe effects on humans and the environment [5,6]. Climate change has had a direct impact on precipitation patterns, affected water and sand transport systems, and had a considerable impact on regional ecological security [7]. Therefore, understanding how vegetation responds to variations in runoff is crucial for determining ecological changes.

Vegetation serves as a link in the material cycle and plays an important role in surface energy conversion, climate regulation, and water transfer [8–10]. Vegetation affects surface runoff through land-based water cycle processes such as precipitation interception, surface evaporation, and soil water infiltration [11,12]. The hydrological cycle in arid zones is extremely vulnerable. Therefore, the response of runoff changes to human activity and climate change is highly sensitive [13,14]. For example, simulations of runoff under different scenarios have shown that the aridity of the Alwand Basin in Iran has increased [15]. Arid locations typically have low population densities. Given the water resource limitations, human activities in arid areas are often concentrated around areas that do have water resources [16]. There are relatively few land-use types in dry zones, with land primarily being used for agriculture or construction [17]. However, agricultural irrigation and urban development require substantial amounts of water, making water resources increasingly scarce [18]. Furthermore, some engineering projects aimed at water conservancy, such as dam closures and cross-basin water diversions, may affect runoff. These projects alter the distribution of natural runoff, thereby affecting hydrological processes [19]. In the Minab River Basin of Iran, the increase in land use has led to a considerable reduction in runoff [20]. When human activities involve water resource development and overuse, water resources are more rapidly depleted [19,21]. As global climate change continues, the shift in precipitation patterns has become most pronounced in arid zones, typified by decreasing rainfall, increasing evaporation, decreasing soil moisture, and exacerbating runoff decline [22,23]. Continuous climate change may lead to the collapse of entire ecosystems in arid areas. Therefore, effective measures need to be taken to strengthen water resource protection and management to address the challenges posed by human activities and climate change.

The Ebinur Lake Basin is part of the "Silk Road Economic Belt", and its soil and water security are closely linked to the economy of China. The ecological environment of the Ebinur Lake Basin is affected by human activity and climate change [24]. Grassland degradation, water scarcity, and land desertification are becoming more prominent in this basin [3]. Grassland deterioration is a critical issue, primarily due to overgrazing and reclamation [25]. These activities have reduced grassland cover and caused vegetation deterioration, which has hastened land desertification [26]. Furthermore, water resources in the Ebinur Lake Basin are under increasing strain from economic expansion and rapid urbanization [27]. Ebinur Lake Basin is a region with scarce hydrological data. There are fewer hydrological stations within the watershed. Obtaining hydrological data is only point data. In this context, many researchers have studied the internal causes of lake area changes, runoff (channel flow) changes, land-use changes, and soil salinization to address increasingly prominent water resource problems [27–29]. Water scarcity in the Ebinur Lake Basin has been linked to a decrease in runoff caused by climate change and an increase in water demand caused by increasing the expansion of cultivated land and plantation [28,30]. Different types of land use have different impacts on vegetation and runoff [19,27,30]. Compared to bare land, surface runoff covered by vegetation is more likely to exhibit gradual characteristics, with runoff time concentrated over a longer period of time rather than a brief peak [19]. The impact of different vegetation on runoff is also different. The runoff of herbaceous plants and forests started significantly later than that of shrubs [31]. To date, the relationship between vegetation and runoff (overland flow) in the Ebinur Lake Basin is not clear. For example, it is not yet known whether the temporal and spatial changes in vegetation and runoff in the sub-basin are the same as those throughout the whole basin and how each vegetation type affects runoff. The spatiotemporal changes of vegetation and runoff (overland flow) are interdependent. Examining the spatial and temporal variations between them can help develop strategies for the sustainable use of watershed water resources. Therefore, ecological and environmental monitoring and scientific research have become important tools for protecting the ecological and environmental security of the Ebinur Lake Basin.

In this study, we used the bivariate spatial autocorrelation method to study the relationship between vegetation and runoff changes from the perspective of sub-watersheds and plant types. This research aimed to investigate three scientific questions: (1) how runoff in the watershed varied in space and time; (2) how the vegetation in the watershed changed over time and space; and (3) whether there is a link between vegetation and runoff in watersheds.

2. Overview of the Study Area

The Ebinur Lake Basin (Figure 1) (44°02′–45°23′ N, 79°53′–83°53′ E) is a component of the Xinjiang Uygur Autonomous Region's Boltara Autonomous Prefecture. The area has a northern temperate continental arid climate. The difference between the daily and annual temperatures of the basin is large, with hot summers and cold winters. For several years, the average annual precipitation has been 116–170 mm. The annual evaporation rate exceeds 1000 mm. The terrain of the watershed is complex, flanked by mountains on three sides, and is a well-known wind outlet in China, with northwest winds dominating in all years [28]. The watershed vegetation is classified into seven types: alpine vegetation, coniferous forests, agricultural fields, grasslands, meadows, shrubs, and deserts. The basin is relatively rich in species, with up to 36 types of national first- and second-class protected animals. There are 79 plant families and 413 plant species with medicinal potential, primarily Chinese wolfberry, ephedra, licorice, and rare plants such as red Mentha [32].

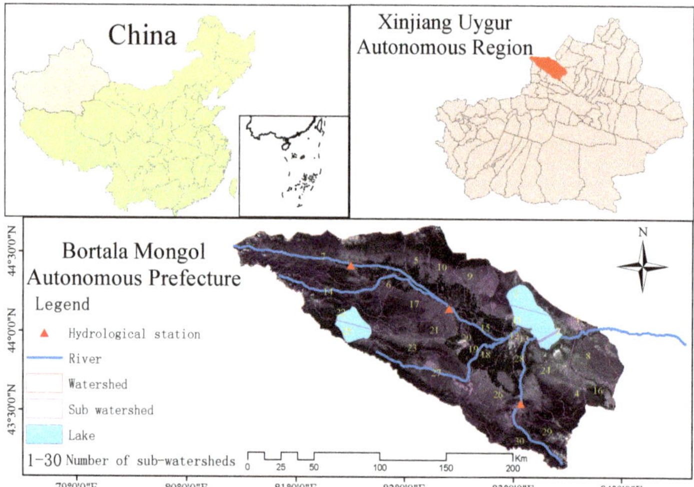

Figure 1. Ebinur Lake Basin.

3. Data and Methods

3.1. Data Sources

The interaction between vegetation and runoff has predominantly been investigated by constructing sample plots [31]. However, to date, actual measured data gathered in the experimental region have been limited. Therefore, we used easily accessible remote sensing data in our study. The water production module of the InVEST model (Figure 2) is a tool used for assessing natural capital management and the sustainable use of water resources [27].

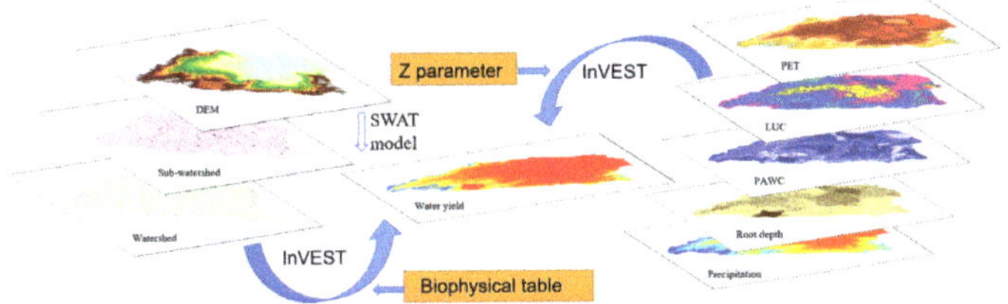

Figure 2. InVEST model water production module.

The water production module was used to evaluate the quantitative and distribution properties of water resources and the impact of factors such as hydrographic conditions, precipitation, and evapotranspiration [33–35]. For this model, data on precipitation, reference evapotranspiration, land use, soil data, soil moisture content, a digital elevation model (DEM), and biophysical tables are required. Table 1 summarizes the data used in the study.

Table 1. Dataset descriptions, processing, usage, and sources used in this study.

Type	Data	Description	Processing and Usage	Source
Hydrological	Surface water volume	Reflects the total amount of surface water in the region during the year	Used to calibrate data and calculate water production modulus	Bortala Autonomous Prefecture Hydrological Bulletin 2000–2020
Remote sensing	Precipitation	Monthly precipitation data with a spatial scale of 1 km resolution	Exploring the spatial distribution differences of precipitation for calibrating InVEST	National Tibetan Plateau Data Center (http://data.tpdc.ac.cn (accessed on 18 April 2023)) [36]
	Potential evapo-transpiration	Monthly potential evapotranspiration dataset with a spatial scale of 1 km	Used for calibrating InVEST	National Tibetan Plateau Data Center (http://data.tpdc.ac.cn (accessed on 18 April 2023)) [36]
	Land use	Landsat images generated through manual visual interpretation	Used for calibrating InVEST	Google Earth Engine Remote Sensing Cloud Computing Platform Download Land Use Classification Maps for Each Year from 2000 to 2020 [37]
	DEM	Digital elevation model with a resolution of 30 m	Calculating watershed boundaries and describing terrain undulation data for calibrating InVEST	National Tibetan Plateau Data Center (http://data.tpdc.ac.cn (accessed on 18 April 2023)) [36]
	NDVI	Maximum NDVI value synthesized over 16 days based on Terra satellite global vegetation index at 250 m	Analyzing the spatiotemporal changes in NDVI and discussing the relationship between runoff and vegetation cover	Google Earth Engine Remote Sensing Cloud Computing Platform Calculate NDVI for Each Year from 2000 to 2020
Soil	Soil data	Contains all attributes of soil (HWSD) Dataset (v1.2)	Calculating soil water content for calibrating InVEST	National Tibetan Plateau Data Center (http://data.tpdc.ac.cn (accessed on 18 April 2023)) [38]
Other data	Biophysical table	Reflects the attributes of soil coverage and LULC, including LULC encoding, plant evapotranspiration coefficient (Kc), and root depth	Used for calibrating InVEST	Literature [25] and InVEST user guide [39]

3.2. Data Analysis

3.2.1. InVEST Water Production Model

Vegetation and runoff data have been used to establish water balance models, Lorenz curves, and Budyko data to explore the relationship between vegetation and runoff [40–42]. However, modeling using these methods only reflects the linear relationship between the two and does not visualize their spatial relationship. Therefore, this study used the water production module of the InVEST model, which is based on the water balance principle, to determine the water production of each raster in the water balance and to obtain the spatial distribution of runoff.

The equations used are as follows:

$$Y_{XJ} = \left(1 - \frac{AET_{XJ}}{P_X}\right) \times P_X,\tag{1}$$

$$\frac{AET_{XJ}}{P_X} = \frac{1 + \omega_X R_{XJ}}{1 + \omega_X R_{XJ} + \frac{1}{R_{XJ}}},\tag{2}$$

$$R_{XJ} = \frac{K_{XJ} \times ET_0}{P_X},\tag{3}$$

$$\omega_X = Z\frac{AWC_X}{P_X},\tag{4}$$

where Y_{XJ} denotes the annual water yield of the study area, AET_{XJ} is the average annual actual evapotranspiration, P_X is the average annual precipitation, and R_{XJ} is the dimensionless drying index obtained from the ratio of potential evapotranspiration to precipitation. ET_0 is the average annual potential evapotranspiration. K_{XJ} is the vegetation evapotranspiration coefficient corresponding to different land-cover types in the raster cell, whose values can be obtained by consulting the data. Z is a seasonal parameter used to characterize the seasonality of precipitation. AWC_X is the available water content of the plant, and its value is determined by soil depth, soil texture, and organic matter content. In this study, the plant available water capacity (PAWC) calculated from soil data was used instead of the AWC_X [43]. The equation used is as follows:

$$\begin{aligned}\text{PAWC} = {}& 54.509 - 0.132\,\text{Sand} - 0.003(\text{Sand})^2 - 0.055\,\text{Silt} - 0.006(\text{Silt})^2 \\ & -0.738\,\text{Clay} + 0.007(\text{Clay})^2 - 2.688c + 0.501(C)^2,\end{aligned}\tag{5}$$

where sand is the soil sand content, silt is the soil powder content, clay is the soil clay content, and C is the soil organic matter content.

3.2.2. Correlation Analysis

Bivariate spatial autocorrelation analysis was used to describe the spatial correlation and dependency characteristics of the two geographical features. Unlike traditional spatial autocorrelation analysis, which considers only one variable, bivariate spatial autocorrelation analysis can more accurately show the spatial relationships between geographical phenomena [44]. The specific equations are as follows:

$$I = \frac{\sum_{a=1}^{c} \sum_{b=1}^{c} W_{ab}(x_a - \bar{x})(y_j - \bar{y})}{N^2 \sum_{a=1}^{c} \sum_{b=1}^{c} W_{\alpha\beta}}\tag{6}$$

where I is the global spatial autocorrelation index, c is the number of research units, W_{ab} is the spatial weight matrix, x_a and y_a are the values of the independent and dependent variables in spatial units a and b, respectively, and N^2 is the variance of all samples. The specific equations are as follows:

$$I_a = z_a \sum_{j=1}^{c} W_{ab} Z_b\tag{7}$$

where I_a represents the local spatial relationship between the independent and dependent variables in study unit a. Z_a and Z_b are the standardized values for the variance of the observation values of study units a and c. The distribution map of local indicators of spatial association (LISA) formed can show the clustering and differentiation characteristics of independent and dependent variables in the local region. The SWAT model is widely used in the field of hydrological research [15,20]. In this study, we only used DEM data to generate watershed boundaries and sub-watershed boundaries through SWAT models. This study refers to the nonparametric linear regression technology and uses Theil–Sen trend analysis and the Mann–Kendall test method to study the spatial and temporal change trend of NDVI and runoff in Ebinur Lake Basin from 2000 to 2020 [15,45]. We used ArcGIS (version 10.2) to create images.

4. Results

4.1. Calibration and Validation of InVEST

The results obtained from the InVEST water production module represented the total amount of water produced in the input catchment. Given that the area of the delineated watershed was not the same as the catchment area of the measured data, the simulated water production values could not be directly compared with the measured surface water volume. The water production modulus was used to compare the differences between the simulated and measured values (Figure 3). The sensitive factor affecting the model was the parameter Z. Therefore, we calibrated the Z parameter based on the water production modulus obtained from the hydrological bulletin of the Bortala Autonomous Prefecture from 2000 to 2020. We used correlation coefficients for several tests between the simulated and measured values to select the optimal Z value, as shown in Table 2. The correlation coefficient between the simulated and measured values was 0.74. The results showed that the larger the Z parameter, the smaller the value of water production. Most of the Z values were in the range of 1–2. When precipitation tended to be stable, close to the multi-year average, the Z value did not change significantly. When the precipitation exceeded the average, Z increased.

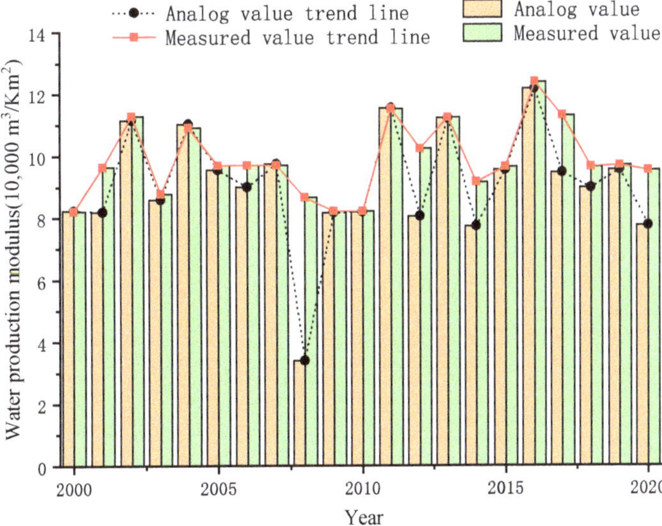

Figure 3. Simulation results and measured values.

Table 2. Z parameters.

Name	Value										
Year	2020	2019	2018	2017	2016	2015	2014	2013	2012	2011	2010
Z	1	1.5	1	1	11	1.7	1	1.6	1	1.7	11
Year	2009	2008	2007	2006	2005	2004	2003	2002	2001	2000	
Z	2	1	1.1	1	1.2	1	1.5	1.6	1	1	

4.2. Spatial Patterns of Interannual Water Content and NDVI

Based on the water production module of the InVEST model, the annual water yield of the basin was estimated for 2000–2020 (Figure 4). Areas with higher water yields were in the upper part of the basin, and smaller areas were in the lower part. The spatial distribution of the water content in the basin was similar to the elevation of the basin. This indicated that the higher the elevation, the higher the water content. The maximum water content from 2000 to 2020 was recorded in 2016, with a maximum value of 993.38 mm. The minimum water content depth was recorded in 2008, with a value of 393.04 mm. The multi-year average water production was 682.938 mm.

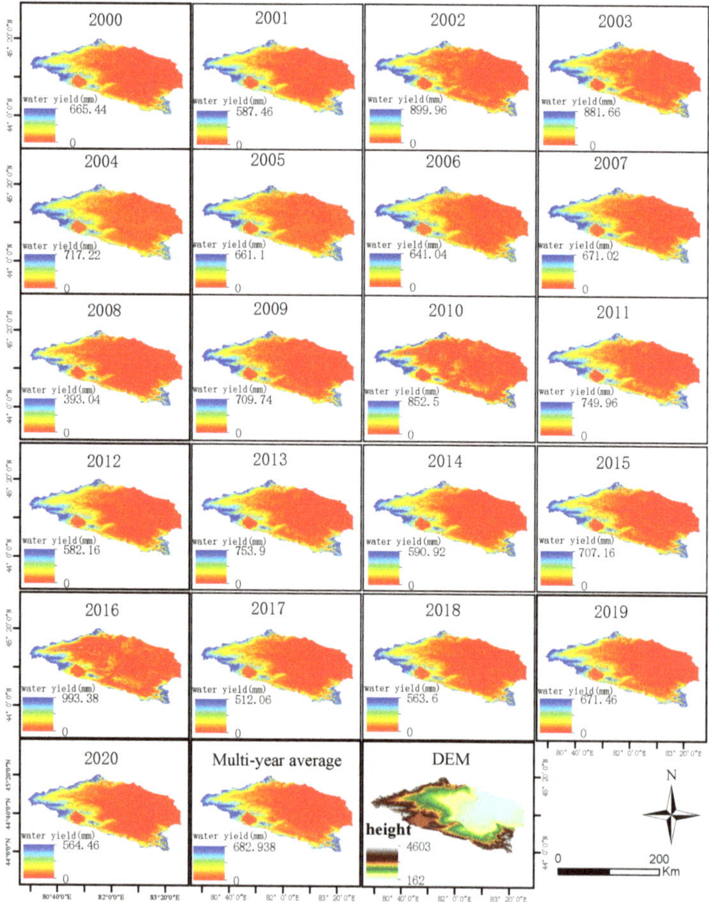

Figure 4. Multi-year average water production and DEM schematic (2000–2020).

The relationship between runoff production depth and elevation was further divided based on the spatial distribution of water production and DEM elevation values from 2000 to 2020 (Figures 4 and 5). The study area was divided into three zones with low, medium, or high values. The elevation of the low-value zone was 162–1250 m, which is in the red area of the spatial distribution of water production (Figure 4). The elevation of the middle-value zone was 1250–3000 m, which is within the red and light-blue zones of the spatial distribution of water production (Figure 4). The high-value area was 3000–4603 m above sea level and was within the light-blue and blue areas of the spatial distribution of water production (Figure 4). There was a saturation point in the high-value area (Figure 5). Its water production capacity tended to saturate at an elevation of 3500 m or more. The water production capacity of the land type was higher because of the presence of snowy mountains above 3500 m in elevation.

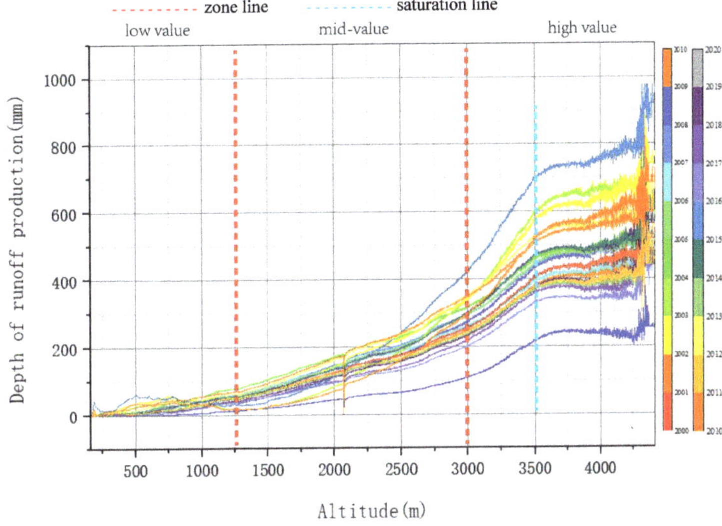

Figure 5. Relationship between runoff production depth and elevation values.

Based on the Google Earth Engine platform, NDVI was downloaded and calculated for each year from 2000 to 2020 (Figure 6). We use red lines to divide the watershed into low value areas, median areas and high value areas based on the trend of the data. The blue line represents the saturation line of runoff at altitude. The maximum NDVI value was 0.6936 in 2003, and the minimum NDVI value was 0.6091 in 2020, with a multi-year average NDVI value of 0.6193. The maximum NDVI values were spatially distributed in the northern and southeastern mountainous areas of the basin, where the main vegetation types were alkaline vegetation and coniferous forests. The minimum NDVI values excluding water bodies were distributed in the marginal areas of the basin. There were large desert areas between the arable land and grasslands, and the NDVI values for desert vegetation were the lowest. The vegetation in the study area was characterized by the distribution of natural plants around it and agricultural fields in the middle.

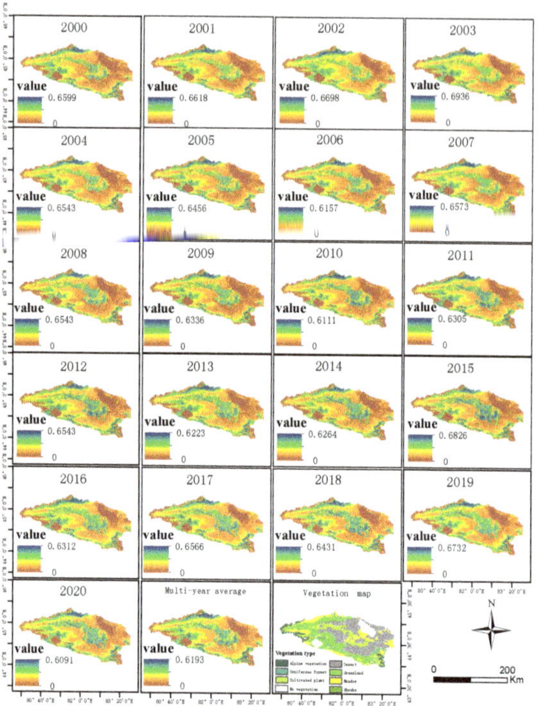

Figure 6. Multi-year average NDVI and vegetation type distribution diagram (2000–2020).

4.3. Interannual Water Content and NDVI Trend Analysis

The temporal and spatial trends of interannual water content and NDVI from 2000 to 2020 were analyzed and tested for significance using the Theil–Sen median trend analysis method (Figure 7a,b). The NDVI values in the middle of the watershed showed a significant increasing trend (Figure 7a). Based on the plant type diagram, cultivated plants showed a significant increase (Figure 5). A significant decrease in NDVI values occurred in the northern part of the watershed. No significant changes were observed in the other areas. In general, the spatial pattern of runoff did not change significantly (Figure 7b). Water content decreased significantly in the central part of the watershed.

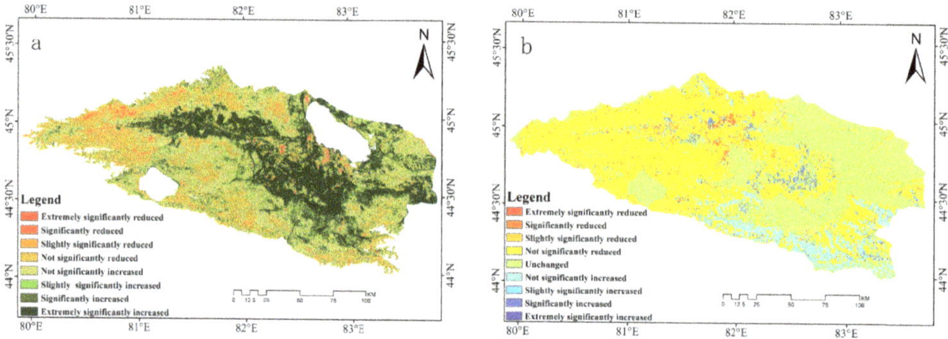

Figure 7. Schematic diagram of NDVI and water production changes from 2000 to 2020: (**a**) trend change in NDVI; (**b**) trend change in runoff.

4.4. Interannual Water Yield and NDVI Autocorrelation Analysis Based on Sub-Basins

A bivariate spatial autocorrelation analysis was performed between the interannual water yield and NDVI values in the sub-basins (Figure 8). The Moran's I was positive from 2000 to 2020, with a range of 0.048–0.263. Its two variables showed a strong positive spatial correlation, that is, the more water yielded, the greater the NDVI value within the sub-basin. Moran's I index in 2002 was the highest at 0.101. The Moran's I index in 2010 was the lowest at 0.048. Most of the 24 sub-watersheds were distributed in quadrants 1, 2, and 3.

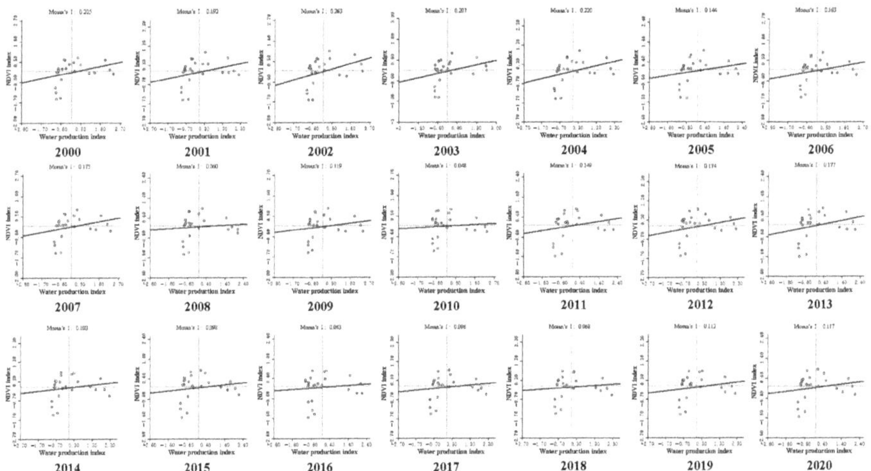

Figure 8. Moran's index based on interannual water yield and NDVI in sub-basins.

Combining the interannual water yield and NDVI (LISA) cluster analysis, high–high clustering was found in sub-basins 3, 5, 8, and 20 (Figure 9). Sub-basin 8 was the area where the maximum NDVI value was recorded, and the maximum water yield was also recorded in that area. Therefore, the spatial autocorrelation between the two was relatively high. The only area with low–high clustering was sub-basin 13, which was in the middle reaches of the basin and had stable low–high clustering of NDVI and water yield over the last 21 years. The low–low clustering areas were sub-basins 2, 6, and 12, which were downstream of the study area. The high–high clustered sub-basins and low–low clustered sub-basins had a contiguous distribution.

4.5. Interannual Water Yield and NDVI Autocorrelation Analysis Based on Vegetation Type

A bivariate spatial autocorrelation analysis was performed between the interannual water yield of the vegetation types and NDVI values (Figure 10). Moran's I index was negative from 2000 to 2020, showing a strong dispersion trend. Moran's I index was the highest in 2008, at −0.116, and the lowest in 2018, at −0.340. Vegetation was divided into seven types: alpine vegetation, coniferous forests, cultivated plants, grasslands, meadows, shrubs, and desert (Figure 5). Given the discrete distribution of vegetation types, they were divided into 162 vegetation areas. The samples were distributed across all four quadrants, mainly in quadrants 2 and 4.

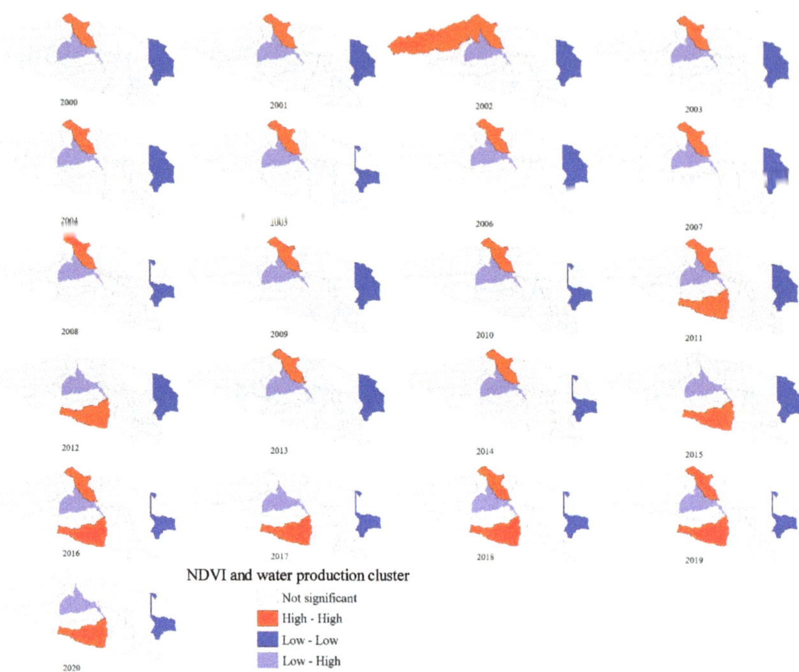

Figure 9. LISA clustering analysis of interannual water production versus NDVI based on sub-basins.

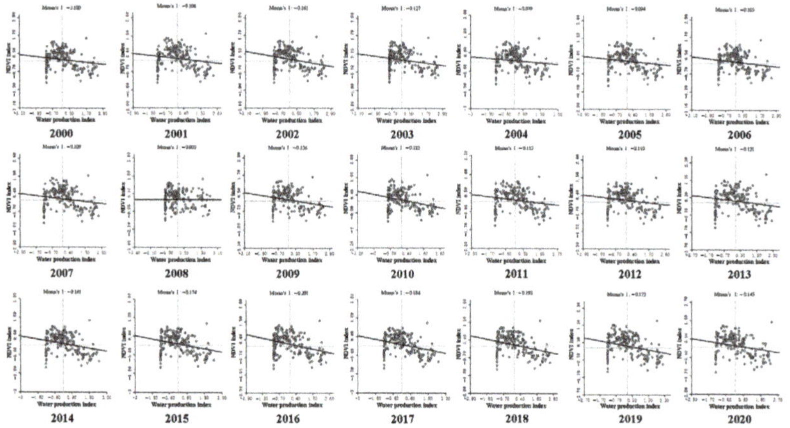

Figure 10. Moran's index of interannual water yield and NDVI based on vegetation type.

Based on the LISA clustering analysis of interannual water yield and NDVI based on vegetation types (Figure 11), the vegetation types with high–high clustering included meadows and grasslands. Meadows in the north were highly clustered for the last 21 years. The year with high–high clustering in grasslands was 2008, with one year of occurrence. The grasslands in other years were generally in a low–high concentration state. Meadows and alpine vegetation exhibited high–low clustering and low–low clustering, respectively, and were located in the south and west of the watershed. The vegetation types with low to high concentrations included grasslands and agricultural fields located in the middle of the watershed.

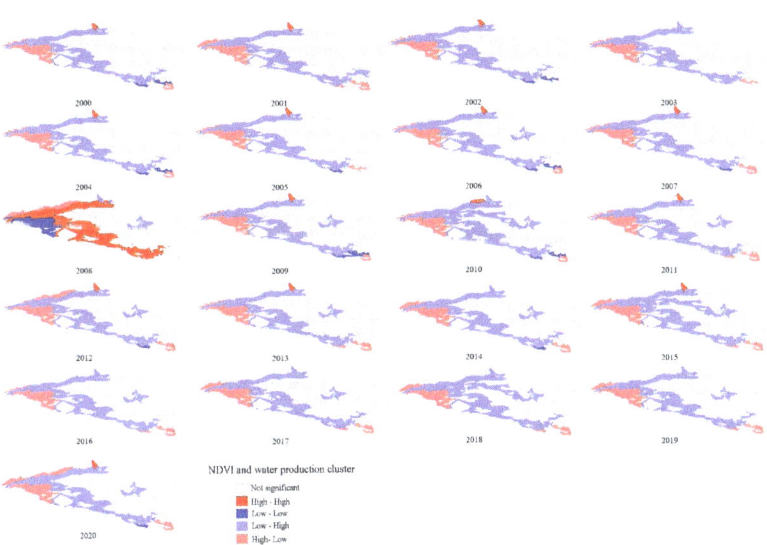

Figure 11. LISA cluster analysis of interannual water yield and NDVI based on vegetation type.

5. Discussion

5.1. Factors Potentially Affecting Changes in Vegetation and Runoff

The changes in vegetation and runoff were influenced by various natural and anthropogenic factors. The current study concluded that water production was positively correlated with precipitation, PAWC, and DEM and negatively correlated with NDVI and PET (Figure 12a). The positive correlation between water production and elevation is the highest. Our results showed that the spatial distribution of water production in the watershed was similar to the elevation of the watershed, with a trend of increased water production at higher elevations. This is because of the presence of snow-capped mountains at higher elevations and iceberg meltwater, which is the main source of water supply in the basin [46,47]. The negative correlation between NDVI and elevation is strongest, as arable land is generally located in plain areas. The proportion of arable land in annual NDVI values is relatively high (Figure 12a). Melting icebergs can provide a large amount of water resources. Climate is an important factor that influences vegetation change and runoff. Different climatic conditions affect vegetation growth and the rate of water evaporation, which has an impact on runoff volume [7,44]. Different hydrological conditions, such as rainfall, evapotranspiration, and soil moisture content, also affect vegetation and runoff [11,12,48]. The higher the effective rainfall, the more suitable the conditions for vegetation growth, and the lower the runoff volume over shorter timescales [48]. The results of this study also showed that precipitation is one of the main causes of regional runoff variability (Figure 12b). There was no significant decrease in runoff in the upstream area, but a change occurred owing to the different amounts of annual iceberg snow melt upstream and precipitation being significantly higher upstream than downstream (Figure 12b). Runoff from the lakes and deserts in the lower reaches did not change. There was no significant increase in runoff at the margins of the basin. However, changes occurred because of variations in snow and ice melt in the high mountains caused by annual differences in precipitation [46,47]. The InVEST model is greatly influenced by precipitation. Among the 21 years analyzed, 2008 had the lowest water production and precipitation. The error between the simulated and measured values was also the largest, indicating that precipitation had the most direct effect on changes in runoff in the study area (Figure 3). The Z parameter is 1 because the lowest adjustable parameter value can only be 1. Topographic conditions are key factors affecting vegetation and runoff. There are differences in vegetation growth conditions and runoff distribution between mountainous and plain areas. The border between mountains and plains is a large area of grasslands and meadows, which grow with a small grass blade area and, therefore, have lower NDVI values, indicating that grasslands in arid areas are more drought tolerant [49].

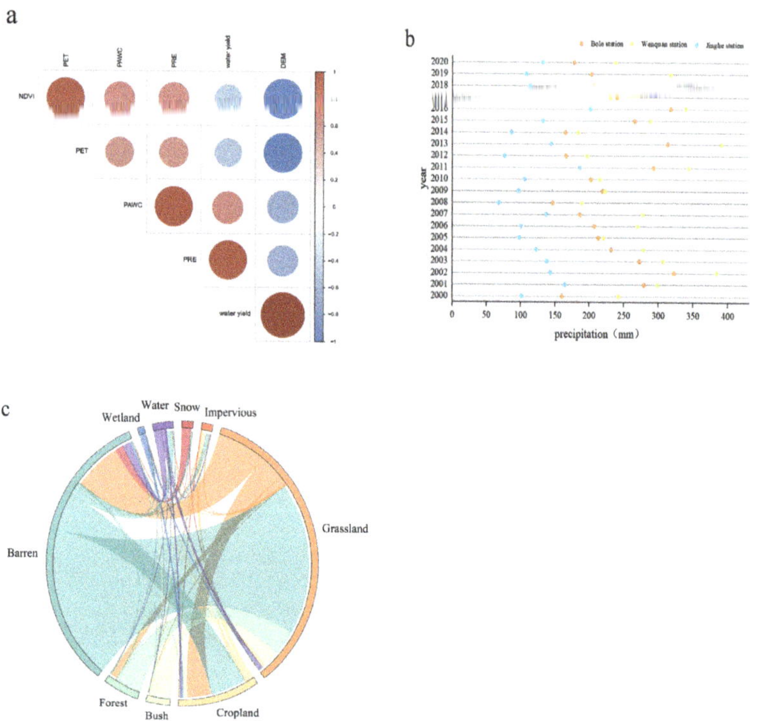

Figure 12. Possible factors affecting changes in vegetation and runoff. (**a**) Correlation analysis of the main factors of runoff; (**b**) precipitation of meteorological stations in the study area; (**c**) land-use change from 2000 to 2020.

Land-use patterns strongly influence vegetation change and runoff. Human activities change the original state of the land and affect vegetation growth, therefore changing the distribution and amount of runoff [19,20,49]. In the last 20 years, all the land-use types in the watershed have changed, with the most significant changes occurring in grasslands and bare ground (Figure 12c). Some grasslands have become bare land, and some have become forests. Some cultivated land has been converted into grassland, which may be due to the implementation of the Grain for Green project [50]. Shrublands have been transformed into grasslands. Areas that usually contain permanent snow/water have been transformed into bare land. Part of the grassland has become bare land, which could have been caused by environmental damage [27]. Overall, the land types in the watershed are improving, with bare land decreasing and grassland increasing. The results of our study showed that the NDVI significantly increased in the watershed for cultivated plants (Figure 6). This may be because the crops being cultivated differ each year and the cultivated area increased (Figure 12c). The basin was found to be richer in cultivated land planted with crops such as corn, wheat, and cotton. Around its perimeter, the NDVI values have significantly changed, and the significant changes may be where the land-use type is changing to croplands. Cultivated plants in the middle of the area also showed a few highly significant decreases, which may have been caused by the conversion of cultivated land to residential land. The NDVI values of grasslands and meadows changed but not significantly. The area where they changed was at the border between the mountains and plains. This may have been from changes in vegetation growth caused by climate change.

5.2. Response of Vegetation Change to Runoff

The vegetation cover affects the entire hydrological cycle [51,52]. At different time scales, changes in vegetation cover can have different effects on indicators such as water production, flow production, and runoff coefficient [40,53]. Changes in vegetation cover may affect rainfall infiltration and evapotranspiration [23,24]. Where significant decreases in NDVI values occur in upper mountainous areas, they may be due to deforestation, leading to soil erosion, which causes a decrease in NDVI values (Figure 7a). The reduction in vegetation during the conversion of forested land to other land types may lead to increased rainfall runoff, which, in turn, affects the availability of water resources [54]. In the present study, excluding extreme years of precipitation, water production had continuity in time when the vegetation did not change substantially. In contrast, changes in vegetation cover also affected soil erosion, particularly in mountainous and hilly areas [55]. Under the influence of global climate change, terrestrial water storage is likely to decrease and increase with drought severity [56]. The vegetation response to water production varied at different spatial scales. At small scales, vegetation can increase soil permeability and water storage capacity through the root system, which can affect groundwater recharge and circulation while at large scales [57]. Changes in vegetation cover may have an impact on water balance and hydrological processes in the watershed [40].

Our study found a positive correlation between the interannual water yield of sub-basins and the overall NDVI and a negative correlation between the interannual water yield of vegetation types and the overall NDVI (Figures 8 and 10). From both perspectives, there were continuity and reproducibility in the local spatial clustering of NDVI values and water yield (Figures 9 and 11). From the sub-basin perspective, high–high clustering existed in sub-basin 8 from 2000 to 2011, and high–high clustering disappeared in 2012, 2015, 2017, and 2020 in region 8 and reoccurred after its disappearance. High–high clustering occurred in sub-basin 20 in 2011. After one year of continuity, the clustering state no longer occurred, before reoccurring after 2015 and persisting until 2020. Sub-basins 2 and 12 had low–low clustering for 21 years. Sub-basin 6 also had low clustering in most years because sub-basins 2, 6, and 12 also had relatively low NDVI in these areas when interannual water production was low. This may indicate that hydrological processes within the watersheds were influenced by the degree of surface vegetation cover and that higher vegetation cover increased soil water retention capacity and, therefore, interannual water production. Under different vegetation types, the high–high meadow aggregation first occurred in 2000, continued in 2001 and 2002, changed in 2003, and was recorded again in 2004. The meadow maintained low–high clustering, except for the extreme year (2008). Over the last hundred years, precipitation has increased in northern North America where vegetation has increased, whereas precipitation has decreased in central North America and almost all of China [14]. Vegetation is particularly abundant at high northern latitudes and in agricultural and afforested regions [58]. Therefore, the mechanisms of influence at different scales need to be considered when studying the responses of spatial and temporal vegetation changes to water production.

In arid zones, substantial amounts of surface water are transported to the atmosphere via transpiration from vegetation or evaporation from the land [59]. Since the 1990s, drought trends have increased in Central Asia owing to insufficient precipitation and increased evapotranspiration [14]. The Ebinur Lake Basin is located in Central Asia and is sensitive to vegetation and runoff changes [25]. From 2000 to 2020, the overall vegetation in the basin showed a greening trend. This was due to an increase in cultivated land area in the basin. Cultivated vegetation requires substantial amounts of water for cultivation, irrigation, and fertilization. In recent decades, soil moisture has significantly decreased at the beginning of the growing season due to the continuous increase in temperature and decrease in precipitation, leading to an increase in agricultural droughts [13,60,61]. Meadows are in the transition zone between high- and low-water production areas, and NDVI values do not vary significantly. Therefore, both values were high and had a high level of aggregation (Figures 4 and 6). Meadows and alpine vegetation have high–low clustering and low–low clustering, respectively, and were located in the south and west of the basin. The two vegetation types differed in their ability to adapt to drought, resulting in relatively low cover at high and low water levels (Figure 11). The low–high clustering vegetation types were grasslands and cultivated plants located in the middle of the watershed. Grasslands had low–high clustering owing to the dominance of dry herbaceous plants in the basin. These are zonal vegetation types under semi-humid and semi-arid climatic conditions. Therefore, when studying the response of spatial and temporal variations in vegetation to runoff in arid zones, the effect of different plant types on runoff needs to be considered.

6. Conclusions

This study investigated the relationship between spatial and temporal changes in vegetation and temporal changes in runoff in the Ebinur Lake Basin. The results showed that the spatial distribution of runoff in the watershed was similar to that of elevation: the higher the elevation, the higher the runoff volume. When the elevation was above 3500 m, the regional runoff capacity tended to saturate. The distribution of vegetation was characterized by natural plants in the surrounding areas and cultivated plants in the middle. The runoff trend in the basin was generally not highly variable. Interannual runoff based on sub-basins showed a positive correlation with the overall NDVI. The interannual runoff based on vegetation type showed a negative correlation with the overall NDVI. There was continuity and reproducibility in the local spatial distribution of clusters of NDVI values and runoff in the watershed.

The response of the spatiotemporal characteristics of vegetation to runoff is a complex process that involves the interaction of multiple factors and mechanisms. Except for extreme years (2008), the spatial distribution of water production in the model watershed is basically similar to the real situation. However, the error for extreme years is still significant, and further research should consider more influencing factors for improvement. This study only focuses on vegetation types. We will select specific vegetation for future research. The results of this study showed the relationship between the spatial distribution of water production and vegetation in the study area, which can provide reference for water resource utilization and land planning in the Ebinur Lake Basin. To achieve rational utilization of water resources, local governments should control the scale of agricultural land.

Author Contributions: Conceptualization, C.Y.; Methodology, Y.W. and G.Y.; Software, C.Y.; Validation, C.Y.; Investigation, B.X.; Resources, Y.T.; Writing—original draft, C.Y. and Y.W.; Writing—review & editing, B.X. and J.Y.; Visualization, C.Y.; Supervision, Y.W.; Funding acquisition, Y.W., G.Y. and H.C. All authors have read and agreed to the published version of the manuscript.

Funding: This research was funded by [the Special project for innovation and development of Shihezi University] grant number [CXFZSK202105, CXFZ202217], [the National Natural Science Foundation of China] grant number [52269006], [Xinjiang Production and Construction Corps] grant number [2021AB021, 2022BC001, 2022DB023], [The Third Xinjiang Scientific Expedition Program] grant number [2021xjkk0804] and [the Program for Youth Innovation and Cultivation of Talents of Shihezi University] grant number [CXPY202223, CXPY202121].

Data Availability Statement: We declare the availability of the data in our manuscript.

Conflicts of Interest: The authors declare no conflict of interest.

References

1. Cui, L.; Shi, J. Temporal and Spatial Response of Vegetation NDVI to Temperature and Precipitation in Eastern China. *J. Geogr. Sci.* **2010**, *20*, 163–176. [CrossRef]
2. Feng, L.; Guo, M.; Wang, W.; Shi, Q.; Guo, W.; Lou, Y.; Zhu, Y.; Yang, H.; Xu, Y. Evaluation of the Effects of Long-Term Natural and Artificial Restoration on Vegetation Characteristics, Soil Properties and Their Coupling Coordinations. *Sci. Total Environ.* **2023**, *884*, 163828. [CrossRef] [PubMed]

3. Chen, Q.; Timmermans, J.; Wen, W.; van Bodegom, P.M. Ecosystems Threatened by Intensified Drought with Divergent Vulnerability. *Remote Sens. Environ.* **2023**, *289*, 113512. [CrossRef]

4. Dai, A.; Zhao, T. Uncertainties in Historical Changes and Future Projections of Drought. Part I: Estimates of Historical Drought Changes. *Clim. Chang.* **2017**, *144*, 519–533. [CrossRef]

5. Gray, L.C.; Zhao, L.; Stillwell, A.S. Impacts of Climate Change on Global Total and Urban Runoff. *J. Hydrol.* **2023**, *620*, 129352. [CrossRef]

6. He, S.; Chen, K.; Liu, Z.; Deng, L. Exploring the Impacts of Climate Change and Human Activities on Future Runoff Variations at the Seasonal Scale. *J. Hydrol.* **2023**, *619*, 129382. [CrossRef]

7. Zhao, D.; Xiong, D.; Zhang, B.; He, K.; Wu, H.; Zhang, W.; Lu, X. Long-Term Response of Runoff and Sediment Load to Spatiotemporally Varied Rainfall in the Lhasa River Basin, Tibetan Plateau. *J. Hydrol.* **2023**, *618*, 129154. [CrossRef]

8. Zhu, S.; Li, Y.; Wei, S.; Wang, C.; Zhang, X.; Jin, X.; Zhou, X.; Shi, X. The Impact of Urban Vegetation Morphology on Urban Building Energy Consumption during Summer and Winter Seasons in Nanjing, China. *Landsc. Urban Plan.* **2022**, *228*, 104576. [CrossRef]

9. Lara, C.; Cazelles, B.; Saldías, G.S.; Flores, R.P.; Paredes, Á.L.; Broitman, B.R. Coupled Biospheric Synchrony of the Coastal Temperate Ecosystem in Northern Patagonia: A Remote Sensing Analysis. *Remote Sens.* **2019**, *11*, 2092. [CrossRef]

10. Xu, H.-L.; Ye, M.; Li, J.-M. Changes in Groundwater Levels and the Response of Natural Vegetation to Transfer of Water to the Lower Reaches of the Tarim River. *J. Environ. Sci.* **2007**, *19*, 1199–1207. [CrossRef]

11. Yang, X.; Long, Y.; Sarkar, B.; Li, Y.; Lü, G.; Ali, A.; Yang, J.; Cao, Y.E. Influence of Soil Microorganisms and Physicochemical Properties on Plant Diversity in an Arid Desert of Western China. *J. For. Res.* **2021**, *32*, 2645–2659. [CrossRef]

12. Chen, J.; Chen, Y.; Wang, K.; Wang, G.; Wu, J.; Zhang, Y. Differences in Soil Water Storage, Consumption, and Use Efficiency of Typical Vegetation Types and Their Responses to Precipitation in the Loess Plateau, China. *Sci. Total Environ.* **2023**, *869*, 161710. [CrossRef] [PubMed]

13. Jiang, J.; Zhou, T. Agricultural Drought over Water-Scarce Central Asia Aggravated by Internal Climate Variability. *Nat. Geosci.* **2023**, *16*, 154–161. [CrossRef]

14. Guo, R.; Deser, C.; Terray, L.; Lehner, F. Human Influence on Winter Precipitation Trends (1921–2015) over North America and Eurasia Revealed by Dynamical Adjustment. *Geophys. Res. Lett.* **2019**, *46*, 3426–3434. [CrossRef]

15. Samavati, A.; Babamiri, O.; Rezai, Y.; Heidarimozaffar, M. Investigating the Effects of Climate Change on Future Hydrological Drought in Mountainous Basins Using SWAT Model Based on CMIP5 Model. *Stoch. Environ. Res. Risk Assess.* **2023**, *37*, 849–875. [CrossRef]

16. Sorg, A.; Bolch, T.; Stoffel, M.; Solomina, O.; Beniston, M. Climate Change Impacts on Glaciers and Runoff in Tien Shan (Central Asia). *Nat. Clim. Chang.* **2012**, *2*, 725–731. [CrossRef]

17. Tran, L.T.; O'Neill, R.V. Detecting the Effects of Land Use/Land Cover on Mean Annual Streamflow in the Upper Mississippi River Basin, USA. *J. Hydrol.* **2013**, *499*, 82–90. [CrossRef]

18. Kuriqi, A.; Pinheiro, A.N.; Sordo-Ward, A.; Garrote, L. Influence of Hydrologically Based Environmental Flow Methods on Flow Alteration and Energy Production in a Run-of-River Hydropower Plant. *J. Clean. Prod.* **2019**, *232*, 1028–1042. [CrossRef]

19. Shi, P.; Li, P.; Li, Z.; Sun, J.; Wang, D.; Min, Z. Effects of Grass Vegetation Coverage and Position on Runoff and Sediment Yields on the Slope of Loess Plateau, China. *Agric. Water Manag.* **2022**, *259*, 107231. [CrossRef]

20. Abbaszadeh, M.; Bazrafshan, O.; Mahdavi, R.; Sardooi, E.R.; Jamshidi, S. Modeling Future Hydrological Characteristics Based on Land Use/Land Cover and Climate Changes Using the SWAT Model. *Water Resour. Manag.* **2023**, *37*, 4177–4194. [CrossRef]

21. Yu, Y.; Zhu, R.; Ma, D.; Liu, D.; Liu, Y.; Gao, Z.; Yin, M.; Bandala, E.R.; Rodrigo-Comino, J. Multiple Surface Runoff and Soil Loss Responses by Sandstone Morphologies to Land-Use and Precipitation Regimes Changes in the Loess Plateau, China. *Catena* **2022**, *217*, 106477. [CrossRef]

22. Wang, B.; Liu, J.; Li, Z.; Morreale, S.J.; Schneider, R.L.; Xu, D.; Lin, X. The Contributions of Root Morphological Characteristics and Soil Property to Soil Infiltration in a Reseeded Desert Steppe. *Catena* **2023**, *225*, 107020. [CrossRef]

23. Pessacg, N.; Flaherty, S.; Brandizi, L.; Solman, S.; Pascual, M. Getting Water Right: A Case Study in Water Yield Modelling Based on Precipitation Data. *Sci. Total Environ.* **2015**, *537*, 225–234. [CrossRef] [PubMed]

24. Sun, Q.; Sun, J.; Baidurela, A.; Li, L.; Hu, X.; Song, T. Ecological Landscape Pattern Changes and Security from 1990 to 2021 in Ebinur Lake Wetland Reserve, China. *Ecol. Indic.* **2022**, *145*, 109648. [CrossRef]

25. Yang, X.D.; Ali, A.; Xu, Y.L.; Jiang, L.M.; Lv, G.H. Soil Moisture and Salinity as Main Drivers of Soil Respiration across Natural Xeromorphic Vegetation and Agricultural Lands in an Arid Desert Region. *Catena* **2019**, *177*, 126–133. [CrossRef]

26. Gbetkom, P.G.; Cretaux, J.F.; Tchilibou, M.; Carret, A.; Delhoume, M.; Bergé-Nguyen, M.; Sylvestre, F. Lake Chad Vegetation Cover and Surface Water Variations in Response to Rainfall Fluctuations under Recent Climate Conditions (2000–2020). *Sci. Total Environ.* **2023**, *857*, 159302. [CrossRef]

27. Wei, Q.; Abudureheman, M.; Halike, A.; Yao, K.; Yao, L.; Tang, H.; Tuheti, B. Temporal and Spatial Variation Analysis of Habitat Quality on the PLUS-InVEST Model for Ebinur Lake Basin, China. *Ecol. Indic.* **2022**, *145*, 109632. [CrossRef]

28. Yao, C.; Wang, Y.; Chen, Y.; Wang, L.; Yao, J.; Xia, B. Meteorological Driving Factors Effecting the Surface Area of Ebinur Lake and Determining Associated Trends and Shifts. *Front. Environ. Sci.* **2022**, *10*, 994260. [CrossRef]

29. Wang, X.; Zhang, F.; Kung, H.-T.; Johnson, V.C. New Methods for Improving the Remote Sensing Estimation of Soil Organic Matter Content (SOMC) in the Ebinur Lake Wetland National Nature Reserve (ELWNNR) in Northwest China. *Remote Sens. Environ.* **2018**, *218*, 104–118. [CrossRef]

30. Zeng, H.; Wu, B.; Zhu, W.; Zhang, N. A Trade-off Method between Environment Restoration and Human Water Consumption: A Case Study in Ebinur Lake. *J. Clean. Prod.* **2019**, *217*, 732–741. [CrossRef]

31. Wen, B.; Duan, G.; Lu, J.; Zhou, R.; Ren, H.; Wen, Z. Response Relationship between Vegetation Structure and Runoff Sediment Yield in the Hilly and Gully Area of the Loess Plateau, China. *Catena* **2023**, *227*, 107107. [CrossRef]

32. Gong, X.; Lü, G. Species Diversity and Dominant Species' Niches of Thermophyte Communities of the Tugai Forest in the Ebinur Basin of Xinjiang, China. *Biodivers. Sci.* **2017**, *25*, 34–43. [CrossRef]

33. Levia, D.F.; Germer, S. A Review of Stemflow Generation Dynamics and Stemflow-Environment Interactions in Forests and Woodlands. *Rev. Geophys.* **2015**, *53*, 673–714. [CrossRef]

34. Wang, Y.; Ye, A.; Peng, D.; Miao, C.; Di, Z.; Gong, W. Spatiotemporal Variations in Water Conservation Function of the Tibetan Plateau under Climate Change Based on InVEST Model. *J. Hydrol. Reg. Stud.* **2022**, *41*, 101064. [CrossRef]

35. Li, M.; Liang, D.; Xia, J.; Song, J.; Cheng, D.; Wu, J.; Cao, Y.; Sun, H.; Li, Q. Evaluation of Water Conservation Function of Danjiang River Basin in Qinling Mountains, China Based on InVEST Model. *J. Environ. Manag.* **2021**, *286*, 112212. [CrossRef]

36. Peng, Z.; Ding, X.; Liu, W.Z.; Li, Z. 1 km monthly temperature and precipitation dataset for China from 1901 to 2017. *Earth Syst. Sci. Data* **2019**, *11*, 1931–1946. [CrossRef]

37. Yang, J.; Huang, X. The 30m Annual Land Cover Dataset and Its Dynamics in China from 1990 to 2019. *Earth Syst. Sci. Data* **2021**, *13*, 3907–3925. [CrossRef]

38. He, Y. *Pan-TPE Soil Map Based on Harmonized World Soil Database (V1.2)*; National Tibetan Plateau Data Center: Lanzhou, China, 2019.

39. Sharp, R.; Douglass, J.; Wolny, S.; Arkema, K.; Bernhardt, J.; Bierbower, W.; Chaumont, N.; Denu, D.; Fisher, D.; Glowinski, K.; et al. *VEST User's Guide*; The Natural Capital Project; Stanford University; University of Minnesota; The Nature Conservancy, and World Wildlife Fund: Gland, Switzerland, 2020. Available online: https://www.researchgate.net/publication/323832082_InVEST_User\T1\textquoterights_Guide (accessed on 16 April 2023). [CrossRef]

40. Fang, Q.; Xin, X.; Guan, T.; Wang, G.; Zhang, S.; Ma, M. Vegetation Patterns Governing the Competitive Relationship between Runoff and Evapotranspiration Using a Novel Water Balance Model at a Semi-Arid Watershed. *Environ. Res.* **2022**, *214*, 113976. [CrossRef]

41. Ye, Q.; Li, Z.; Duan, L.; Xu, X. Science of the Total Environment Decoupling the in Fl Uence of Vegetation and Climate on Intra-Annual Variability in Runoff in Karst Watersheds. *Sci. Total Environ.* **2022**, *824*, 153874. [CrossRef]

42. Luo, Y.; Yang, Y.; Yang, D.; Zhang, S. Quantifying the Impact of Vegetation Changes on Global Terrestrial Runoff Using the Budyko Framework. *J. Hydrol.* **2020**, *590*, 125389. [CrossRef]

43. Zhou, W. Distribution of Available Soil Water Capacity in China. *J. Geogr. Sci.* **2005**, *15*, 3. [CrossRef]

44. Zhuang, Q.; Zhou, Z.; Liu, S.; Wright, D.B.; Tavares Araruna Júnior, J.; Makhinov, A.N.; Makhinova, A.F. Bivariate Rainfall Frequency Analysis in an Urban Watershed: Combining Copula Theory with Stochastic Storm Transposition. *J. Hydrol.* **2022**, *615*, 128648. [CrossRef]

45. Fernandes, R.; Leblanc, S.G. Parametric (Modified Least Squares) and Non-Parametric (Theil–Sen) Linear Regressions for Predicting Biophysical Parameters in the Presence of Measurement Errors. *Remote Sens. Environ.* **2005**, *95*, 303–316. [CrossRef]

46. Kim, J.K.; Onda, Y.; Kim, M.S.; Yang, D.Y. Plot-Scale Study of Surface Runoff on Well-Covered Forest Floors under Different Canopy Species. *Quat. Int.* **2014**, *344*, 75–85. [CrossRef]

47. Wang, L.; Li, Z.; Wang, F.; Li, H.; Wang, P. Glacier Changes from 1964 to 2004 in the Jinghe River Basin, Tien Shan. *Cold Reg. Sci. Technol.* **2014**, *102*, 78–83. [CrossRef]

48. Yao, J.; Hu, W.; Chen, Y.; Huo, W.; Zhao, Y.; Mao, W.; Yang, Q. Hydro-Climatic Changes and Their Impacts on Vegetation in Xinjiang, Central Asia. *Sci. Total Environ.* **2019**, *660*, 724–732. [CrossRef] [PubMed]

49. Wei, X.; He, W.; Zhou, Y.; Ju, W.; Xiao, J.; Li, X.; Liu, Y.; Xu, S.; Bi, W.; Zhang, X.; et al. Global Assessment of Lagged and Cumulative Effects of Drought on Grassland Gross Primary Production. *Ecol. Indic.* **2022**, *136*, 108646. [CrossRef]

50. Lou, Y.; Wang, W.; Guo, M.; Guo, W.; Kang, H.; Feng, L.; Zhu, Y.; Yang, H. Vegetation Affects Gully Headcut Erosion Processes by Regulating Runoff Hydrodynamics in the Loess Tableland Region. *J. Hydrol.* **2023**, *616*, 128769. [CrossRef]

51. Zhao, Y.; Miao, Y.; Li, Y.; Fang, Y.; Zhao, J.; Wang, X.; An, C. Non-Linear Response of Mid-Latitude Asian Dryland Vegetation to Holocene Climate Fluctuations. *Catena* **2022**, *213*, 106212. [CrossRef]

52. Crockford, R.H.; Richardson, D.P. Partitioning of Rainfall in a Eucalypt Forest and Pine Plantation in Southeastern Australia: IV the Relationship of Interception and Canopy Storage Capacity, the Interception of These Forests, and the Effect on Interception of Thinning the Pine Plantation. *Hydrol. Process.* **1990**, *4*, 169–188. [CrossRef]

53. Efon, E.; Ngongang, R.D.; Meukaleuni, C.; Wandjie, B.B.S.; Zebaze, S.; Lenouo, A.; Valipour, M. Monthly, Seasonal, and Annual Variations of Precipitation and Runoff Over West and Central Africa Using Remote Sensing and Climate Reanalysis. *Earth Syst. Environ.* **2023**, *7*, 67–82. [CrossRef]

54. Zhang, P.; Cai, Y.; He, Y.; Xie, Y.; Zhang, X.; Li, Z. Changes of Vegetational Cover and the Induced Impacts on Hydrological Processes under Climate Change for a High-Diversity Watershed of South China. *J. Environ. Manag.* **2022**, *322*, 115963. [CrossRef] [PubMed]

55. Long, Y.; Yang, X.; Cao, Y.; Lv, G.; Li, Y.; Pan, Y.; Yan, K.; Liu, Y. Relationship between Soil Fungi and Seedling Density in the Vicinity of Adult Conspecifics in an Arid Desert Forest. *Forests* **2021**, *12*, 92. [CrossRef]
56. Pokhrel, Y.; Felfelani, F.; Satoh, Y.; Boulange, J.; Burek, P.; Gädeke, A.; Gerten, D.; Gosling, S.N.; Grillakis, M.; Gudmundsson, L.; et al. Global Terrestrial Water Storage and Drought Severity under Climate Change. *Nat. Clim. Chang.* **2021**, *11*, 226–233. [CrossRef]
57. Appels, W.M.; Bogaart, P.W.; van der Zee, S.E.A.T.M. Influence of Spatial Variations of Microtopography and Infiltration on Surface Runoff and Field Scale Hydrological Connectivity. *Adv. Water Resour.* **2011**, *34*, 303–313. [CrossRef]
58. Berdugo, M.; Gaitan, J.J.; Delgado-Baquerizo, M.; Crowther, T.W.; Dakos, V. Prevalence and Drivers of Abrupt Vegetation Shifts in Global Drylands. *Proc. Natl. Acad. Sci. USA* **2022**, *119*, e2123393119. [CrossRef] [PubMed]
59. Li, S.; Jing, H.; Yuan, Q.; Yue, L.; Li, T. Investigating the Spatio-Temporal Variation of Vegetation Water Content in the Western United States by Blending GNSS-IR, AMSR-E, and AMSR2 Observables Using Machine Learning Methods. *Sci. Remote Sens.* **2022**, *6*, 100061. [CrossRef]
60. Yang, G.; Li, F.; Chen, D.; He, X.; Xue, L.; Long, A. Assessment of Changes in Oasis Scale and Water Management in the Arid Manas River Basin, North Western China. *Sci. Total Environ.* **2019**, *691*, 506–515. [CrossRef]
61. Yang, X.D.; Anwar, E.; Zhou, J.; He, D.; Gao, Y.C.; Lv, G.H.; Cao, Y.E. Higher Association and Integration among Functional Traits in Small Tree than Shrub in Resisting Drought Stress in an Arid Desert. *Environ. Exp. Bot.* **2022**, *201*, 104993. [CrossRef]

Disclaimer/Publisher's Note: The statements, opinions and data contained in all publications are solely those of the individual author(s) and contributor(s) and not of MDPI and/or the editor(s). MDPI and/or the editor(s) disclaim responsibility for any injury to people or property resulting from any ideas, methods, instructions or products referred to in the content.

Article

Evaluating the Stand Structure, Carbon Sequestration, Oxygen Release Function, and Carbon Sink Value of Three Artificial Shrubs alongside the Tarim Desert Highway

Lin Li [1,2,3], Abudoukeremujiang Zayiti [1,2,3] and Xuemin He [1,2,3,*]

1   College of Ecology and Environment, Xinjiang University, Urumqi 830017, China;
    107552103474@stu.xju.edu.cn (L.L.); krimjan411@xju.edu.cn (A.Z.)
2   Key Laboratory of Oasis Ecology of Education Ministry, Xinjiang University, Urumqi 830017, China
3   Xinjiang Jinghe Observation and Research Station of Temperate Desert Ecosystem, Ministry of Education,
    Jinghe 833300, China
*   Correspondence: hxm@xju.edu.cn

Abstract: Currently, the ecological problems caused by the greenhouse effect are growing more serious, and implementing carbon sequestration methods is an effective way to address them. Arid and semi-arid desert areas have tremendous potential as carbon sinks, and artificial forests in these areas play an important role in absorbing and sequestering carbon dioxide. This study selected three main species of artificial protective trees along the Tarim Desert Highway—*Haloxylon ammodendron* (C.A.Mey.) Bunge, *Calligonum mongolicum* Turcz. and *Tamarix chinensis* Lour.—and evaluated them for their carbon sequestration, oxygen release capacity, and economic benefits using *Pn* (net photosynthetic rate) and biomass methods. The results showed that the average daily *Pn* value and carbon sequestration and oxygen release per unit leaf area of *T. chinensis* were significantly higher than those of *H. ammodendron* and *C. mongolicum* ($p < 0.05$). The total carbon storage of the three shelterbelts was 15.41×10^4 t, and the carbon storage of *H. ammodendron* was significantly higher than that of *C. mongolicum* and *T. chinensis* ($p < 0.05$). According to the net photosynthetic rate method, the annual carbon sequestration and oxygen release of the shelter forest is 6.13×10^4 t a^{-1}, and the transaction price is CNY 13.73 million a^{-1}. The total amount of carbon sequestration and oxygen release of the shelter forest obtained via the biomass method is 97.61×10^4 t, and the transaction price is CNY 218.77 million. This study conducted research on the carbon sequestration capacity of protective forests along the Tarim Desert Highway located in an extremely arid region. It highlights the significant contribution of these protective forests in terms of carbon storage, playing a crucial role in promoting ecological restoration and sustainable development in arid areas. Additionally, this study provides a scientific basis for estimating carbon storage and promoting the sustainable management of artificial forests in arid desert regions.

Keywords: arid area; artificial shelter forest; photosynthesis; biomass; carbon storage; carbon trading

Citation: Li, L.; Zayiti, A.; He, X. Evaluating the Stand Structure, Carbon Sequestration, Oxygen Release Function, and Carbon Sink Value of Three Artificial Shrubs alongside the Tarim Desert Highway. *Forests* **2023**, *14*, 2137. https://doi.org/10.3390/f14112137

Academic Editor: Ricardo Ruiz-Peinado

Received: 20 September 2023
Revised: 21 October 2023
Accepted: 24 October 2023
Published: 26 October 2023

Copyright: © 2023 by the authors. Licensee MDPI, Basel, Switzerland. This article is an open access article distributed under the terms and conditions of the Creative Commons Attribution (CC BY) license (https://creativecommons.org/licenses/by/4.0/).

1. Introduction

According to the assessment report by The Intergovernmental Panel on Climate Change (IPCC), excessive emissions of greenhouse gases have had a significant impact on human survival and societal development [1–4]. The mitigation of climate change has become an urgent global challenge, and the concentration of carbon dioxide (CO_2) in the atmosphere has a significant impact on the climate [5]. As a major carbon emitter, China will strive to achieve a "carbon peak" by 2030 and strive to achieve "carbon neutrality" (referred to as the "double carbon target") by 2060 [6]. Enhancing forestry carbon sink capacity is an important path to achieve the "dual carbon" goal, and the utilization of forestry carbon sinks has gradually risen to the national strategic level [7]. Plantation ecosystems have gradually become an important component of forestry carbon sequestration [8]. In the past three decades, natural forests have only sequestrated nearly the same amount of carbon as plantation forests despite the fact that the area of natural forests in China is four times that of plantation forests [9]. Fang et al. [10] showed that the increase in plantation area was the main reason for the increase in forest carbon sequestration in China. China has the largest planted forest area in the world, and planted forests will play an irreplaceable role in mitigating the impact of climate change [11]. In the arid regions of northwest China, the main approach to improving the ecological environment was through the establishment of artificial forests [12]. Artificial forests provide multiple benefits, including preventing desertification, improving land conditions, and enriching species diversity in arid areas, thus playing a crucial role in forest ecosystems [13]. With the increasing demand for mitigating climate change, the area of artificial forests is expected to further expand [14]. Therefore, strengthening research on carbon sequestration in artificial forests is of great significance for enhancing the carbon storage capacity of forests in arid regions.

As a carbon pool, forests play an important role in increasing sink area, mainly reflected in the function of carbon sequestration and oxygen release [15,16]. At present, the research methods used to examine forest vegetation carbon sequestration mainly include the biomass method and the *Pn* (net photosynthetic rate) method. The biomass method is used to estimate carbon sequestration by measuring the dry matter mass accumulated by the community per unit of time and per unit area, which are important indicators of ecosystem function and productivity [17]. The advantage of this method is that the technology is relatively mature, but the disadvantage is that it has a certain destructive effect on vegetation. The *Pn* method is used to obtain the instantaneous photosynthesis and respiration rate per unit leaf area of plants and then obtain the carbon sequestration per unit area. The advantage of this method is that it can conduct multi-factor quantitative research on photosynthesis and respiration rate, but the disadvantage is that it is more complicated to perform.

In recent decades, vegetation restoration projects have been widely adopted in many countries and regions with the aim of achieving economic, ecological, and climate change mitigation goals [18,19]. Moreover, the international academic community has paid extensive attention to the ecological benefits of desert vegetation construction [20,21]. In arid areas of China, artificial desert vegetation, especially scrub, plays a very important role in carbon sequestration [22]. In addition, the area of shrublands in the "Three northern regions" has increased significantly, and shrublands are characterized by *H. ammodendron* (C.A.Mey.) Bunge plays an important role in forestry carbon sinks [23]. However, concerning the potential carbon sequestration and role of sand-fixation vegetation plantation in arid areas, there is still a lack of detailed and measured data verification [24].

The Tarim Desert Highway is located in the Tarim Basin, which starts from Luntai in the north and reaches Minfeng in the south. The total length of the desert highway is 562 km, including 443 km through the desert, which is the longest continuous highway through a desert in the world. In June 2003, the state approved the construction of the Tarim Desert Highway Protection Forest Ecological Project. In 2005, the Tarim Desert Highway shelterbelt project was completed, with a total length of 436 km, an overall width of 72~78 m, and a total area of 3128 ha [25]. Since 2 June 2022, newly built photovoltaic power stations have been scattered across the ecological protection forest belt, achieving zero carbon emissions across the road, and the Tarim Desert Highway has become the first zero-carbon desert highway in China. At present, most studies on carbon sequestration using vegetation focus on highly productive ecosystems such as forests (mainly trees) [26,27], grasslands, wetlands and farmland, while relatively few studies have centered on carbon sequestration in desert shrublands [28,29]. In particular, there is less research on the capacity and value of carbon sequestration and oxygen release of artificial shelterbelts in desert areas. In addition, what is the existing carbon storage of artificial shelterbelts on desert roads? How significant is the carbon benefit of vegetation? These are pressing scientific questions. Therefore, based on previous research, this study selected *H. ammodendron* (C.A.Mey.) Bunge, *C. mongolicum* Turcz. and *T. chinensis* Lour. from the artificial protective forest alongside the Taklamakan

Desert Highway as research subjects. By combining the *Pn* (net photosynthetic rate) method and biomass method, a comprehensive study was conducted on the *Pn* and biomass of these three shrubs in the protective forest. Subsequently, the carbon sequestration and oxygen release capacity, carbon storage, and economic value were estimated, providing insights into the methodology of studying artificial shrub carbon sinks in arid regions.

2. Materials and Methods

2.1. Overview of the Study Area

The Tarim Desert Highway is located between 37°–42° N and 82°–85° E, running north–south through the Taklimakan Desert. The landscape of the Taklimakan Desert is dominated by composite sand ridges with high mobility, and the terrain along the highway is high in the south and low in the north, with a relative elevation difference of less than 80 m. The main soil type along the highway is aeolian sand soil, with a coarse texture and little accumulation of organic matter. Precipitation along the desert highway is unevenly distributed throughout the year, with annual precipitation ranging from 11 mm to 50 mm, and potential evapotranspiration as high as 3638.6 mm [30]. The average annual temperature is 12.7 °C, the maximum extreme temperature is 45.6 °C, and the minimum extreme temperature is −22.2 °C. The cumulative annual sunshine duration is 2854.2 h, the annual average wind speed is 2.5 m s^{-1}, the maximum instantaneous wind speed is 24 m s^{-1}, the annual sand wind duration is 550~800 h, and floating dust and sand are very frequent [31]. The desert highway protective forest lacks surface runoff along its route, but it has a large amount of underground water storage, with water quality ranging from 2.6 to 30 g L^{-1} in terms of mineralization [32]. The tree species selected for the desert highway shelterbelt mainly included *H. ammodendron* (C.A.Mey.) Bunge, *C. mongolicum* Turcz. and *T. chinensis* Lour. These species have strong adaptability, wind erosion resistance, drought resistance, and sand burial tolerance. The entire Tarim Desert Highway protective forest was irrigated using drip irrigation. The irrigation cycle was set to 15 days with a water volume of 30 L m^{-2} per irrigation. During the winter season (November to February of the following year), irrigation was not conducted. Groundwater was utilized through on-site extraction.

2.2. Quadrat Setup

The experiment utilized the shelterbelt of the desert highway, which spans a total length of 436 km, as a natural sample belt. Ten sample plots were designated along the length of the road section, labeled S1–S10 from north to south, each measuring 2500 m^2 (50 m × 50 m), with an interval of 45 km between the adjacent plots. Measurements were taken for the plant height, base diameter and crown of three plant species within the sample plots. Additionally, the quantity of vegetation was recorded. The area of unplanted vegetation within the sample plots was also measured. A 10 m × 10 m quadrat was established in each plot and three individuals of similar size and with healthy growth were randomly selected from the species *H. ammodendron*, *C. mongolicum* and *T. chinensis* in each quadrat as the photosynthetic sample plants, for a total of 90 plants.

2.3. Carbon Sequestration and Oxygen Release by Net Photosynthetic Rate Method

The test was conducted in June 2023. The *Pn* value of the photosynthetic sample plants in the 10 quadrats was measured using a portable photosynthetic measurement system LI-6400XT(LI-COR, Lincoln, NE, USA), and a standard leaf chamber of 2 cm × 3 cm was selected. The measurement time was 8:00–18:00 (Beijing time) daily at 2 h intervals. Each plant measurement was repeated three times, and five instantaneous values were recorded in each measurement. Finally, the average values of three different plants in each quadrat were calculated, respectively.

According to the diurnal variation curve of the *Pn* value of each plant, the net assimilation of the tested plants on the day of determination was calculated using the simple integration method [33]. The calculation formula is as follows:

$$P = \sum_{i=1}^{j}[(P_{i+1} + P_i)/2 \times (t_{i+1} - t_i) \times 3600/1000] \tag{1}$$

where P is the total net assimilation per unit leaf area measured daily (mmol m^{-2} d^{-1}). P$_i$ is the instantaneous photosynthesis rate at the initial measurement point. P$_{i+1}$ is the instantaneous photosynthesis rate (μmol m^{-2} s^{-1}) at the measuring point i + 1. t$_i$ is the instantaneous time of the initial measurement point. t$_{i+1}$ is the time of t + 1 measuring point (h). j is the number of tests. In this study, 3600 signifies 3600 s/h and 1000 signifies that 1 mmol is equal to 1000 μmol.

Generally, the nocturnal dark respiration of plants is calculated based on 20% of the assimilation rate during the day. Therefore, the daily net carbon sequestration and oxygen release per unit leaf area are calculated according to Formulas (2) and (3):

$$W_{CO_2} = P \times (1 - 0.2) \times 44/1000 \tag{2}$$

where W$_{CO_2}$ is the daily net fixed CO_2 mass per unit leaf area (g m^{-2} d^{-1}). Here, 44 is the molar mass of CO_2 (g mol^{-1}).

$$W_{O_2} = P \times (1 - 0.2) \times 32/1000 \tag{3}$$

where W$_{O_2}$ is the daily net O_2 mass released per unit leaf area (g m^{-2} d^{-1}). Here, 32 is the molar mass of O_2 (g mol^{-1}).

The calculation of carbon sequestration and oxygen release by photosynthetic rate method adopts Formulas (4) and (5) [34]:

$$C_{CO_2} = W_{CO_2} \times S \times N \tag{4}$$

where C$_{CO_2}$ is annual carbon sequestration (t). Here, S is vegetation area (ha). Here, N is the total number of days in the growing season (from May to September, 150 days).

$$C_{O_2} = W_{O_2} \times S \times N \tag{5}$$

where C$_{O_2}$ is annual oxygen release (t). Here, S is vegetation area (ha). Here, N is the total number of days in the growing season (from May to September, 150 days).

2.4. Carbon Sequestration and Oxygen Release by Biomass Method

Among the 10 quadrats, we randomly selected 12 shrubs of *H. ammodendron*, *C. mongolicum* and *T. chinensis* (7 for biomass model construction and 5 for the verification of the biomass model), totaling 36 shrubs as samples. We recorded the plant height, diameter at breast height, and crown width of plants within the quadrats. For the aboveground portions of the plants, we used the total harvest method to collect them and subsequently separate them into different organs (roots, stems, branches, and leaves). For the underground portions of plants, we employed the complete excavation method for manual digging [35]. However, shrub roots were typically deep and intricately complex, making the complete excavation of the root system highly challenging. Therefore, in practical operations, we chose a threshold of 2 mm in root diameter as the criterion for excavation and determined the average area for collecting roots based on this standard. The decision to adopt this threshold was based on the fact that fine roots had small diameters and were difficult to access while also contributing little to the overall biomass and representing a very low proportion of the total root system [36]. Samples were brought back to the laboratory and dried at a temperature of 72 °C for 48 h [37]. Then, the dry weight of each part of the plant was calculated and added together to obtain the total biomass of the whole plant (Table S1).

Through the relationship between carbon storage and biomass (biomass multiplied by carbon content), the vegetation carbon storage of the artificial shrub community is estimated. The average carbon content of shrub forests recommended by the IPCC was 0.50 [38].

$$Q = W \times 0.5 \tag{6}$$

where Q is carbon storage (t). W is the whole-plant biomass (t). Here, 0.5 was the carbon content.

In this study, the carbon storage density was calculated according to Formula (7):

$$N = Q/A \tag{7}$$

where N is the carbon storage density (t ha^{-1}). A is the green area (ha).

The calculation of carbon sequestration and oxygen release by biomass method adopts Formulas (8) and (9) [39]:

$$G_C = Q \times (44/12) \tag{8}$$

where G_C is the amount of carbon sequestration (t). Here, 44 is the molar mass of CO_2 (g mol^{-1}). Here, 12 is the molar mass of C (g mol^{-1}).

$$G_O = Q \times (32/12) \tag{9}$$

where G_O is the released oxygen (t). Here, 32 is the molar mass of O_2 (g mol^{-1}). Here, 12 is the molar mass of C (g mol^{-1}).

2.5. Establishment and Testing of Three Shrub Biomass Models

The measured growth factors (diameter at breast height D, plant height H, crown average diameter and their combination factors (crown area $C = \pi \times L_1 \times L_2/4$, L_1 and L_2 are, respectively, east–west and north–south crown diameters, plant volume $V = CH = \pi \times L_1 \times L_2 \times H/4$, and plant cross-sectional area LH, $L = (L_1 + L_2)/2$)) were used as the independent variables when fitting the biomass model. Through screening the best-fitting variable, the model was selected for fitting according to previous studies on the whole-plant biomass of three shrubs [40,41] (Table 1). In the biomass model, the judgment coefficient (R^2), standard error (*SEE*) and the significance level of the regression test were used to evaluate the advantages and disadvantages of the equation. Statistically, the equation with the largest judgment coefficient R^2, the smallest *SEE* and the significant regression ($p < 0.001$) is considered the best, indicating that the fitting accuracy of the equation is the highest.

Table 1. Function types and expressions.

Function Type	Expression
Primary function	$W = a + bX$
Quadratic function	$W = a + bX + cX^2$
Logarithmic function	$W = a + b\ln X$
Power function	$W = aX^b$
Exponential function	$W = ae^{bX}$

Note: *W* represents biomass, a, b, and c are equation fitting parameters, e is the base of the natural logarithm, and X is the independent variable (D, H, C, LH, CH, etc.).

To verify the accuracy of the optimal model, the average relative error (RMA) and total relative error (RS) of the predicted value and the measured value were used to test the accuracy. At the same time, partially predicted and measured values of three shrub species were selected for regression analysis. Among them:

$$RMA = \frac{1}{N} \sum \left| \frac{X_i - \overline{X}_i}{\overline{X}_i} \right| \tag{10}$$

$$RS = \frac{\sum X_i - \sum \overline{X}_i}{\sum \overline{X}_i} \tag{11}$$

where X_i represents the measured value. $\overline{X}_i$ represents the predicted value. N represents the number of samples.

2.6. Estimation of Carbon Sequestration and Oxygen Release Value

For the valuation of carbon sequestration and oxygen release, we used the market price method—that is, the amount of carbon sequestration was multiplied by the price of carbon sequestration, and the amount of oxygen released was multiplied by the price of oxygen production—and then the cost of carbon sequestration and oxygen production in the study area was calculated. The carbon sequestration price was CNY 53.30 t^{-1}, and the average transaction price of the carbon emission quota (CEA) in the national carbon market from July 2021 to August 2023 was issued by the Shanghai Environmental Energy Exchange (https://www.cneeex.com/, accessed on 15 September 2023). Oxygen price was determined according to the relevant information released by Zhuochuang Information (https://www.sci99.com/, accessed on 10 September 2023), and the average transaction price of oxygen in the first half of 2023 was CNY 459 t^{-1}.

According to the relevant provisions of the latest national standard, "Standard for Evaluation of Forest Ecosystem Service Functions (GB/T38582-2020)" [42], the value of carbon sequestration and oxygen release functions was calculated using Formulas (12) and (13):

$$U_C = G_C \times C_C \tag{12}$$

where U_C is the annual carbon sequestration price of the stand (CNY a^{-1}). G_C is the annual carbon sequestration mass of the stand (t a^{-1}). C_C is the price of carbon dioxide (CNY t^{-1}).

$$U_O = G_O \times C_O \tag{13}$$

where U_O is the annual oxygen release price of the stand (CNY a^{-1}). G_O is the annual oxygen release mass of the stand (t a^{-1}). C_O is the oxygen price (CNY t^{-1}).

2.7. Data Processing

Data were collated and tabulated using Microsoft Office 2016 (Microsoft, Washington, DC, USA). The least significant difference (LSD) method of using the one-way ANOVA (95% confidence interval) was used for analysis and comparison, and $p > 0.05$ was considered as indicating no significant difference. Using the systematic clustering method and square Euclidean distance as basic options, the carbon sequestration and oxygen release capacity of plant per unit leaf area and daily average *Pn* value were measured using various squares for cluster analysis. Linear and nonlinear regression methods were used to simulate the equations of vegetation characteristic factors and total plant biomass, and T-tests were used to screen the equations. SPSS 24.0 (IBM, Armonk, NY, USA) was used for data processing, and Origin 2022 (OriginLab, Northampton, MA, USA) was used for drawing pictures and fitting regression models.

3. Results

3.1. Quadrat Vegetation Characteristics

Based on the survey of desert highway quadrats, The green area of vegetation accounted for 81.55% of the total forested area (2550.88 ha). The average coverage of the three species was 70%, and the vegetation area was 1785.62 ha. The proportion of three shrubs in the vegetation area was as follows: *H. ammodendron* was 72.32% (1291.36 ha), *C. mongolicum* was 18.38% (328.20 ha) and *T. chinensis* was 9.30% (166.06 ha). The sample plots vegetation frequency was: *H. ammodendron* (1008 trees), *C. mongolicum* (330 trees), and *T. chinensis* (118 trees). The vegetation relative frequency (Table S1) was: *H. ammodendron* (69.23%), *C. mongolicum* (22.69%), and *T. chinensis* (8.08%). The vegetation density was: *H. ammoden-*

dron stood at 0.403 trees m^{-2} (12.6 million trees), *C. mongolicum* stood at 0.1319 trees m^{-2} (4.13 million trees), and *T. chinensis* stood at 0.047 trees m^{-2} (1.47 million trees).

3.2. Carbon Sequestration and Oxygen Release Capacity under Photosynthesis

3.2.1. Comparative Analysis of Carbon Sequestration and Oxygen Release in the Quadrat

The change trend of the daily net assimilation of the 10 quadrats showed an "M" pattern. Among them, the maximum values of the three types of data appear in quadrat 9, the minimum values of daily net assimilation and net carbon sequestration and oxygen release appear in quadrat 1, and the minimum values of daily average *Pn* value appear in quadrat 6. The difference between the maximum and minimum values is as follows: the daily net assimilation is 110.56 mmol m^{-2} d^{-1}, the annual carbon sequestration is 0.59 kg m^{-2} a^{-1}, the annual oxygen release is 0.43 kg m^{-2} a^{-1}, and the daily average *Pn* value is 2.9 µmol m^{-2} s^{-1} (Table S3, Figure 1).

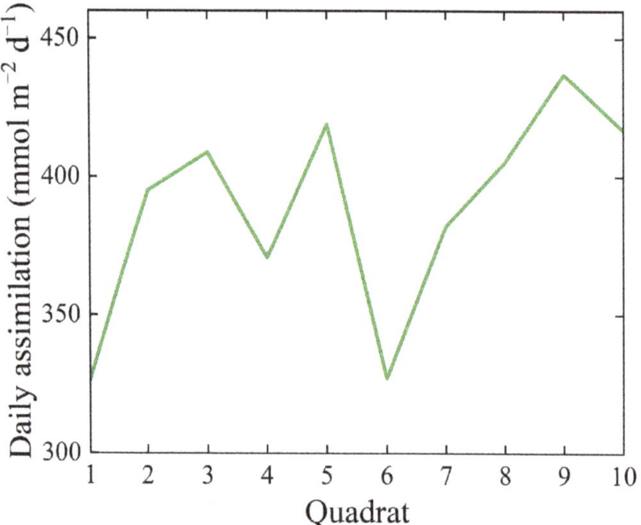

Figure 1. Daily assimilation of photosynthesis in various quadrats.

According to the annual net carbon sequestration and oxygen release of the 10 quadrats, they can be divided into three categories (Table S3, Figure 2), and there are significant differences among the three categories quadrats ($p < 0.05$). According to the annual net carbon sequestration and oxygen release, the inter-class order is as follows: the first category (higher carbon sequestration and oxygen release per unit leaf area) is seen in quadrat 2, quadrat 3, quadrat 5, quadrat 8, quadrat 9 and quadrat 10. The second category (moderate carbon sequestration and oxygen release per unit leaf area) is seen in quadrat 4 and quadrat 7. The third category (low carbon sequestration and oxygen release per unit leaf area) is seen in quadrat 1 and quadrat 6.

In addition, the carbon sequestration/oxygen release per unit leaf area and the daily average *Pn* value of vegetation also showed this trend; that is, the first category > the second category > the third category, and the differences among the three categories were obvious ($p < 0.05$) (Figure 3A,B).

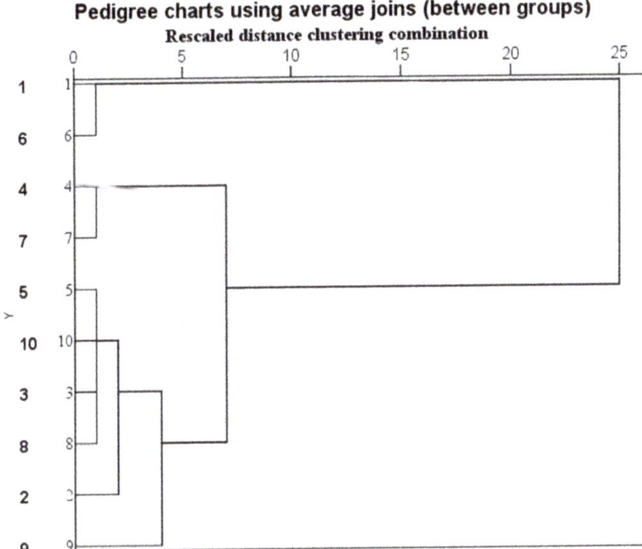

Figure 2. Cluster analysis results of carbon sequestration and oxygen release capacity of 10 quadrats.

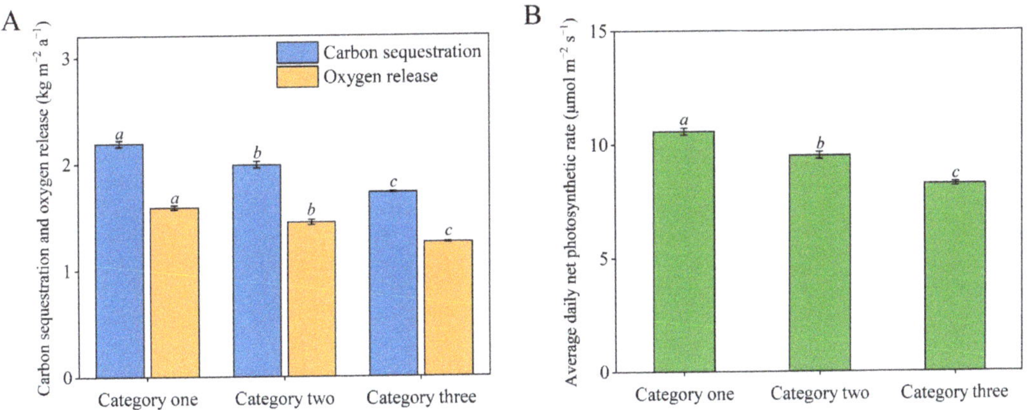

Figure 3. Carbon sequestration and oxygen release capacity of three kinds of quadrats (**A**). Average daily net photosynthetic rate value of three kinds of quadrats (**B**). Different lowercase letters indicate significant differences ($p < 0.05$).

3.2.2. Comparative Analysis of Carbon Sequestration and Oxygen Release Capacity of Different Plants in Shelterbelt by Net Photosynthetic Rate Method

The *Pn* value of the three shelterbelt plants showed an obvious "single peak" curve, and the peaks appeared at around 12:00, among which the peak net photosynthetic rate (*Pn*max) of *T. chinensis* was the highest (Figure 4A). At the same time, the daily average *Pn* value and carbon sequestration and oxygen release per unit leaf area of *T. chinensis* were significantly higher than those of *H. ammodendron* and *C. mongolicum* ($p < 0.05$), but the difference between *H. ammodendron* and *C. mongolicum* was not significant ($p > 0.05$) (Figure 4B,C).

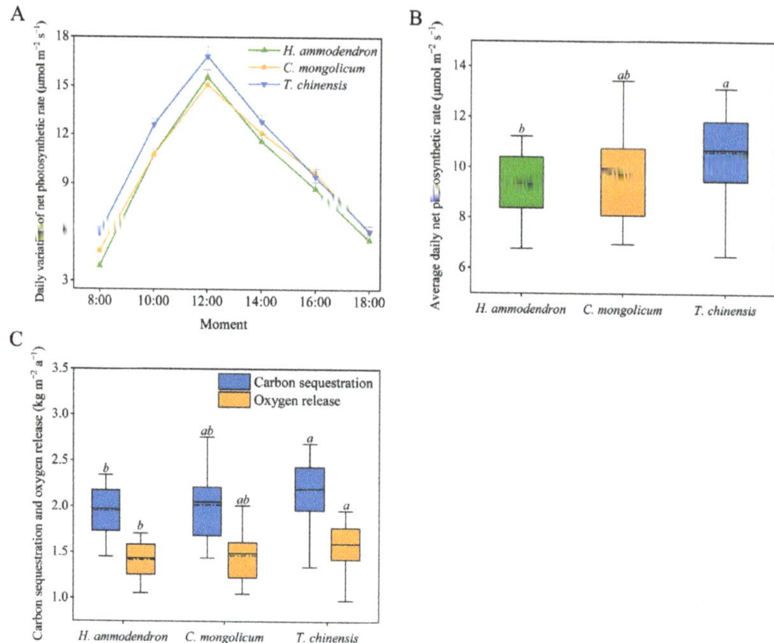

Figure 4. Daily variation of net photosynthetic rate value of three plants (**A**). Average daily net photosynthetic rate value of three plants (**B**). Carbon sequestration and oxygen release capacity of three plants (**C**). Different lowercase letters indicate significant differences ($p < 0.05$).

3.2.3. Annual Carbon Sequestration and Oxygen Release via the Net Photosynthetic Rate Method

Although *T. chinensis* had the highest annual carbon sequestrations and oxygen releases per unit leaf surface, the annual carbon sequestrations and oxygen releases per unit plant of *H. ammodendron* were higher than those of *C. mongolicum* and *T. chinensis* (Table 2).

Table 2. Annual carbon sequestration and oxygen release of three kinds of shelterbelts on the Tarim Desert Highway.

Species of Trees	Vegetation Area (ha)	Carbon Sequestration		Oxygen Release	
		Carbon Sequestration per Unit Leaf Surface (kg m^{-2} a^{-1})	Annual Carbon Sequestration (10^4 t a^{-1})	Annual Oxygen Release per Leaf Surface (kg m^{-2} a^{-1})	Annual Oxygen Release (10^4 t a^{-1})
H. ammodendron	1291.36	1.95	2.52	1.42	1.83
C. mongolicum	328.20	2.02	0.66	1.47	0.48
T. chinensis	166.06	2.20	0.37	1.60	0.27
Total	1785.62		3.55		2.58

3.3. Carbon Sequestration and Oxygen Release Capacity under Biomass Method

3.3.1. Biomass Model Construction

According to the measured data concerning the shrubs (Table 3), the model was established, and the results showed that the optimal biomass models of the three shrubs all adopted the power function form (Table 4, Figure 5). In the biomass equations of *H. ammodendron* and *T. chinensis*, plant volume (V) is the best predictive variable. In the biomass equation of *C. mongolicum*, the longitudinal sectional area (LH) is the best predictive variable. The fitting degree of these biomass models is very high; the R^2 value ranges from 0.9414 to 0.9641, and the *SEE* ranges from 0.578 to 0.908, which shows that the fitting effect of the models is good. By testing the optimal models, it is found that the p values of all biomass estimation models are less than 0.001, which meets the requirements of an extremely significant level test. This means that the predictive variables of each model have a good explanation for the dependent variables.

Table 3. Shrub characteristics for building biomass regression models.

Species of Trees	Sample Size	Diameter at Breast Height (cm)	Plant Height (cm)	Crown Average Diameter (m)	Crown Area (m²)	Measured Total Biomass (kg plant⁻¹)
H. ammodendron	12	5.70 ± 0.53	181.83 ± 7.06	1.95 ± 0.94	3.05 ± 0.29	14.62 ± 1.17
C. mongolicum	12	2.57 ± 0.26	158.12 ± 6.73	1.58 ± 0.11	2.07 ± 0.29	12.33 ± 0.88
T. chinensis	12	2.63 ± 0.32	193.30 ± 6.93	1.64 ± 0.07	2.15 ± 0.18	7.95 ± 0.55

Note: Data = mean ± standard error.

Table 4. Three shrub biomass regression models.

Species of Trees	Independent Variable	Optimum Model	Argument a	Argument b	R^2	*SEE*	P
H. ammodendron	CH	$W = aX^b$	2.7982	0.9736	0.9641	0.908	<0.001
C. mongolicum	LH	$W = aX^b$	6.0349	0.7696	0.9540	0.827	<0.001
T. chinensis	CH	$W = aX^b$	2.8663	0.7265	0.9414	0.578	<0.001

Note: W is whole-plant biomass (kg), C is crown area (m²), L is crown average diameter (m), H is plant height (m), R^2 is discriminant coefficient, *SEE* is estimation standard error, P is significant level of regression test, a is the coefficient in the formula and b is the exponential term in the formula.

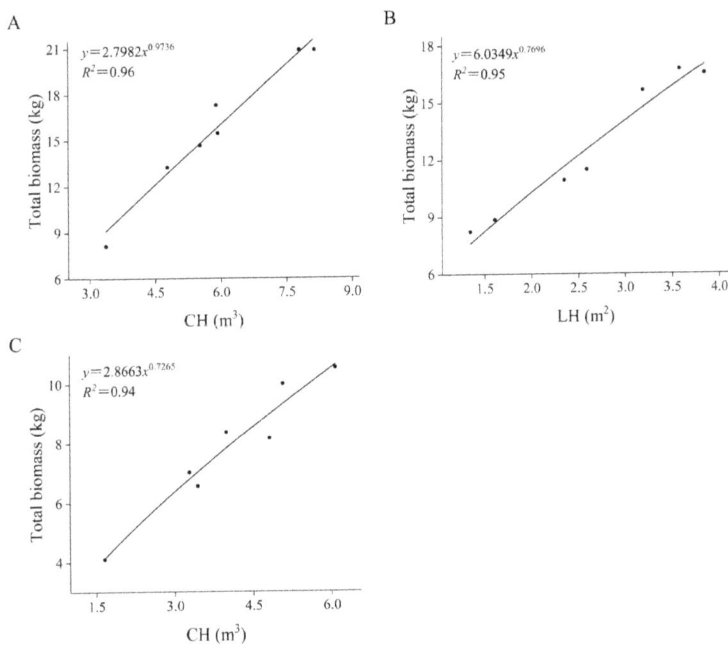

Figure 5. Scatter diagram of *H. ammodendron* biomass regression model (**A**). Scatter diagram of *C. mongolicum* biomass regression model (**B**). Scatter diagram of *T. chinensis* biomass regression model (**C**).

3.3.2. Verification of Biomass Model

Regression analysis between the measured data and the predicted values of shrubs show that the total relative error (RS) of the fitting model is between −2.49% and 1.47%, and the average relative error (RMA) is between 5.40% and 7.20%, which shows that the fitting accuracy of the whole-plant biomass of the three shrubs is high (Table 5). At the same time, there is a good correlation between the predicted values and the measured values, and both values meet the requirements of a 95% confidence interval (Figure 6). The fitting rate is between 0.90 and 0.95, which shows that the model is effective. These results show that regression analysis can accurately predict the whole-plant biomass of three shrubs, and the model effect is ideal.

Table 5. Accuracy test of three kinds of shrub biomass regression models.

Species of Trees	Biomass	Total Relative Error/RS (%)	Mean Relative Error/RMA (%)
H. ammodendron		1.27	7.20
C. mongolicum	W	1.47	5.40
T. chinensis		−2.49	6.60

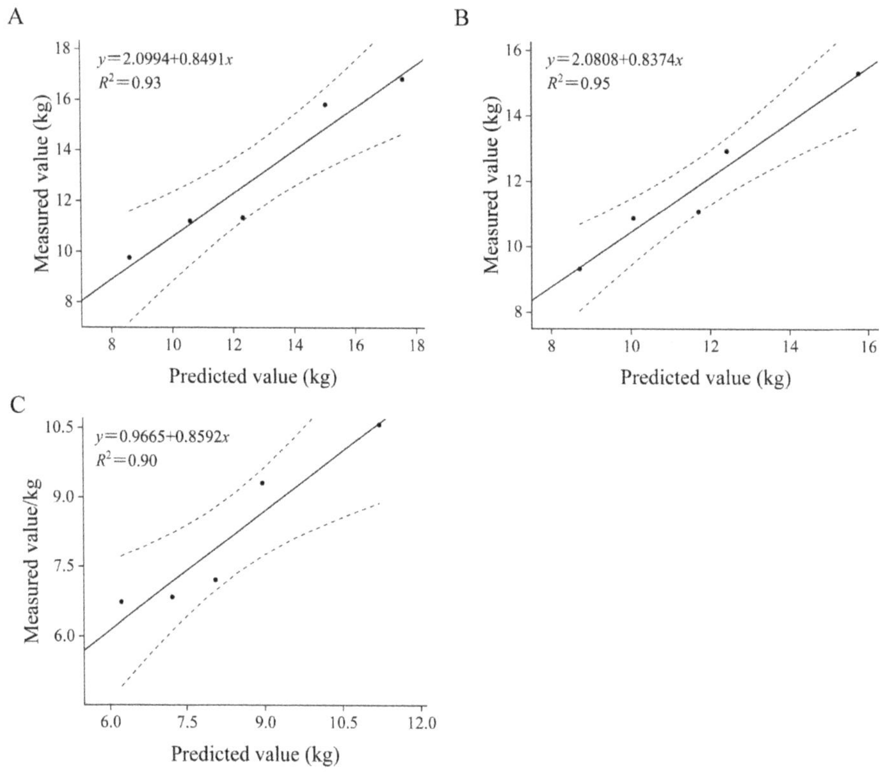

Figure 6. Comparison of measured and predicted biomass of *H. ammodendron* (**A**), *C. mongolicum* (**B**), *T. chinensis* (**C**). The dashed lines in the figure represent the upper and lower 95% confidence intervals.

3.3.3. Total Carbon Sequestration and Oxygen Release Using the Biomass Method

The average biomass of the three shrubs on the desert highway ranged from 8.72 to 19.62 kg/plant (Table 6). The order of the three shelterbelt plants in terms of total number of plants, total biomass, carbon storage, total carbon sequestration and oxygen release is *H. ammodendron* > *C. mongolicum* > *T. chinensis*. The carbon storage densities of *H. ammodendron*, *C. mongolicum* and *T. chinensis* are 67.00 t ha^{-1}, 51.40 t ha^{-1} and 26.98 t ha^{-1}, respectively. In addition, the total carbon sequestration and oxygen release of *H. ammodendron* is 5.13 times that of *C. mongolicum* and 19.29 times that of *T. chinensis* (Table 7).

Table 6. Characteristics of shrubs within the quadrat.

Species of Trees	Diameter at Breast Height (cm)	Plant Height (cm)	Crown Average Diameter (m)	Crown Area (m^2)	Predicted Biomass (kg plant^{-1})
H. ammodendron	4.08 ± 0.15	178.61 ± 5.13	2.19 ± 0.05	4.13 ± 0.22	19.62
C. mongolicum	2.50 ± 0.22	156.68 ± 7.43	1.50 ± 0.09	1.96 ± 0.21	11.67
T. chinensis	2.73 ± 0.25	181.92 ± 10.51	1.69 ± 0.11	2.54 ± 0.40	8.72

Note: Data = mean ± standard error.

Table 7. Total carbon sequestration and oxygen release of three shelterbelt plants on the Tarim Desert Highway.

Species of Trees	Biomass (kg Plant^{-1})	Number of Plants (Million)	Total Biomass (10^4 t)	Carbon Content (kg kg^{-1})	Carbon Stock (10^4 t)	Carbon Storage Density (t ha^{-1})	Total Carbon Sequestration (10^4 t)	Total Oxygen Release (10^4 t)
H. ammodendron	19.62	12.6	24.72		12.36	67.00	45.32	32.96
C. mongolicum	11.67	4.13	4.82	0.5	2.41	51.40	8.84	6.43
T. chinensis	8.72	1.47	1.28		0.64	26.98	2.35	1.71
Total		18.20			15.41	60.41	56.51	41.10

3.4. Carbon Sequestration and Oxygen Release Benefits

The total annual value of carbon sequestration and oxygen release in the whole shelter forest is CNY 13.73 million, of which the total annual value of carbon sequestration is CNY 1.88 million, and the total annual value of oxygen release is CNY 11.85 million. At the same time, the total value of carbon sequestration and oxygen release of the whole shelter forest is CNY 218.77 million, of which the total value of carbon sequestration is CNY 30.12 million, and the total value of oxygen release is CNY 188.65 million. In addition, in terms of three kinds of shelter forest plants, the performance order of carbon sequestration and oxygen release value is *H. ammodendron* > *C. mongolicum* > *T. chinensis* (Table 8).

Table 8. Estimation of carbon sequestration and oxygen release value of three kinds of shelterbelts on the Tarim Desert Highway.

Species of Trees	Annual Price of *Pn* Method (Million CNY a^{-1})			Total Value of Biomass Method (Million CNY)		
	Carbon Sequestration	Oxygen Release	Carbon Sequestration and Oxygen Release	Carbon Sequestration	Oxygen Release	Carbon Sequestration and Oxygen Release
H. ammodendron	1.34	8.42	9.76	24.16	151.29	175.45
C. mongolicum	0.35	2.21	2.56	4.71	29.51	34.22
T. chinensis	0.19	1.22	1.41	1.25	7.85	9.10
Total	1.88	11.85	13.73	30.12	188.65	218.77

4. Discussion

The results of the quadrat survey showed that the three dominant shrub species, with *H. ammodendron* (C.A.Mey.) Bunge as the main species, had an average coverage of 70%. The total number of individuals among the three shrub species exceeded 12.8 million trees, indicating that progress had been made in the greening work along the desert road. The continued presence of protective forests helped reduce soil erosion and sandstorms, thereby maintaining ecological stability in the area [43]. Additionally, in terms of vegetation frequency, density, and canopy width, *H. ammodendron* far surpassed *C. mongolicum* Turcz. and *T. chinensis* Lour. The vegetation density of *H. ammodendron* was three times that of *C. mongolicum* and eight times that of *T. chinensis*, reflecting its superior status as a dominant species in the protective forest. Plants increased their ability to acquire resources by optimizing the utilization of ecological space [44]. *H. ammodendron* had stronger salt and alkali resistance, deeper root depth, and lower transpiration water consumption compared to the other two plant species [45,46]. As a result, *H. ammodendron* demonstrated a higher capacity for survival under arid conditions. These findings provided important data for understanding the distribution of vegetation and the ecological environment along the desert road and contributed to the development of corresponding protective and management measures.

After conducting actual observations in the study area, we found that the number of deaths among the three types of shrubs was extremely limited, and their proportion within the overall plant population was quite low. This may have been due to the fact that these drought-resistant plants could grow well under conditions where there was underground water irrigation. Additionally, the overall age of the protective forest was relatively young, and it had not yet entered into senescence [47]. Therefore, taking into account the actual circumstances, the biomass of the deceased plant parts was not evaluated in this study. Additionally, plant residues such as fallen leaves and branches were easily dispersed by strong winds, and the soil conditions of the desert road limited the decomposition rate of plant residues and microbial activity [48]. In this study, the carbon release caused by the deterioration of biomass was negligible. Furthermore, due to various constraints such as the growth environment and water supply, the total amount of existing plant residues in the desert road protection forest belt was very low [49]. Meanwhile, due to restricted organic matter inputs, high temperatures caused the rapid evaporation of soil moisture, and strong winds would damage the soil environment. All these factors resulted in significantly lower soil organic carbon content in the protective forests along the desert road compared to the shrub forests in the national and surrounding areas [50]. The quantity of organic carbon in the soil, when compared to the carbon content of the plants themselves, could be disregarded. Therefore, in our research, we only calculated the carbon content contained within the plants themselves along the desert road, while neglecting the carbon stored in litter and soil.

This study used two methods to investigate the carbon sequestration and oxygen release capacity of plants. Among them, the photosynthesis rate method is considered a potential approach as it infers the carbon storage of plants by measuring their photosynthesis rate. This method focuses more on the metabolic activity and biochemical processes of plants to infer the carbon storage within plants [34]. In contrast, the biomass method is regarded as an observational approach that directly measures the plant's biomass to estimate its carbon storage [51]. This method is based on the relationship between biomass and carbon content and places more emphasis on the observation and measurement of the plant's external manifestations. This study found that the carbon sequestration calculated based on the biomass method (56.51×10^4 t) was slightly lower than that calculated based on the photosynthetic rate method (63.9×10^4 t). Possibly because this study did not consider the biomass of dead plants and litter, it led to slightly lower carbon storage estimates under the biomass method compared to the photosynthesis rate method. The photosynthetic rate method focuses on the process and rate of photosynthesis, while the biomass method emphasizes the carbon sequestration capacity and cumulative effects of organisms. They can complement and verify each other in studying the process of carbon sequestration, helping us to comprehensively understand the mechanisms and effects of carbon sequestration.

The *Pn* value of three shelterbelt plants showed a "single peak" curve in one day; that is, it reached the highest peak at noon (12:00 p.m.). In terms of carbon sequestration capacity per unit leaf area, *T. chinensis* has the highest carbon sequestration capacity. However, from the point of view of carbon sequestration capacity per plant, *H. ammodendron* showed the strongest capacity, followed by *T. chinensis*. Jia et al. [52] analyzed the carbon sequestration capacity of *T. chinensis*, *H. ammodendron* and *C. mongolicum* in the desert area on the southern edge of the Zhungeer Basin. They measured the carbon sequestration capacity by comparing the size of the *Pn*max value and concluded that *T. chinensis* was the first-choice tree species. If the carbon sequestration level is considered from the perspective of the *Pn*max value, this is consistent with the conclusion of this study. However, only using this value to measure the carbon sequestration capacity of vegetation may overestimate the overall carbon sequestration capacity of vegetation. This is because the *Pn* value and the fixed amount of CO_2 per unit leaf area reflect the photosynthetic carbon sequestration capacity of shrub leaves, while the carbon sequestration capacity per plant comprehensively considers the carbon sequestration capacity of individual shrubs [53]. Therefore, it is suggested to keep *H. ammodendron* as the dominant species and appropriately increase the planting area of *T. chinensis* in order to improve the carbon sequestration and oxygen release capacity of the whole shelterbelt. This study found that there were differences in the total leaf area per plant of the three shelterbelts, which resulted in different rules of fixed CO_2 per unit leaf area and carbon sequestration per plant. This is similar to the conclusion made by Dang et al. [54] and Li et al. [55] to compare the carbon sequestration capacity of different plants by combining the fixed amount of CO_2 per unit leaf area and the leaf area index. Therefore, the fixed amount of CO_2 per unit leaf area and the leaf area index should be considered comprehensively when evaluating the carbon sequestration capacity of plants.

The large-scale construction of artificial forests in arid regions not only provides windbreaks, sand stabilizers, and landscape enhancers but also contributes to the increase in carbon storage in arid and semi-arid ecosystems [13,19]. This study found that the average carbon storage density in the shelter forest vegetation of the Tarim Desert Highway was 60.41 t ha^{-1}, which exceeded the value reported in other arid areas. Compared with *H. ammodendron* in arid desert areas [56,57], the carbon storage density of *H. ammodendron* on desert roads is about 13 times higher. Compared with *T. chinensis* plants in arid desert areas [58], the carbon storage density of *T. chinensis* plants on desert roads is 18.8 times higher. Compared with *T. chinensis* plants in coastal areas [59], the carbon storage density of *T. chinensis* plants on desert roads is also 5.3 times higher. Therefore, compared with plants in similar arid areas, desert highway artificial shelterbelts have higher carbon sequestration potential. This shows that plantation has great potential in carbon sequestration. Therefore, we should not only pay attention to the shelter forest along the desert highway but also realize that the carbon sequestration potential of other plantations mainly composed of these species cannot be ignored. This discovery is of great significance for us to evaluate and protect the carbon sequestration value of the plantation.

Estimating biomass carbon is crucial for understanding changes in carbon density and facilitating assessments of carbon management [60,61]. The power function model is selected as the best-fitting model for the three vegetation biomass-estimation models, but the corresponding dependent variables are different. Zhao [62] constructed an equation for estimating the biomass of *H. ammodendron* and *T. chinensis* in desert areas by using the plant height H and diameter at breast height D, but the selection of its dependent variables was not completely consistent with this paper. This is because the growth of desert plants is closely influenced by climate, soil, and precipitation conditions, so even the biomass model of the same species in different environments will be different [63]. In addition, Zhao et al. [64] and Yang et al. [65] think that the power function model is the most suitable model for estimating shrub biomass, which is consistent with our research conclusions. However, the optimal biomass model chosen by Tang et al. [66] in their research on *C. mongolicum* and *T. chinensis* in the northern margin of the Kubuqi Desert and by Wei et al. [67] in their research on *H. ammodendron* and *C. mongolicum* in the desert oasis transitional zone of

Minqin is not a power function model, which is different from our research results to some extent. This difference is probably due to the extreme climatic conditions in desert highway areas, which leads to differences in plant morphological characteristics [68]. Therefore, when biomass estimation is carried out in arid areas, each biomass-estimation model has problems and limitations in applicable areas. In practical application, we should choose a suitable model according to the specific situation. Furthermore, in order to address the difficulties in excavating deeply rooted shrub roots caused by extreme weather conditions and minimize damage to the local ecological environment, a small number of shrubs were used for complete excavation in this study [69]. Seven samples were used for model fitting, and five samples were used for model validation. During the model construction phase, it typically required a larger amount of sample data to capture the variability between plants. In contrast, model accuracy validation only required an independent validation dataset to assess the predictive ability of the model [70]. Moreover, through strict sampling methods and randomization, a certain number of standard shrubs were selected to ensure that the chosen plant samples were representative and could infer the overall population.

As the first zero-carbon highway in China, the Tarim Desert Highway has greater value in terms of its carbon sequestration and oxygen release function due to shelter forest vegetation. According to the estimation using the *Pn* method, the value of carbon sequestration and oxygen release of the three types of shelter forests is CNY 247.14 million (18 years old). Compared with this, the total value of carbon sequestration and oxygen release calculated using the biomass method is CNY 218.77 million. It can be seen that the value obtained when using the *Pn* method is higher than that when using the biomass method. The *Pn* method is mainly used to study plant carbon sinks at the micro-scale, and it is studied at various intermediate points, producing more detailed results [71]. The biomass law is applied to the macro-scale study of plant carbon sinks, mainly considering the first and last time points, and is suitable for the study of larger-scale areas [72]. At present, there is no mainstream method to calculate the carbon sequestration and oxygen release capacity of vegetation with both accuracy and convenience, so future development needs to be prioritized according to actual needs, and a more reasonable and simple testing method should be established. From the perspective of economic evaluation, due to the differences in theory and evaluation methods, the evaluation results may be different when studying shrubs, especially those growing in arid areas. At present, there is no systematic methodology, which leads to some errors in the calculation results. However, with the continuous improvement of the low-carbon economy theory and the maturity of the carbon-trading mechanism, the results of the economic benefit evaluation of ecological forests will be enriched.

5. Conclusions

Vegetation alongside desert highways exhibits overall fluctuations in carbon sequestration processes, showing a characteristic "M"-shaped variation pattern in the north-south direction. The diurnal variations in the *Pn* value of the three plant species all exhibit a "unimodal" curve. Among them, the *Pn* value of *T. chinensis* Lour. was the strongest, while the individual carbon sequestration capacity of *H. ammodendron* (C.A.Mey.) Bunge was more remarkable. It is suggested that, while maintaining *H. ammodendron* (C.A.Mey.) Bunge as the dominant species, introducing *T. chinensis* Lour. for planting in moderation would provide greater carbon sink benefits to the protective forest belt alongside desert highways. In addition, according to the 18-year growth cycle of desert highway vegetation, based on the *Pn* method, the accumulated carbon sequestration and oxygen release value is CNY 247.14 million. Based on the biomass method, the accumulated carbon sequestration and oxygen release value is CNY 218.77 million.

Studying the carbon storage and dynamic changes of artificial sand-fixation vegetation ecosystems is beneficial for establishing a deeper understanding and conducting a quantitative evaluation of the carbon sequestration potential of artificial sand-fixation vegetation.

This holds significant practical and strategic importance in accurately estimating land carbon stocks at regional, and even national, scales.

Supplementary Materials: The following supporting information can be downloaded at: https://www.mdpi.com/article/10.3390/f14112137/s1, Table S1: The measured total biomass of the whole plant; Table S2: Quadrat vegetation characteristics; Table S3: Daily assimilation amount per unit leaf area, annual carbon sequestration, annual oxygen release and average daily net photosynthetic rate.

Author Contributions: Investigation, L.L.; conceptualization, L.L. and X.H.; methodology, L.L.; software, L.L.; writing—original draft, L.L.; writing—review and editing, L.L.; writing—review and editing, A.Z.; supervision, X.H.; funding acquisition, X.H. All authors have read and agreed to the published version of the manuscript.

Funding: This work was funded by the determination of carbon sink in the sand-fixing forest belt in Luntai–Minfeng section of the desert highway. Funder: Xuemin He. Funding number: 202205140011.

Institutional Review Board Statement: Not applicable.

Informed Consent Statement: Not applicable.

Data Availability Statement: Not applicable.

Conflicts of Interest: The authors declare no conflict of interest.

References

1. Forzieri, G.; Alkama, R.; Miralles, D.G.; Cescatti, A. Satellites reveal contrasting responses of regional climate to the widespread greening of earth. *Science* **2017**, *356*, 1180–1184. [CrossRef] [PubMed]
2. Sohngen, B.; Tian, X. Global climate change impacts on forests and markets. *For. Policy Econ.* **2016**, *72*, 18–26. [CrossRef]
3. Zhang, S.; Bai, X.; Zhao, C.; Tan, Q.; Xi, H. Global CO_2 Consumption by silicate rock chemical weathering: Its past and future. *Earth's Future* **2021**, *9*, e1938E–e2020E. [CrossRef]
4. Smith, P.; Porter, J.R. Bioenergy in the IPCC assessments. *GCB Bioenergy* **2018**, *10*, 428–431. [CrossRef]
5. Yao, L.; Tan, S.; Xu, Z. Towards carbon neutrality: What has been done and what needs to be done for carbon emission reduction? *Environ. Sci. Pollut. Res. Int.* **2023**, *30*, 20570–20589. [CrossRef]
6. Li, H.; Qin, Q. Challenges for China's carbon emissions peaking in 2030: A decomposition and decoupling analysis. *J. Clean. Prod.* **2019**, *207*, 857–865. [CrossRef]
7. Lin, B.; Ge, J. Carbon sinks and output of China's forestry sector: An ecological economic development perspective. *Sci. Total Environ.* **2019**, *655*, 1169–1180. [CrossRef]
8. Hoque, M.Z.; Cui, S.; Islam, I.; Xu, L.; Ding, S. Dynamics of plantation forest development and ecosystem carbon storage change in coastal Bangladesh. *Ecol. Indic.* **2021**, *130*, 107954. [CrossRef]
9. Guo, Z.D.; Hu, H.F.; Li, P.; Li, N.Y.; Fang, J.Y. Spatio-Temporal changes in biomass carbon sinks in China's forests from 1977 to 2008. *Sci. China Life Sci.* **2013**, *56*, 661–671. [CrossRef]
10. Nghiem, N. Optimal rotation age for carbon sequestration and biodiversity conservation in Vietnam. *Forest Policy Econ.* **2014**, *2014*, 56–64. [CrossRef]
11. Fang, J.Y.; Chen, A.P.; Peng, C.H.; Zhao, S.Q.; Ci, L. Changes in forest biomass carbon storage in China between 1949 and 1998. *Science* **2001**, *292*, 2320–2322. [CrossRef] [PubMed]
12. He, Y.J.; Han, X.R.; Wang, X.P.; Wang, L.Q.; Liang, T. Long-term ecological effects of two artificial forests on soil properties and quality in The Eastern Qinghai-Tibet Plateau. *Sci. Total Environ.* **2021**, *796*, 148986. [CrossRef] [PubMed]
13. Qiu, D.D.; Zhu, G.F.; Lin, X.R.; Jiao, Y.Y.; Lu, S.Y.; Liu, J.T.; Liu, J.W.; Zhang, W.H.; Ye, L.L.; Li, R.; et al. Dissipation and movement of soil water in artificial forest in arid oasis areas: Cognition based on stable isotopes. *CATENA* **2023**, *228*, 107178. [CrossRef]
14. Jörgensen, K.; Granath, G.; Lindahl, B.D.; Strengbom, J. Forest management to increase carbon sequestration in boreal pinus sylvestris forests. *Plant Soil* **2021**, *466*, 165–178. [CrossRef]
15. Cook-Patton, S.C.; Leavitt, S.M.; Gibbs, D.; Harris, N.L.; Lister, K.; Anderson-Teixeira, K.J.; Briggs, R.D.; Chazdon, R.L.; Crowther, T.W.; Ellis, P.W.; et al. Mapping carbon accumulation potential from global natural forest regrowth. *Nature* **2020**, *585*, 545–550. [CrossRef] [PubMed]
16. Kumar, A.; Kumar, M.; Pandey, R.; Yu, Z.G.; Cabral-Pinto, M. Forest soil nutrient stocks along altitudinal range of Uttarakhand Himalayas: An aid to nature based climate solutions. *CATENA* **2021**, *207*, 105667. [CrossRef]
17. Ye, T.; Zhang, Y.M. Biomass allocation patterns and allometric relationships of six ephemeroid species in Junggar Basin, China. *Acta Prataculturae Sin.* **2014**, *23*, 38–48.
18. Walker, W.S.; Gorelik, S.R.; Cook-Patton, S.C.; Baccini, A.; Farina, M.K.; Solvik, K.K.; Ellis, P.W.; Sanderman, J.; Houghton, R.A.; Leavitt, S.M.; et al. The global potential for increased storage of carbon on land. *Pnas* **2022**, *119*, e2111312119. [CrossRef]

19. Yang, Y.; Liu, L.; Zhang, P.P.; Wu, F.; Wang, Y.Q.; Xu, C.; Zhang, L.K.; An, S.S.; Kuzyakov, Y.K. Large-scale ecosystem carbon stocks and their driving factors across Loess Plateau. *Carb Neutrality* **2023**, *2*, 5. [CrossRef]

20. Niu, J.P.; Yang, K.; Tang, Z.; Wang, Y. Relationships between soil crust development and soil properties in the Desert Region of North China. *Sustainability* **2017**, *9*, 725. [CrossRef]

21. Liu, X.P.; Zhang, W.J.; Cao, J.S.; Yang, B.; Cai, Y.J. Carbon sequestration of plantation in Beijing-Tianjin sand source areas. *J. Mt. Sci.* **2018**, *15*, 2148–2158. [CrossRef]

22. Xu, M.; Cao, C.; Tong, Q.; Li, Z.; Zhang, H.; He, Q.; Gao, M.; Zhao, J.; Zheng, S.; Chen, W.; et al. Remote sensing based shrub above-ground biomass and carbon storage mapping in Mu Us desert, China. *Sci. China Technol. Sci* **2010**, *53*, 176–183. [CrossRef]

23. Chen, G.L. Photosynthetic Characteristics and Carbon Sequestration and Oxygen Release Capacity of Typical Plants in Helan Mountain. Master's Thesis, Ningxia University, Yinchuan, China, 2021.

24. Zhang, W.; Li, Y.Q.; Zhao, X.Y.; Zhang, T.H.; Ma, Q.L.; Tang, J.N.; Feng, J.; Su, N. Effects of Caragana microphylla plantations on organic carbon sequestration in total and labile soil organic carbon fractions in the Horqin Sandy Land, northern China. *Arid. Land* **2017**, *9*, 688–700. [CrossRef]

25. Liu, J. The Dynamic of Soil Water Shelterbelt Plants Photosynthetic and Stem Flow under Saline drip irrigation in Taklamakan Desert. Ph.D. Thesis, Northwest Agriculture and Forestry University, Xianyang, China, 2020.

26. Moradi, A.; Shabanian, N. Sacred groves: A model of Zagros forests for carbon sequestration and climate change mitigation. *Environ. Conserv.* **2023**, *50*, 163–168. [CrossRef]

27. Devi, N.B.; Lepcha, N.T. Carbon sink and source function of Eastern Himalayan forests: Implications of change in climate and biotic variables. *Environ. Monit. Assess.* **2023**, *195*, 843. [CrossRef]

28. Barkhordarian, A.; Bowman, K.W.; Cressie, N.; Jewell, J.; Liu, J. Emergent constraints on tropical atmospheric aridity—Carbon feedbacks and the future of carbon sequestration. *Environ. Res. Lett.* **2021**, *16*, 114008. [CrossRef]

29. Gao, Y.; Liu, L.; Ma, S.; Zhou, Y.; Jia, R.; Li, X.; Yang, H.; Wang, B. Vegetation restoration in dryland with shrub serves as a carbon sink: Evidence from a 13-year observation at the Tengger Desert of Northern China. *Land Degrad. Develop.* **2023**, 1–12. [CrossRef]

30. An, Z.; Zhang, K.; Tan, L.; Niu, Q.; Wang, T. Mechanisms responsible for sand hazards along Desert Highways and their control: A case study of the Wuhai–Maqin Highway in the Tengger Desert, Northwest China. *Front. Environ. Sci.* **2022**, *10*, 878778. [CrossRef]

31. Ding, X.Y.; Zhou, Z.B.; Xu, X.W.; Lei, J.Q.; Lu, J.J.; Ma, X.X. Three-dimension temporal and spatial dynamics of soil water for the artificial vegetation in the center of Taklimakan Desert under saline water drip-irrigation. *Chin. J. Appl. Ecol.* **2015**, *26*, 2600–2608.

32. Zhang, X.M.; Wang, Y.D.; Xu, X.W.; Jiaqiang, L. Biomass, composition and dynamics of litterfall in Taklimakan Desert highway shelterbelt. *J. Desert Res.* **2017**, *37*, 6.

33. Wang, S.C.; Luo, Y.Y.; Wang, S.S.; Zhang, H.W. Study on carbon reserve and carbon sequestration and oxygen release function of four artificial shrub communities in the south foot of Langshan Mountain. *Shandong Agric. Sci.* **2022**, *54*, 116–122.

34. Fu, D.F.; Bu, B.; Wu, J.G.; Singh, R.P. Investigation on the carbon sequestration capacity of vegetation along a heavy traffic load expressway. *J. Environ. Manag.* **2019**, *241*, 549–557. [CrossRef] [PubMed]

35. Xu, Y.; Deng, L. Effects of replanted on stand growth and species diversity. *J. Est Environ.* **2023**, *43*, 329–336.

36. Gou, J.H.; Li, C.J.; Zeng, F.J.; Zhang, B.; Liu, B.; Gou, Z.C. Relationship between root bio-mass distribution and soil moisture, nutrient for two desert plant species. *Arid. Zone Res.* **2016**, *33*, 1.

37. Ahmad, A.; Liu, Q.J.; Nizami, S.M.; Mannan, A.; Saeed, S. Carbon emission from deforestation, forest degradation and wood harvest in the temperate region of Hindukush Himalaya, Pakistan between 1994 and 2016. *Land Use Policy* **2018**, *78*, 781–790. [CrossRef]

38. Ma, X.W.; Xiong, K.N.; Zhang, Y.; Lai, J.L.; Zhang, S.H.; Ji, C.Z. Research progresses and prospects of carbon storage in forest ecosystems. *J. Northwest For. Univ.* **2019**, *34*, 62–72.

39. Li, M. Carbon stock and sink economic values of forest ecosystem in the forest industry region of Heilongjiang Province, China. *J. For. Res.* **2022**, *33*, 875–882. [CrossRef]

40. Chen, W.; Zhao, J.; Cao, C.; Tian, H. Shrub biomass estimation in semi-arid sandland ecosystem based on remote sensing technology. *Glob. Ecol. Conserv.* **2018**, *16*, e00479. [CrossRef]

41. Chen, J.L.; Jiang, X.; Zhou, X.L.; Pang, X.A. Studies on aboveground biomass model of *Tamarix Ramosissima* at the upper reaches of the Tarim River. *Xinjiang Agric. Sci.* **2014**, *51*, 1893–1899.

42. GB/T 38582-2020; State Forestry and Grassland Administration. Specification for Functional Assessment of Forest Ecosystem Services. State Administration for Market Regulation. National Standardization Administration Commission: Beijing, China, 2020.

43. Luo, Y.; Gong, Y. α Diversity of desert shrub communities and Its relationship with Climatic factors in Xinjiang. *Forests* **2023**, *14*, 178. [CrossRef]

44. Shan, L.S.; Li, Y.; Zhang, X.M.; Wang, H. Effects of different irrigation regimes on characteristics of transpiring water-consumption of three desert species. *Acta Ecol. Sin.* **2012**, *32*, 5692–5702. [CrossRef]

45. Wei, Y.J. Study on Ecological Benefits and Optimal Configuration of Jilantai Salt Lake Protection System. Ph.D. Thesis, Inner Mongolia Agricultural University, Hohhot, China, 2022.

46. Li, W.; Wang, W.Q.; Sun, R.M.; Li, M.K.; Liu, H.W.; Shi, Y.F.; Zhu, D.D.; Li, J.Y.; Ma, L.; Fu, S.L. Influence of nitrogen addition on the functional diversity and biomass of fine roots in warm-temperate and subtropical forests. *For. Ecol. Manag.* **2023**, *545*, 121309. [CrossRef]

47. Wang, X.Y.; Ma, Q.L.; Wang, Y.L. Carbon benefits evaluation of the artificial shelter forest in the Shiyanghe River Basin. *J. Desert Res.* **2020**, *40*, 197–205.
48. Meng, T.G.; Wu, L.Y.; Zhang, S.L.; Xu, Y.Y.; Li, X.; Zhang, J.G. Vertical distribution of soil dissolved carbon and its influencing factors in the artificial shelterbelt irrigated with saline water in an Extreme Drought Desert. *Huan Jing Ke Xue Huanjing Kexue* **2020**, *41*, 1950–1959. [PubMed]
49. Zou, Y.Y.; Jin, Z.Z.; Zhang, D.D.; Li, S.Y.; Xu, X.W. Study on litter standing biomass and soil properties under different forest in shelterbelt along the Tarim Desert Highway. *Chin. J. Soil Sci.* **2015**, *46*, 656–663.
50. Zhang, Q.; Zhang, J.G.; Wang, L.M., Ding, X X.; Ma, A.S.; Zhang, H.; Li, L.M. Vertical distribution of soil organic and inorganic carbon in the Taklimakan Desert Highway shelterbelt drip-irrigated with different mineralization water. *J. Northwest For. Univ.* **2019**, *34*, 1–7.
51. Liang, Y.; Gustafson, E.J.; He, H.S.; Serra-Diaz, J.M.; Duveneck, M.J.; Thompson, J.R. What is the role of disturbance in catalyzing spatial shifts in forest composition and tree species biomass under climate change? *Glob. Change Biol.* **2023**, *29*, 1160–1177. [CrossRef]
52. Saini, D. Screening and evaluation of candidate trees for terrestrial carbon storage in regions with high air pollution and water stress. *Carbon Manag.* **2017**, *8*, 445–456. [CrossRef]
53. Yu, D.; Han, S. Ecosystem service status and changes of degraded natural reserves–A study from the Changbai Mountain Natural Reserve, China. *Ecosyst. Serv.* **2016**, *20*, 56–65. [CrossRef]
54. Tang, X.H.; Meng, Z.J.; Gao, Y.; Wang, J.; Zhang, P.; Liu, B. Photosynthetic carbon sequestration capacity of five natural desert shrubs in west Ordos region. *J. Arid. Land Resour. Environ.* **2017**, *31*, 128–135.
55. Li, Z.G.; Zhu, Q.; Li, J. A comparison of photosynthetic carbon sequestration of four shrubs in Ningxia. *Pratacultural. Sci.* **2012**, *29*, 352–357.
56. Zhang, D.M. Study Carbon Budget of Desert Shrubs in Alashan Desert Region. Master's Thesis, Inner Mongolia Agricultural University, Hohhot, China, 2012.
57. Niu, P.X. Study on Biomass and Carbon Storage of *Haloxylon ammodendron* Community in Gurbantunggut Desert. Master's Thesis, Shihezi University, Shihezi, China, 2015.
58. Zheng, Z.H.; Ma, C.X.; Ma, J.L.; Li, J.Y.; Lei, S.X.; Liu, C.; Li, H. Analysis of carbon sequestration and energy producing efficiency of four shrub species. *Hubei Agric. Sci.* **2011**, *50*, 4633–4635+4643.
59. Fen, X.H.; Zhang, X.M.; Liu, X.J.; Cheng, R.M.; Sun, H.R. Growth dynamics of *Tamarix chinensis* plantations in heavy-saline coastal lands and related ecological effects. *Chin. J. Eco. Agric.* **2013**, *21*, 1233–1240.
60. Ali, F.; Khan, N.; Abd_Allah, E.F.; Ahmad, A. Species Diversity, Growing Stock Variables and Carbon Mitigation Potential in the Phytocoenosis of *Monotheca buxifolia* Forests along Altitudinal Gradient across Pakistan. *Appl. Sci.* **2022**, *12*, 1292. [CrossRef]
61. Li, J.Q.; Chen, Q.B.; Li, Z.; Peng, B.X.; Zhang, J.L.; Xing, X.X.; Zhao, B.Y.; Song, D.H. Distribution and altitudinal patterns of carbon and nitrogen storage in various forest ecosystems in the central Yunnan Plateau, China. *Sci. Rep.* **2021**, *11*, 6269. [CrossRef] [PubMed]
62. Zhao, C.Y.; Song, Y.D.; Wang, Y.C.; Jaing, P.A. Estimation of aboveground biomass of desert plants. *Chin. J. Appl. Ecol.* **2004**, *15*, 49–52.
63. Yasseen, B.T.; Al-Thani, R.F. Endophytes and halophytes to remediate industrial wastewater and saline soils: Perspectives from Qatar. *Plants* **2022**, *11*, 1497. [CrossRef]
64. Zhao, M.Y.; Sun, W.; Luo, Y.K.; Liang, C.Z.; Li, Z.Y.; Shen, H.T.; Niu, X.X.; Zheng, C.Y.; Hu, H.F.; Ma, W.H. Models for estimating the biomass of 26 temperate shrub species in Inner Mongolia, China. *Arid. Zone Res.* **2019**, *36*, 1219–1228.
65. Yang, H.T.; Li, X.R.; Wang, X.R.; Jia, R.L.; Liu, L.C.; Gao, Y.H.; Li, G. Biomass prediction model of four shrub species at southeast edge of the Tengger Desert. *J. Desert Res.* **2013**, *33*, 1699–1704.
66. Tang, X.H.; Gao, H.; Yu, Y.; Meng, Z.J.; Liu, Y.; Wang, S.; Wu, H.; Ding, Y.L. The biomass estimation models for eight desert shrub species in northern edge of the Hobq Desert. *J. Arid. Land Resour. Environ.* **2016**, *30*, 168–174.
67. Wei, X.P.; Zhao, C.M.; Wang, G.X.; Chen, B.M.; Chen, D.L. Estimation of above-and below-ground biomass of dominant desert plant species in an oasis-desert ecotone of Minqin, China. *Chin. J. Plant Ecol.* **2005**, *29*, 12–17.
68. Liu, D.J.; Zhang, C.; Ogaya, R.; Estiarte, M.; Peñuelas, J.; Valencia, E. Effects of decadal experimental drought and climate extremes on vegetation growth in Mediterranean forests and shrublands. *J. Veg. Sci.* **2020**, *31*, 768–779. [CrossRef]
69. Yang, Q. Stability of Typical Desert Plant Populations, Communities and Ecosystems Across Water-Salt Gradient. Master's Thesis, Xinjiang University, Urumqi, China, 2019.
70. Yang, H.T.; Wang, Z.R.; Tan, H.J.; Gao, Y.H. Allometric models for estimating shrub biomass in desert grassland in northern China. *Arid. Land Res. Manag.* **2017**, *31*, 283–300. [CrossRef]
71. Friedlingstein, P.; O'Sullivan, M.; Jones, M.W.; Andrew, R.M.; Gregor, L.; Hauck, J.; Le Quéré, C.; Luijkx, I.T.; Olsen, A.; Peters, G.P.; et al. Global carbon budget 2022. *Earth Syst. Sci. Data* **2022**, *14*, 4811–4900. [CrossRef]
72. Lahn, B. A history of the global carbon budget. *WIREs Clim. Change* **2020**, *11*, e636. [CrossRef]

Disclaimer/Publisher's Note: The statements, opinions and data contained in all publications are solely those of the individual author(s) and contributor(s) and not of MDPI and/or the editor(s). MDPI and/or the editor(s) disclaim responsibility for any injury to people or property resulting from any ideas, methods, instructions or products referred to in the content.

MDPI

St. Alban-Anlage 66

4052 Basel

Switzerland

www.mdpi.com

Forests Editorial Office

E-mail: forests@mdpi.com

www.mdpi.com/journal/forests

Disclaimer/Publisher's Note: The statements, opinions and data contained in all publications are solely those of the individual author(s) and contributor(s) and not of MDPI and/or the editor(s). MDPI and/or the editor(s) disclaim responsibility for any injury to people or property resulting from any ideas, methods, instructions or products referred to in the content.